Concepts in Viral Pathogenesis

Concepts in Viral Pathogenesis

Edited by
Abner Louis Notkins and
Michael B. A. Oldstone

With 26 Figures

Springer-Verlag
New York Berlin Heidelberg Tokyo

ABNER LOUIS NOTKINS, M.D.
National Institutes of Health, Bethesda, Maryland 20205, U.S.A.

MICHAEL B. A. OLDSTONE, M.D.
Department of Immunopathology, Scripps Clinic and Research
Foundation, La Jolla, California 92037, U.S.A.

Cover photograph: Computer graphic model of antibody interaction with viral capsid by Arthur J. Olson, Ph.D., Research Institute of Scripps Clinic. The model uses x-ray crystallographic coordinates for the independently solved structures of an intact immunoglobulin (Silverton, E. W., Navia, M. A., and Davies, D. R. *Proc. Natl. Acad. Sci. U.S.A.*, Vol. 74, 5140–5144) and an intact spherical virus particle (Olson, A. J., Bricogne, G., and Harrison, S. C. *J. Mol. Bio.* (1983) 171, 61–93). Although the interaction depicted is purely hypothetical, it shows the size relationship between antibody and viral particle. Software used to produce the image: GRAMPS, graphics language developed by T. J. O'Donnell and A. J. Olson; and GRANNY, molecular modeling package developed by M. L. Connelly and A. J. Olson.

Library of Congress Cataloging in Publication Data
Main Entry under title:
Concepts in viral pathogenesis.
  Bibliography: p.
  1. Host-virus relationships. 2. Virus diseases
I. Notkins, Abner Louis. II. Oldstone, Michael B. A. [DNLM: 1. Viruses—Pathogenicity. QW 160 C744]
QR482.C65   1984   616'.0194   84-5416

© 1984 by Springer-Verlag New York Inc.
Softcover reprint of the hardcover 1st edition 1984
All rights reserved. No part of this book may be translated or reproduced in any form without written permission from Springer-Verlag, 175 Fifth Avenue, New York, New York 10010, U.S.A.
The use of general descriptive names, trade names, trademarks, etc., in this publication, even if the former are not especially identified, is not to be taken as a sign that such names, as understood by the Trade Marks and Merchandise Marks Act, may accordingly be used freely by anyone.

In recognition of the authors' work undertaken as part of their official duties as U.S. Government employees, the following chapters are not covered by copyright: 14, 23, 24, 25, 30, 34, 39, 43, 47, and 51. Reproduction of these chapters in whole or in part for any purpose of the United States Government is permitted.

Typeset by Bi-Comp, Incorporated, York, Pennsylvania.
Printed and bound by Halliday Lithograph, West Hanover, Massachusetts.

9 8 7 6 5 4 3 2 1

ISBN-13: 978-1-4612-9756-7   e-ISBN-13: 978-1-4612-5250-4
DOI:10.1007/ 978-1-4612-5250-4

Preface

The current proliferation of scientific information makes it difficult for even the most diligent reader to keep up with the latest developments in his/her own field, let alone other areas of interest. Review articles are one solution, but they too have become so voluminous and detailed that they often defeat the purpose for which they were intended.

We have attempted to ease this problem by using a different format. In this volume on *Concepts in Viral Pathogenesis*, we have assembled a series of mini-reviews/editorials, 1,000 to 2,000 words in length. Each is a pithy distillation of the state-of-the-art with emphasis on current thinking and unifying concepts rather than a compendium of the literature. The 53 articles, all written by active workers in their respective fields, are organized systematically so that the book will provide busy investigators, teachers and students a conceptual core of up-to-date information in a very brief and easily readable form. In addition, the authors have attempted to identify unresolved problems and point to future directions.

The articles in this book concentrate on the mechanisms by which viruses cause disease. The book begins with chapters on immune and non-immune mechanisms that control infection and then turns to virus tropism, mutation and diversity as factors in causing tissue injury and disease. Later chapters deal with virus persistence, integration, reactivation and interaction with the immune system. Although the emphasis in the book is on concepts and mechanisms rather than a description of individual viral infections, selected viruses that best exemplify important or novel virus–host interactions and that are under active investigation were singled out for further discussion. The book concludes with a series of chapters on new approaches for controlling viral diseases.

Bethesda, Maryland ABNER LOUIS NOTKINS, M.D.
La Jolla, California MICHAEL B. A. OLDSTONE, M.D.

Contents

Host Control of Viral Spread

Early Non-immune Control

1. Interferon and Viral Pathogenesis
   G. JOHN STANTON AND S. BARON 3

2. Natural Killer Cells in Viral Infection
   PAOLO CASALI AND GIORGIO TRINCHIERI 11

3. The Role of the Complement System in Host Defense Against Virus Diseases
   NEIL R. COOPER 20

4. Proteolytic Cleavage and Virus Pathogenesis
   ANDREAS SCHEID AND PURNELL W. CHOPPIN 26

Immune Control—Humoral

5. Mechanisms of Virus Neutralization
   BENJAMIN MANDEL 32

6. Antibody- and Complement-Dependent Lysis of Virus-Infected Cells
   J. G. P. SISSONS... 39

Immune Control—Cellular

7. The Inflammatory Response to Acute Viral Infections
   DIANE E. GRIFFIN 46

8. MHC Restriction and Cytotoxic T Lymphocytes
   P. C. DOHERTY AND R. M. ZINKERNAGEL 53

9. Antibody-Dependent Cellular Cytotoxicity (ADCC) Mediated by Human Killer Lymphocytes (K-Cells)
   PETER PERLMANN 58

10. Cloning of Functional T Lymphocytes
    JEAN-CHARLES CEROTTINI AND H. ROBSON MACDONALD 65

Genetic Control

11. Host Genes that Influence Susceptibility to Viral Diseases
    MARGO A. BRINTON, KENNETH J. BLANK, AND
    NEAL NATHANSON 71

12. Host Genes Controlling the Immune Response
    HUGH O. MCDEVITT 79

13. On the Role of Recombinant Retroviruses in Murine Leukemia
    JOHN H. ELDER... 86

Host Range and Tissue Tropism

14. Viral Receptors: Expression, Regulation and Relationship to Infectivity
    PATRICK R. MCCLINTOCK AND ABNER LOUIS NOTKINS......... 97

15. Viral Genes and Tissue Tropism
    BERNARD N. FIELDS....................................... 102

16. Utilization of Host Proteins as Virus Receptors
    GREGORY H. TIGNOR, ABIGAIL L. SMITH, AND ROBERT E. SHOPE   109

17. Host Range and Tissue Tropisms: Antibody-Dependent Mechanisms
    JAMES S. PORTERFIELD AND M. JANE CARDOSA 117

18. Enveloped Virus Maturation at Restricted Membrane Domains
    RICHARD W. COMPANS 123

19. Viruses and Differentiation: The Molecular Basis of Viral Tissue Tropisms
    ARNOLD J. LEVINE 130

Virus Maturation and Diversity

20. Continuum of Change in RNA Virus Genomes
    JOHN J. HOLLAND 137

21. Reassortment Continuum
    PETER PALESE 144

22. Immune Selection of Virus Variants
    JANICE E. CLEMENTS AND OPENDRA NARAYAN 152

23. Antigenic Variants of Viruses and Their Relevance to Clinical Disease
    BELLUR S. PRABHAKAR AND ABNER LOUIS NOTKINS 158

Virus Persistence

24. Unique Interactions of Retroviruses with Eukaryotic Cells
    S. R. TRONICK AND S. A. AARONSON 165

25. Molecular Biology of Herpes Simplex Virus Latency
    EDOUARD M. CANTIN, ALVARO PUGA, AND
    ABNER LOUIS NOTKINS 172

26. Cellular Oncogenes and Pathogenesis of Cancer
    ROBERT A. WEINBERG 178

27. Antibody Initiates Virus Persistence: Immune Modulation and Measles Virus Infection
    ROBERT S. FUJINAMI AND MICHAEL B. A. OLDSTONE 187

28. Ovarian Infection and Transovarial Transmission of Viruses in Insects
    LEON ROSEN 194

Virus-Interaction with the Immune System

29. Virus-Induced Immune Complex Formation and Disease: Definition, Regulation, Importance
    MICHAEL B. A. OLDSTONE 201

30. Virus-Induced Autoimmunity
    ABNER LOUIS NOTKINS, TAKASHI ONODERA, AND
    BELLUR S. PRABHAKAR 210

31. The Role of Immunospecific Receptors in Retrovirus-Induced Lymphomagenesis
    M. S. MCGRATH AND I. L. WEISSMAN 216

32. Viruses as Regulators of Delayed Hypersensitivity T-Cell and Suppressor T-Cell Function
    A. A. NASH 225

33. Mechanisms and Biological Implications of Virus-Induced Polyclonal B-Cell Activation
    RAFI AHMED AND MICHAEL B. A. OLDSTONE 231

Evolving Concepts in Viral Pathogenesis Illustrated by Selected Animal Models

34. Virus-Induced Diabetes Mellitus
    ABNER LOUIS NOTKINS AND JI-WON YOON 241

35. Herpesvirus-Induced Atherosclerosis
    CATHERINE G. FABRICANT 248

36. Retrovirus-Induced Arthritis
    TRAVIS C. MCGUIRE 254

37. Virus-Induced Demyelination
    PETER W. LAMPERT AND MOSES RODRIGUEZ 260

38. Virus Can Alter Cell Function Without Causing Cell Pathology: Disordered Function Leads to Imbalance of Homeostasis and Disease
    MICHAEL B. A. OLDSTONE 269

Evolving Concepts in Viral Pathogenesis Illustrated by Selected Diseases in Humans

39. Chronic Leukemia
 P. S. Sarin and R. C. Gallo 279

40. Hepatitis B Virus Diseases
 William S. Robinson 288

41. Herpesviruses and Cancer
 Fred Rapp and Mary K. Howett 300

42. Epithelial Cell Interactions of the Epstein–Barr Virus
 Joseph S. Pagano 307

43. Viral Gastroenteritis
 Albert Z. Kapikian 315

44. Hemorrhagic Fever Viruses
 C. J. Peters and K. M. Johnson 325

45. Marburg and Ebola Viruses: New Agents on the Frontiers of Virology
 Michael J. Buchmeier 338

46. Rabies
 Hilary Koprowski 344

47. Unconventional Viruses
 D. Carleton Gajdusek 350

Control of Viral Diseases

48. Antibodies to Synthetic Peptide Immunogens as Probes for Virus Protein Expression and Function
 Thomas M. Shinnick, J. Gregor Sutcliffe, and Richard A. Lerner 361

49. Recombinant DNA Vaccines
 Laurence A. Lasky and John F. Obijeski 366

50. Monoclonal Antibodies
 Walter Gerhard and Hilary Koprowski 376

51. Antiviral Drugs
    GEORGE J. GALASSO 382

52. The Use of Interferons in the Control of Viral Diseases
    ANTHONY L. CUNNINGHAM AND THOMAS C. MERIGAN 389

53. Insect Viruses as Pesticides
    THOMAS W. TINSLEY 398

Index 405

Contributors

S. A. AARONSON, National Cancer Institute, Building 37, Room 1A07 Bethesda, MD 20205, USA

RAFI AHMED, Department of Immunology, Scripps Clinic and Research Foundation, La Jolla, CA 92037, USA

S. BARON, Department of Microbiology, The University of Texas Medical Branch, Galveston, TX 77550, USA

KENNETH J. BLANK, Department of Pathology, School of Medicine, University of Pennsylvania, Philadelphia, PA 19104, USA

MARGO A. BRINTON, The Wistar Institute of Anatomy and Biology, Philadelphia, PA 19104, USA

MICHAEL J. BUCHMEIER, Department of Immunology, Scripps Clinic and Research Foundation, La Jolla, CA 92037, USA

EDOUARD M. CANTIN, NIDR, NIH, Building 30, Room 121, Bethesda, MD 20205, USA

M. JANE CARDOSA, Sir William Dunn School of Pathology, Oxford OX1 3RE, England

PAOLO CASALI, Department of Immunology, Scripps Clinic and Research Foundation, La Jolla, CA 92037, USA

JEAN-CHARLES CEROTTINI, Ludwig Institute for Cancer Research, Lausanne Branch, CH-1066 Epalinges, Switzerland

PURNELL W. CHOPPIN, The Rockefeller University, New York, NY 10021, USA

JANICE E. CLEMENTS, The Johns Hopkins University School of Medicine, Baltimore, MD 21205, USA

RICHARD W. COMPANS, Department of Microbiology, University of Alabama, University Station, Birmingham, AL 35294, USA

NEIL R. COOPER, Department of Immunology, Scripps Clinic and Research Foundation, La Jolla, CA 92037, USA

ANTHONY L. CUNNINGHAM, School of Medicine, Stanford University, Stanford, CA 94305, USA

P. C. DOHERTY, Department of Experimental Pathology, The John Curtin School of Medical Research, The Australian National University, Canberra City, A.C.T. 2601, Australia

JOHN H. ELDER, Department of Immunology, Scripps Clinic and Research Foundation, La Jolla, CA 92037, USA

CATHERINE G. FABRICANT, New York State College of Veterinary Medicine, Cornell University, Ithaca, NY 14853, USA

BERNARD N. FIELDS, Microbiology and Molecular Genetics Department, Harvard Medical School, Boston, MA 02115, USA

ROBERT S. FUJINAMI, Department of Immunology, Scripps Clinic and Research Foundation, La Jolla, CA 92037, USA

D. CARLETON GAJDUSEK, NINCDS, NIH, Building 36, Room 5B25, Bethesda, MD 20205, USA

GEORGE J. GALASSO, NIAID, Building WB, Room 750, Bethesda, MD 20205, USA

R. C. GALLO, Laboratory of Tumor Cell Biology, NCI, NIH, Building 37, Room 6B04, Bethesda, MD 20205, USA

WALTER GERHARD, The Wistar Institute of Anatomy and Biology, Philadelphia, PA 19104, USA

DIANE E. GRIFFIN, The Johns Hopkins University School of Medicine, Baltimore, MD 21205, USA

JOHN J. HOLLAND, Department of Biology, University of California, San Diego, La Jolla, CA 92093, USA

MARY K. HOWETT, Department of Microbiology and Cancer Research Center, The Pennsylvania State University College of Medicine, Hershey, PA 17033, USA

K. M. JOHNSON, Program Hazardous Viruses, Department of the Army, U.S. Army Medical Research Institute of Infectious Diseases, Fort Detrick, Frederick, MD 21701, USA

ALBERT Z. KAPIKIAN, Laboratory of Infectious Diseases, NIAID, NIH, Building 7, Room 103, Bethesda, MD 20205, USA

HILARY KOPROWSKI, The Wistar Institute of Anatomy and Biology, Philadelphia, PA 19104, USA

PETER W. LAMPERT, Department of Pathology, School of Medicine, University of California, San Diego, La Jolla, CA 92093, USA

LAURENCE A. LASKY, Genentech, Inc., South San Francisco, CA 94068, USA

RICHARD A. LERNER, Department of Immunology, Scripps Clinic and Research Foundation, La Jolla, CA 92037, USA

ARNOLD J. LEVINE, Department of Microbiology, School of Medicine, Health Sciences Center, State University of New York at Stony Brook, Stony Brook, NY 11794, USA

BENJAMIN MANDEL, 1848 E. 24th St., Brooklyn, NY 11229, USA

PATRICK R. MCCLINTOCK, NIDR, NIH, Building 30, Room 121, Bethesda, MD 20205, USA

HUGH O. MCDEVITT, School of Medicine, Stanford University, Stanford, CA 94305, USA

H. ROBSON MACDONALD, Ludwig Institute for Cancer Research, Lausanne Branch, 1066 Epalinges, Switzerland

M. S. McGrath, Department of Pathology, Stanford University Medical Center, Stanford, CA 94305, USA

Travis C. McGuire, Department of Veterinary Microbiology and Pathology, Washington State University, Pullman, WA 99164, USA

Thomas C. Merigan, Department of Medicine, Stanford University School of Medicine, Stanford, CA 94305, USA

Opendra Narayan, The Johns Hopkins University School of Medicine, Baltimore, MD 21205, USA

A. A. Nash, Department of Pathology, University of Cambridge, Cambridge CB2 1QP, England

Neal Nathanson, Department of Microbiology, School of Medicine, University of Pennsylvania, Philadelphia, PA 19104, USA

Abner Louis Notkins, NIDR, NIH, Building 30, Room 121, Bethesda, MD 20205, USA

John F. Obijeski, Genentech, Inc., South San Francisco, CA 94068, USA

Michael B. A. Oldstone, Department of Immunology, Scripps Clinic and Research Foundation, La Jolla, CA 92037, USA

Takashi Onodera, NIDR, NIH, Building 30, Room 121, Bethesda, MD 20205, USA

Joseph S. Pagano, Cancer Research Center, School of Medicine, University of North Carolina, Chapel Hill, NC 27514, USA

Peter Palese, Department of Microbiology, The Mt. Sinai Medical Center, New York, NY 10029, USA

Peter Perlmann, The University of Stockholm, The Wenner-Gren Institute, Norrfullsgatan 16, S-113 45, Stockholm, Sweden

C. J. Peters, Department of the Army, U.S. Army Medical Research Institute of Infectious Diseases, Fort Detrick, Frederick, MD 21701, USA

James S. Porterfield, Sir William Dunn School of Pathology, University of Oxford, Oxford OX1 3RE, England

Bellur S. Prabhakar, NIDR, NIH, Building 30, Room 121, Bethesda, MD 20205, USA

ALVARO PUGA, NIDR, NIH, Building 30, Room 121, Bethesda, MD 20205, USA

FRED RAPP, The Milton S. Hershey Medical Center, The Pennsylvania State University, Hershey, PA 17033, USA

WILLIAM S. ROBINSON, School of Medicine, Stanford University, Stanford, CA 94305, USA

MOSES RODRIGUEZ, Mayo Clinic and Research Foundation, Rochester, MN 55901, USA

LEON ROSEN, Pacific Biomedical Research Center, University of Hawaii at Manoa, Leahi Hospital, Honolulu, Hawaii 96816, USA

P. S. SARIN, Laboratory of Tumor Cell Biology, National Cancer Institute, Bethesda, MD 20205, USA

ANDREAS SCHEID, The Rockefeller University, New York, NY 10021, USA

THOMAS M. SHINNICK, Department of Immunology, Scripps Clinic and Research Foundation, La Jolla, CA 92037, USA

ROBERT E. SHOPE, Yale Arbovirus Research Unit, New Haven, CT 06510, USA

J. G. P. SISSONS, University of London, Royal Postgraduate Medical School, Hammersmith Hospital, London W12 OHS, England

ABIGAIL L. SMITH, Yale Arbovirus Research Unit, New Haven, CT 06501, USA

G. JOHN STANTON, Department of Microbiology, The University of Texas Medical Branch, Galveston, TX 77550, USA

J. GREGOR SUTCLIFFE, Department of Immunology, Scripps Clinic and Research Foundation, La Jolla, CA 92037, USA

GREGORY H. TIGNOR, Yale Arbovirus Research Unit, New Haven, CT 06510, USA

THOMAS W. TINSLEY, Natural Environment Research Council, Institute of Virology, Oxford OX1 3SR, England

GIORGIO TRINCHIERI, The Wistar Institute of Anatomy and Biology, Philadelphia, PA 19104, USA

S. R. TRONICK, Laboratory of Cellular and Molecular Biology, National Cancer Institute, Building 37, Room 1A07, Bethesda, MD 20205, USA

ROBERT A. WEINBERG, Center for Cancer Research, Massachusetts Institute of Technology, Cambridge, MA 02139, USA

I. L. WEISSMAN, Department of Pathology, Stanford University Medical Center, Stanford, CA 94305, USA

JI-WON YOON, NIDR, NIH, Building 30, Room 121, Bethesda, MD 20205, USA

R. M. ZINKERNAGEL, Institute of Pathology, University of Zurich, Universitatsspital, 8091 Zurich, Switzerland

Host Control of Viral Spread

Early Non-immune Control

1. Interferon and Viral Pathogenesis
   G. John Stanton and S. Baron 3

2. Natural Killer Cells in Viral Infection
   Paolo Casali and Giorgio Trinchieri 11

3. The Role of the Complement System in Host Defense Against Virus Diseases
   Neil R. Cooper 20

4. Proteolytic Cleavage and Virus Pathogenesis
   Andreas Scheid and Purnell W. Choppin 26

Immune Control—Humoral

5. Mechanisms of Virus Neutralization
   Benjamin Mandel 32

6. Antibody- and Complement-Dependent Lysis of Virus-Infected Cells
   J. G. P. Sissons... 39

Immune Control—Cellular

7. The Inflammatory Response to Acute Viral Infections
   Diane E. Griffin 46

8. MHC Restriction and Cytotoxic T Lymphocytes
   P. C. DOHERTY AND R. M. ZINKERNAGEL 53

9. Antibody-Dependent Cellular Cytotoxicity (ADCC) Mediated by
   Human Killer Lymphocytes (K-Cells)
   PETER PERLMANN 58

10. Cloning of Functional T Lymphocytes
    JEAN-CHARLES CERTOTTINI AND H. ROBSON MACDONALD 65

Genetic Control

11. Host Genes that Influence Susceptibility to Viral Diseases
    MARGO A. BRINTON, KENNETH J. BLANK,
    AND NEAL NATHANSON.. 71

12. Host Genes Controlling the Immune Response
    HUGH O. MCDEVITT 79

13. On the Role of Recombinant Retroviruses in Murine Leukemia
    JOHN H. ELDER... 86

Early Non-immune Control

CHAPTER 1
Interferon and Viral Pathogenesis

G. John Stanton and S. Baron*

This chapter presents in an editorial manner the main concepts regarding the defensive role of interferons (IFNs) during virus infections. The chapter begins with a description of the background and individual mechanisms of action of IFNs and ends with an integrated hypothesis concerning how the individual mechanisms may interact during an *in vivo* infection. The format calls primarily for citing reviews in which more specific references can be found [1–12].

Background

Interferon (IFN) was discovered 25 years ago by Isaacs and Lindenmann, who showed that fluids from virus-infected cell cultures contained a protein that could protect cells against a wide variety of viruses. Since then, at least three antigenically different proteins or glycoproteins have been found in humans and mice to possess the functional characteristics of IFNs but have major differences in antigenicity and amino acid sequences. These IFNs have been named alpha, beta, and gamma and they all react with cells to induce various degrees of antiviral activity along with other biological functions; including antitumor action, immunoregulatory action, cell growth inhibition, alteration of cell membranes, macrophage activation, enhancement of cytotoxicity of lymphocytes, influence on subsequent production of interferon, and hormone-like activation of cells.

The biochemical changes known to occur in cells treated with IFNs are listed in Fig. 1. The relationships between the biochemical and functional changes occurring in cells exposed to IFNs are not known. The principal

---
* University of Texas, Galveston, Texas.

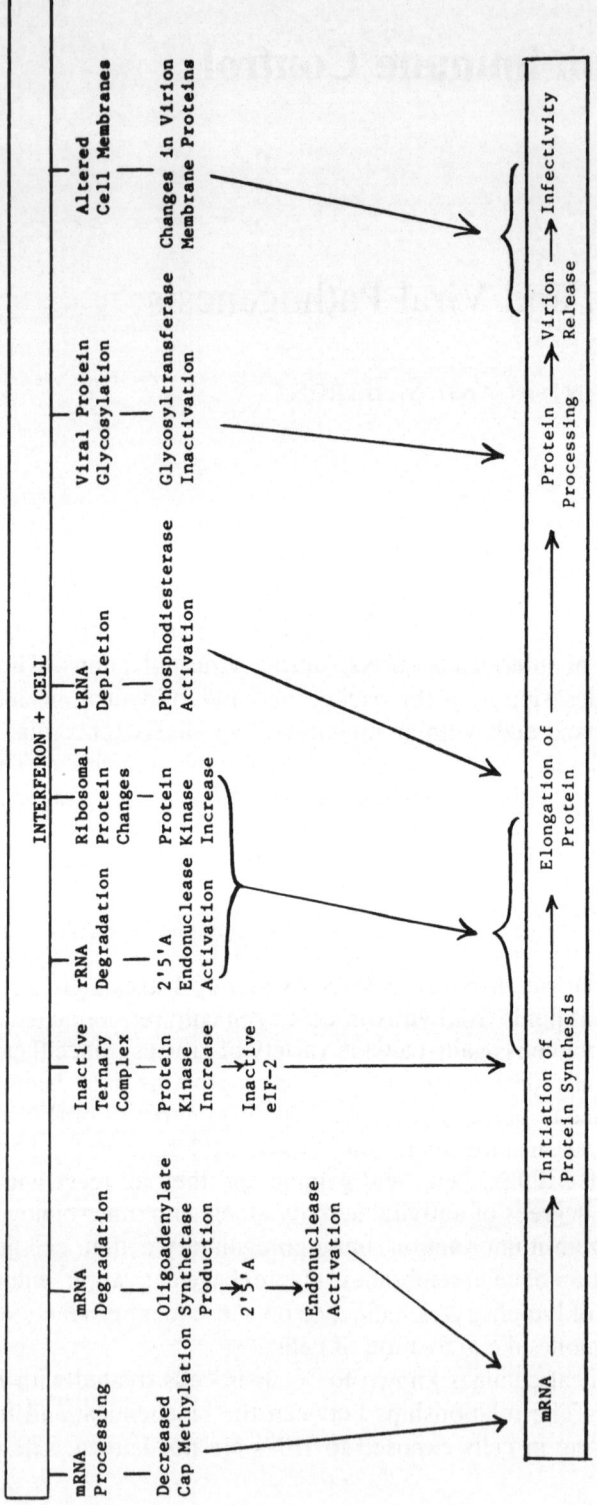

Fig. 1. Possible biochemical mechanisms by which the IFN-activated cell (top) may inhibit virus replication (bottom) [12].

function of IFNs that is discussed in this chapter is their antiviral activity, especially as it relates to the pathogenesis of viral infections.

Overview of Interferon and Pathogenesis

Before discussing the IFN defense mechanism in more depth, it may be worthwhile to recall the factors that help determine the effectiveness of IFN during viral infections. Some important variables include: (1) the sensitivity of the virus to IFN, (2) the quantity of IFN produced by the infected cell, and (3) the degree of antiviral activity capable of being induced in a cell. Variables that determine virus sensitivity to IFN are incompletely understood but are being studied actively. The ability of a particular virus to induce IFN may vary considerably and sometimes depends on the type of cell induced and the rapidity and degree to which the virus shuts down host–cell macromolecular synthesis. Since IFN can impede the inhibitory effect of virus on host–cell macromolecular synthesis, viruses that do not ordinarily induce much IFN can induce more IFN in cells previously exposed to IFN. Mechanisms explaining the level of antiviral activity inducible in a cell by IFNs are not completely understood, however, the number of IFN receptors present on the plasma membrane is thought to be a determining factor.

The ability of IFNs to inhibit viruses *in vivo* is more complex than the degree of inhibition measurable in cell culture because different types, subtypes, and mixtures of IFNs present in tissues during different phases of infection can provide marked variations in degrees of inhibition. In addition, the kinetics of each variable may differ with the virus, the host, and the host tissue.

An additional factor that determines the effectiveness of inhibition by IFN is the presence of a high concentration of virus in a tissue, which often enables the virus to overcome the antiviral effect of IFNs. Recent evidence *in vitro* suggests that the production of specific antibody to virus may help overcome this effect by reducing virus levels [13]. Thus, early-appearing antiviral antibodies may act synergistically with the IFN system to control virus infections.

The Interferon System

Types of Interferon Inducible by Virus. Whether IFNα, β, and/or γ is induced by a virus depends on the virus and cell interactions involved [14]. Viruses can induce HuIFNα by several mechanisms: (1) by infection and direct induction of leukocytes (T, B, null and/or macrophages [Mϕ]), and/or less directly (2) by infection of any nucleated cell and incorporation of viral antigens into its plasma membrane followed by interaction and induction of B or null lymphocytes, and (3) adsorption of virus or virus components to membranes of cells followed by interaction of these cells with B or null cells.

Viruses may also act as B- or T-cell mitogens or antigens and induce the formation of IFNα or γ. HuIFNβ production primarily results from virus infection of epithelial or fibroblastic cells. IFNγ is synthesized along with other lymphokines following the interaction of virus antigens with sensitized T lymphocytes and possibly some sets of null cells [15]. In the mouse system, MuIFNα and β are produced together by fibroblasts, epithelial cells, or leukocytes following interaction with virus. Thus, depending on the type of cells involved in the infection, one or more types of IFN may be produced at the same or different times. This may be very important because mixtures of IFNγ with α or β have greatly potentiated activities [16].

Interferon Induction by Virus. The exact nature of the interaction of virus or virus antigens with host cells that leads to induction of IFN is not known. In the case of infectious virus (Fig. 2), the virion penetrates into the cytoplasm and releases its nucleic acid, which can be either DNA or RNA. The intracellular presence or replication of viral nucleic acids is associated with the derepression of genes that code for IFN. For IFNα, a dozen or more genes may be derepressed and give rise to antigenically related IFNs that are thought to vary somewhat in their functions. The newly synthesized and

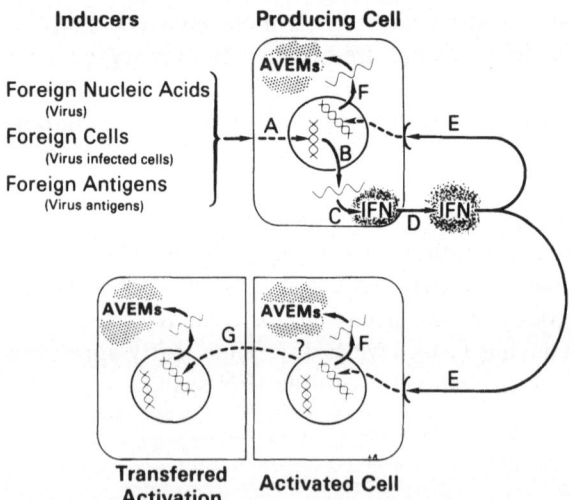

Fig. 2. Cellular events of the induction, production, and action of interferon (IFN). Inducers of IFN react with cells to derepress the IFN gene(s) (A). This leads to the production of mRNA for IFN (B). The mRNA is translated into the IFN protein (C) that is secreted into the extracellular fluid (D) where it reacts with the membrane receptors of cells (E). The IFN-stimulated cells derepress genes (F) for antiviral effector molecules (AVEMs) that are required to establish antiviral resistance eventually, and other cell changes (see Table 1). The activated cells also stimulate contacted cells (G) by a still unknown mechanism to produce AVEMs. (From Baron *et al.* [12].)

sometimes glycosylated (β and γ are glycosylated, while α has little or no carbohydrate) IFN polypeptides are quickly secreted into the extracellular fluid. Once the IFN is released, it must interact with specific receptors on the cell secreting it or on neighboring cells to induce antiviral activity. IFNs are sometimes capable of reducing virus yields in the cells initially infected and producing the IFN; however, they are more effective in inhibiting virus replication in neighboring and distal cells as: (1) levels of IFN in the cellular environment increase, and (2) time of exposure of cells to IFN increases. During the time when IFNs are being synthesized, virus is also replicating and eventually leaves the cell and infects adjacent cells. In many infections, IFN is produced within a few hours and thus usually precedes or appears simultaneously with virus release. IFN synthesis is eventually diminished by both a feedback mechanism and elimination of the viral inducer. Induction and synthesis of IFN constitutes the first part of the IFN system.

Induction of the Antiviral State by IFN. After the IFN reacts with a specific receptor on a cell membrane, an induction process, probably involving secondary cell "signal molecules," is initiated (Fig. 2). The signal molecule induces the synthesis of antiviral effector molecules (AVEMs) by a derepressional event. (We will use the term AVEMs to refer to those molecules that will eventually be important in the development of the antiviral state.) The newly transcribed mRNAs are then translated into AVEMs. The functions of the AVEMs that result in the development of the antiviral state are not fully understood, but probably include IFN-induced enzymes and other molecules shown in Fig. 1. An intriguing and unresolved aspect of the AVEMs' mechanism of action is their ability to inhibit viral mRNA translation significantly more than cell mRNA translation. The level of antiviral activity maintained in cells by the AVEMs appears to be dependent upon the concentration of IFN interacting with the cell membrane receptors. Control may also be mediated by an inhibitor of IFN action that is produced with IFNγ and possibly other IFNs [17].

Interferon-Induced Transfer of Antiviral Activity. In addition to the classic pathway for IFN induction of antiviral activity in cell populations, an antiviral state can be transferred to neighboring cells not exposed to IFN [18]. Effector cells treated with IFN and cocultivated with untreated allogeneic or xenogeneic recipient cells are able to transfer antiviral activity to the recipient cells that were not exposed to IFN (Fig. 2). The development of this antiviral activity requires cell-to-cell contact. Since an IFN-treated cell can transfer resistance before the mRNAs for the AVEMs are produced, the transfer molecules probably induce synthesis of AVEMs in the recipient cell. Although the process is initiated by IFNs, it does not require their continued presence. It has also been shown in cocultures that the cell type reacting most rapidly to IFN governs the rate of development of antiviral activity in the entire culture. Furthermore, as few as 10% of the cells in a population control the overall sensitivity of the population to IFN. Thus, transfer amplifies the IFN system by recruiting slower-responding cells to develop antiviral activity more rapidly and by inducing cells that are unresponsive or not even exposed to IFN to develop antiviral states. The overall importance of the transfer process in controlling virus infections is not known; however, it is reasonable to hypothesize that this could be a mechanism of spread of antiviral activity, especially in avascular tissues (e.g., surfaces of respiratory tract, gastrointestinal tract, and eye) and in tissues accessible to mobile leukocytes (e.g., areas of infection and inflammation).

IFN—An Important Defensive Mechanism during Viral Infections. The IFN system appears to be the earliest of the known host-defense mechanisms. It becomes operative within hours of infection. Its importance is suggested by the large numbers of subclinical virus infections that occur and the significant exacerbation of virus infections in animals when the IFN defense is deleted, as in the young embryo or by treatment with antibodies to IFN. The overall evidence for the importance of IFNs in viral infections is given in Table 1 [19].

How IFNs May Operate to Control Virus Infections: A Model. Following implantation of virus at the portal of entry (e.g., epithelium of the upper respiratory tract), newly formed virus and IFNβ begin to diffuse radially. The locally produced IFN in extremely high concentration [20] reacts with receptors on adjacent cells to make them more resistant to virus and also primes neighboring and distal cells to make higher levels of IFN following infection by virus. At the same time, the IFN-induced transfer mechanism begins operating to enhance the level of antiviral activity rapidly in low-responding cells at the implantation site and areas where IFN cannot readily diffuse, such as underlying cells in epithelial tissue. If the virus reaches the

Table 1. Evidence that interferons play a defensive role during viral infections.

Type of evidence	Example
Time correlations	Interferon produced just before and during arrest of virus replication
Place correlations	Interferon produced at the site of its action
Transfer	Exogenous interferon controls viral infections
Absence	Deletion of interferon in cell culture, eggs, and animals increases severity of viral infections
Sufficiency of quantity	Concentrations of interferon produced *in vivo* are sufficient to account for defensive action
Specificity of probes	Antibody to interferon; broad antiviral activity of both infected organ and interferon; characterization of interferon produced during infection
Reproducibility	Confirmation in many animal species and humans

lymphatic vessels, lymph nodes, or the blood stream, leukocytes may become infected and produce IFNα. IFNα may also be produced at the local site of infection as soon as leukocytes begin to migrate into the area. IFNα diffuses more effectively into the blood stream than IFNβ and γ. The circulating IFN can protect distant target organs. If the virus stimulates certain null cells, is mitogenic for lymphocytes, or is present later when sensitized T cells develop, IFNγ will be produced. Since IFNγ can potentiate the various activities of IFNα or β, the whole system becomes amplified, i.e., potentiation of AVEMs and other host-defense mechanisms (natural killer cells, macrophages, ADCC, etc.). Lymphocytes migrating out of the site of local infection and those exposed to IFNs in the spleen, lymph nodes, blood, and/or lymphatics can also induce antiviral activity throughout the body by the transfer mechanism, even in the absence of circulating IFN. If the virus continues to replicate and spread, it may still reach target organs. At this later stage, the IFN defense becomes fully activated: (1) most target organs have been primed by circulating IFNs to produce higher levels of IFN; (2) T lymphocytes have been sensitized that will produce IFNγ when they interact with viral antigens; (3) in target organs where IFN may not readily diffuse, some level of antiviral activity may exist from transfer of antiviral activity by migrating lymphocytes; and (4) the conditions are optimal for the presence of mixtures of IFNs that would result in a potentiated antiviral state. Nevertheless, in some cases virus may do damage to target organs even in the presence of high levels of host-defense factors, i.e., IFN, antibody, cytotoxic T cells, ADCC, Mφ, and other cytokines. Fortunately because of the IFN system, these instances are rare.

In summary, IFNs serve important defensive roles both early and later by their direct antiviral effects on cells, potentiation of antiviral effects of IFN mixtures, induction of the transfer system, and activation of host lymphocytes to destroy virus-infected cells.

References
1. Isaacs A (1963) Interferon. Adv Virus Res 10:1
2. Baron S (1963) Mechanism of recovery from viral infection. Adv Virus Res 10:39
3. Ho M (1964) Identification and "induction" of interferon. Bact Rev 28:367
4. Glasgow LA (1965) Interferon: a review. J Pediat 67:104
5. Wagner RR (1965) Interferon. Am J Med 38:726
6. Baron S (1970) The biological significance of the interferon system. Arch Intern Med 126:84
7. Glasgow LA (1970) Interrelationship of interferon and immunity. J Gen Physiol 56:212
8. Baron S (1973) The defensive and biological roles of the interferon system. *In* Finter NB (ed) Interferons and Interferon Inducers, Elsevier, Amsterdam, The Netherlands, p 267
9. Baron S (1967) Host defenses during virus infection. *In* Waterson AP, Heath RB (eds) Modern Trends in Medical Virology. Butterworth and Co., London, p 77
10. Stewart WE (1979) The interferon system. Springer-Verlag, New York

11. Stanton GJ, Johnson HM, Baron S (1978) The role of interferon in virus infections and antibody formation. *In* Ioachim HL (ed) Pathobiology Annual, Vol 8. Raven Press, New York, p 285
12. Baron S, Dianzani F, Stanton GJ (eds) (1982) Texas Reports on Biology and Medicine, Vol 41. The Interferon System: A Review to 1982, Part I and II. The University of Texas Medical Branch, Galveston, Texas, p 1
13. Langford MP, Stanton GJ (1982) Synergistic antiviral effects of interferon plus antibody against acute hemorrhagic conjunctivitis viruses, herpes virus and adenovirus. *In* Baron S, Dianzani F, Stanton GJ (eds) Texas Reports on Biology and Medicine, Vol 41. The Interferon System: A Review to 1982, Part II. The University of Texas Medical Branch, Galveston, Texas, p 538
14. Stanton, GJ, Langford MP, Weigent DA (1982) Cell types involved in production of interferon by leukocytes. *In* Baron S, Dianzani F, Stanton GJ (eds) Texas Reports on Biology and Medicine, Vol 41. The Interferon System: A Review to 1982, Part I. The University of Texas Medical Branch, Galveston, Texas, p 84
15. Herberman RB, Ortaldo JR, Timonen T, Reynolds CW, Djeu JY, Pestka S, Stanton J (1982) Interferon and natural killer (NK) cells. *In* Baron S, Dianzani F, Stanton GJ (eds) Texax Reports on Biology and Medicine, Vol 41. The Interferon System: A Review to 1982, Part II. The University of Texas Medical Branch, Galveston, Texas, p 590
16. Schwarz LA, Fleischmann WR Jr. (1982) Potentiation of interferon action. *In* Baron S, Dianzani F, Stanton GJ (eds) Texas Reports on Biology and Medicine, Vol 41. The Interferon System: A Review to 1982, Part I. The University of Texas Medical Branch, Galveston, Texas, p 298
17. Lefkowitz EJ, Fleischmann WR Jr. (1982) Control of the interferon system: An inhibitor of interferon action. *In* Baron S, Dianzani F, Stanton GJ (eds) Texas Reports on Biology and Medicine, Vol 41. The Interferon System: A Review to 1982, Part I. The University of Texas Medical Branch, Galveston, Texas, p 317
18. Blalock JE, Baron S, Johnson HM, Stanton GJ (1982) Transmission of IFN-induced activities by cell to cell communication. *In* Baron S, Dianzani F, Stanton GJ (eds) Texas Reports on Biology and Medicine, Vol 41. The Interferon System: A Review to 1982, Part I. The University of Texas Medical Branch, Galveston, Texas, p 344
19. Baron S (1979) The interferon system. American Society for Microbiology. Reprinted from ASM News, p 158
20. Dianzani F, Baron S, Levy HB, Gullino P, Viano I, Zucca M (1975) Studies on the action of human interferon under physiologic conditions. *In* Proc Symposium on Clinical Use of Interferon. Yugoslav Academy of Sciences and Arts, Zagreb, p 93

CHAPTER 2
Natural Killer Cells in Viral Infection

PAOLO CASALI* AND GIORGIO TRINCHIERI†

Mechanisms of Defense in Viral Infection

In human and experimental animals, the outcome of a virus infection is determined by the infecting agent's virulence and the host's ability to limit virus spread. Both specific immune response and natural mechanisms of defense act in limiting infection. The efferent arm of the specific immune response may limit virus growth by various effector mechanisms: virus inactivation by antibodies, complement-mediated lysis of enveloped viral particles or virus-infected cells coated with appropriate antibody, lysis of infected cells by antibody-dependent cytotoxic cells or specific cytotoxic T lymphocytes (CTL). T cells have virtually no detectable spontaneous cytotoxic activity. Generation of CTL requires activation of T cells by antigen presented on accessory cells such as macrophages. Usually, a lag period of 7 to 10 days is required before T cells develop peak primary reactivity, which rapidly abates. An accelerated memory response of T cells occurs in 2 to 5 days upon reexposure to the antigen. Antibody-dependent cell-mediated cytotoxicity (ADCC) can be mediated by different kinds of leukocytes bearing receptors of the Fc portion of IgG molecules, including monocytes, polymorphonuclear leukocytes (PMN), and a subset of lymphocytes, without the requirement of prior activation. ADCC, however, as well as complement-mediated lysis of infected cells, requires the presence of antibodies specific for structures expressed at the surface of the infected cells. Because of the requirement for specific recognition molecules, ADCC and complement-dependent lysis take place with a delay after virus infection, comparable to

---

* Scripps Clinic and Research Foundation, La Jolla, California.
† The Wistar Institute, Philadelphia, Pennsylvania.

that of specific CTL. It is conceivable that nonspecific mechanisms of defense play a relevant role against viral infections when specific immunological memory is absent (primary infection) and/or in early phases of infection when the immune response is not yet fully expressed. The mechanisms of natural resistance are characterized by their ability to intervene rapidly either because they do not require preactivation or because they can be induced very rapidly. Although these mechanisms appear to operate in nonspecific fashion, they are effectively regulated *in vivo* to act selectively against pathogenic organisms or cells. Natural resistance is mediated by soluble factors such as the complement system (e.g., opsonization of particles by deposition of C3b molecules on the cell surface upon activation of the alternate pathway), interferons (IFN) and other cytokines, and also by cellular effectors such as macrophages, PMN, and natural killer (NK) cells. Activated macrophages and monocytes and, to some extent, PMN are cytotoxic or cytostatic against a variety of tumoral and virus-infected target cells. Their interaction with target cells does not depend upon specific antigenic recognition and they can kill a broad range of syngeneic, allogeneic, and xenogeneic target cells. Macrophage activity is enhanced by certain lymphokines and microbial products. Among early cellular defense mechanisms, NK cell activity would likely play a primary role. Indeed, cells infected with a variety of viruses are efficiently lysed by an NK mechanism *in vitro* and augmented NK function parallels the kinetics of viral elimination in acutely infected experimental animals [1-4]. Like phagocytic and PMN cells, NK cells are capable of lysing various target cells, including virus-infected, neoplastic, and normal cells, without prior exposure to specific antigens *in vivo*.

Natural Killer Cells and Their Activity

Cells with NK activity are nonadherent and nonphagocytic and have been studied in human, rat, mouse, and other animal species [5-7]. In addition to their spontaneous cytotoxicity, NK cells express surface receptors for the Fc portion of IgG and act as effector cells in ADCC systems. NK cells and the lymphocytes responsible for ADCC activity (killer [k] cells) are identical or largely overlapping subsets. Most or all cells with NK activity in humans, rats, and as recently shown in mice have a characteristic morphology and are defined as large granular lymphocytes (LGL) [8]. LGL in humans represent 5% to 15% of the peripheral blood lymphocytes (PBL) and are medium- to large-sized lymphocytes characterized by a high cytoplasm-to-nucleus ratio, an indented nucleus, and a variable number of distinct azurophilic granules. Ultrastructural characteristics of LGL are: an extended Golgi apparatus; numerous "coated" vesicles diffused through the cytoplasm or, more often, concentrated in the nuclear notch; and granules containing a matrix of variable density. LGL can be identified among Giemsa-stained lymphoid cells prepared on cytocentrifuged slides and can be highly enriched by Percoll gradient centrifugation. Most LGL are endowed with NK activity and virtually all NK activity in PBL can be attributed to LGL.

Lineage, Differentiation, and Phenotypic Markers of NK Cells

Although NK cells appear to represent a morphologically homogeneous cell subset, their specific hematopoietic lineage has not been unambiguously identified. LGL could represent a stage of differentiation of either the myelomonocytic lineage, or, according to the major prevailing hypothesis, of the T-cell lineage, or they could be terminally differentiated cells of a distinct lineage. It has also been proposed that LGL are a heterogeneous group of cells, some of myelomonocytic and others of lymphocytic origin. Human NK cells (and LGL) share surface markers of both T cell and myelomonocytic cells. The large majority of NK cells express the antigen recognized by the monoclonal antibody OKM1 (and others) that also reacts with all mature PMN and monocytes, and recognizes the receptor for third component of complement. All NK cells also react with B73.1, a monoclonal antibody that detects the human Fc receptor for aggregated immunoglobulins [9]. This receptor is also present, though at much lower density, on neutrophilic PMN. However, NK cells do not express several other antigens that are present on both mature and immature myelomonocytic cells. On the other hand, most NK cells bear the receptor for sheep erythrocytes, as detected by several monoclonal antibodies (e.g., OKT11). Other markers of mature T cells (e.g., OKT1, OKT3, OKT4) are not present in NK cells, whereas OKT10, a monoclonal antibody reactive with thymocytes and activated T cells, reacts with most NK cells. The antigen recognized by OKT8 antibody, and present at high density in thymocytes and in the cytotoxic/suppressor T-cell subset, is present at low density in about 50% of NK cells. A recently described antibody, HNK-1, reacts with a variable proportion of NK cells and also with a proportion of OKT3-, OKT8-positive cells that have been tentatively identified by some authors as NK precursors [10]. Mouse NK cells share the Thy1, Qa4, Qa5, Ly5, and asialo-GM$_1$ markers with cells of the T lineage, but lack the Lyt1 and Lyt2 markers. Like T cells, both human and murine NK cells can be grown *in vitro* in the presence of T-cell growth factor (interleukin 2, IL-2). In general, the analysis of NK clones has been hampered by the ability of activated T cells to mimic NK-cell activity. Further unambiguous identification of NK-cell clones has been difficult. Putative NK-cell clones, described by their activity, bear various surface markers, such as HLA-DR antigens or those recognized by OKT3 and OKT4 in human cells, and by Lyt2 in murine cells, that are not present on NK cells of peripheral blood or spleen [6,11].

Very little is known about the differentiation and ontogenesis of NK cells.

Specific NK-cell markers are lacking and it is difficult to identify immature cells on the basis of cytotoxic function. In the mouse, NK-cell activity is absent at birth. In human cord blood, cells with the morphology and cell surface phenotype of NK cells are present in normal numbers, although their cytotoxic ability is quite low. NK cells have been shown to differentiate in the bone marrow with no requirement for processing through the thymus. The data regarding *in vivo* turnover of NK cells are conflicting and vary according to the experimental system used, with half-lives ranging from a few days to 2 to 3 weeks. The alloantigen NK1.1 has been used to study NK ontogeny in the mouse, and it has been possible to demonstrate different stages of differentiation of NK cells (based on the presence of the antigen and acquisition of the ability to mediate cytotoxicity). IFN is the most important differentiation-inducing factor. However, these studies are also hampered by the fact that anti-NK1.1 antisera reacts with a greater proportion of mouse cells than can be accounted for by NK activity.

Mechanisms of NK-Cell-Mediated Lysis

Incubation of PBLs with a variety of virus-infected cells, with transformed lymphoblastoid cells, or with tumor cells results in significant lysis of these target cells within 4 to 12 hours and also in high levels of antiviral activity in the culture supernatant. The viral inhibitor produced in these cultures has been identified as IFN and it has been shown that lymphocytes are responsible for its production [1,2,12,13]. Among nonacutely infected cells, most of the tumor-derived human cell lines and lines transformed with EBV or murine sarcoma virus are able to induce IFN production in lymphocytes (induction might in some cases be attributed to mycoplasma contaminants). By contrast, normal fibroblasts and most of the cell lines transformed with simian virus 40 or with adenovirus 5 are unable to induce IFN. The human IFN produced has characteristics of leukocyte IFN type I or IFNα, although some antiviral activity is due to immune IFN or IFNγ. The same IFN mixture is observed with virus-infected leukocytes. IFN efficiently enhances the cytotoxic activity of NK cells. This enhancing effect can be readily demonstrated and quantitated by preincubating lymphocytes in the presence of IFN and then testing these lymphocytes in a cytotoxicity assay, using target cells that are unable to induce IFN. All three known types of human IFN (i.e., fibroblast (β), leukocyte type I (α), and leukocyte type II (γ) IFN), increase human NK-cell activity similarly. Neither B cells nor T cells acquire cytotoxic activity when exposed to IFN. Among human LGL, positive self regulation of NK activity by production of IFN *in vitro* upon infection with influenza and herpes simplex viruses has been proposed [14]. In the presence of virus, IFN production has been attributed solely to LGL, without participation of other cell types. However, in other studies, the producers of IFN have been identified as HLA-DR positive, Fc-receptor positive, nonadherent, nonphagocytic cells that might only copurify with LGL on Percoll gradients. IFN production by monocytes can also be detected, albeit at lower levels, after coculture with viruses for 18 hours, but no cytotoxicity is induced. Based on experiments involving treatment of purified LGL with IFN, it has been shown that IFN affects NK activity in at least three different ways: (1) by increasing the number of LGL able to bind to the target cells and the proportion of cytotoxic cells within the LGL population bound to the target cells; (2) by accelerating the kinetics of lysis; and (3) by increasing the recycling ability of active NK cells [15]. Enhancement of NK activity happens in the absence of extracellular IFN. Using human fibroblasts (non-IFN inducers) as target cells, it has been demonstrated recently that virus-infected cells and purified viral glycoproteins efficiently enhance NK-cell cytotoxic activity by a mechanism independent of IFN release into the extracellular fluid [16]. This IFN-independent NK activity is exerted earlier than the IFN-associated cytotoxic activity and is completely abrogated by $F(ab')_2$ fragments of antibodies to viral glycoproteins [17].

The ability of IFN to induce blast formation, DNA synthesis, and possibly proliferation of NK cells has been demonstrated *in vivo* in mice [18]. Because IFN *in vitro* induces expression of IL-2 receptors, it is possible that the *in vivo* effect of IFN on NK-cell proliferation is mediated by IL-2. IL-2 itself can indirectly enhance NK activity, and the effects of IFN and IL-2 are synergistic. Thus, in the case of virus infection, the IFN produced *in vivo* can induce the NK system at various levels, including the differentiation of NK precursors, proliferation of mature NK cells and enhancement of its cytotoxic activity. The effect of IFN pretreatment *in vitro* of target cells is paradoxically opposite to that obtained by pretreating effector cells. Normal cells (nontumoral, noninfected) become refractory to NK activity following incubation with IFN [2,13]. Conversely, virus-infected cells and most tumor cells when treated with IFN are efficiently killed by NK cells, similar to their untreated counterparts. These *in vitro* findings might have important implications *in vivo*, as the NK cells might be rendered selective for virus-infected cells.

Information about the recognition of target cells by NK cells is limited to the observation that a number of different target cells bind NK cells to varying degrees. However, the mechanism by which NK cells kill their targets has been partially clarified. Morphologic and metabolic inhibitor studies indicate that lysis is mediated by vesicular secretion. Drugs that block vesicular secretion in other cells inhibit NK-cell killing without affecting the binding of the effector to the target cell. Degranulating agents both deplete the granules from LGL and inhibit killing [19]. Thus, NK-cell killing probably proceeds by: (1) binding of the effector cell to the target, (2) polarization of granules to the part of the effector cell in contact with the target, (3) Ca^{2+}-dependent degranulation and release of the NK cytotoxic molecules that bind to the target surface, and (4) lysis of target cells by the released cytotoxic factor. Electron microscopic study of killing by cloned NK cells reveals insertion of tubular complexes in the membrane of target cell. These tubular formations are products of NK cells. Based on ultrastructural images, it appears possible that these tubules arise by circular polymerization of monomeric precursors [20]. Thus, the formation of NK-cell lesions may be analogous to complement where it was shown that component C9 upon circular polymerization forms tubules that, when bound to membranes, appear as membrane lesions.

Role of NK Cells in Virus Infection

The role of NK cells *in vivo* in man, awaits clear definition. Most cell activity *in vivo* by virus infections has been done in mice. Concurrent with the observation of virus-specific H-2-restricted cytotoxic T cells in virus infection [21,22], cells with nonspecific cytotoxic activity were discovered in virus-infected mice [4,23,24]. *In vivo* induction of putative NK cells was demonstrated first in mice acutely infected with lymphocytic choriomeningitis virus (LCMV) and ectromelia virus prior to specific T-cell response. NK activity is greatly enhanced in various viral infections of mice, including those by cytomegalovirus, lactic dehydrogenase, mumps, Moloney sarcoma, mouse hepatitis, Newcastle disease, Pichinde, Semliki forest, vaccinia, and vesicular stomatitis viruses [3]. A typical paradigm of the kinetics of NK activity in relation to IFN production and generation of immune-specific response in acute LCMV infection of adult mouse is seen in Fig. 1. The induction of NK-cell activity in LCMV-infected mice directly parallels the level of virus-induced IFN and elimination of infectious virus particles. Similar NK enhancement can be induced in uninfected mice by injection of IFN, and IFN-dependent NK activation *in vivo* can be abrogated by antibodies to IFN. In humans, *in vivo* enhancement of NK activity occurs following administration of IFNα. Activated NK cells are present in humans during acute viral infection, including infections with Epstein–Barr, measles, mumps, and cytomegaloviruses. Although more data are needed to define the role of NK cells in human viral infection, recent findings suggest a correlation between fatal CMV infection and failure to develop NK-cell activity in immunosuppressed bone marrow-transplanted patients [25]. The role of NK cells is currently being defined in other viral infections associated with immunosuppression. Viruses themselves can affect lymphocyte functions. Recently, evidence has been provided that some human pathogenic viruses turn off cytotoxic activity of human NK cells *in vitro* [26]. This may prove important in negative modulation of the host's defense against viral infections *in vivo*.

In experimental animals, tumorigenesis is reduced or inhibited during viral infection and viral infection enhances NK activity and IFN production. Indeed, NK cells and IFN play an essential role in the control of tumor outgrowth. Further, cloned NK cells reject bone marrow grafts, inhibit im-

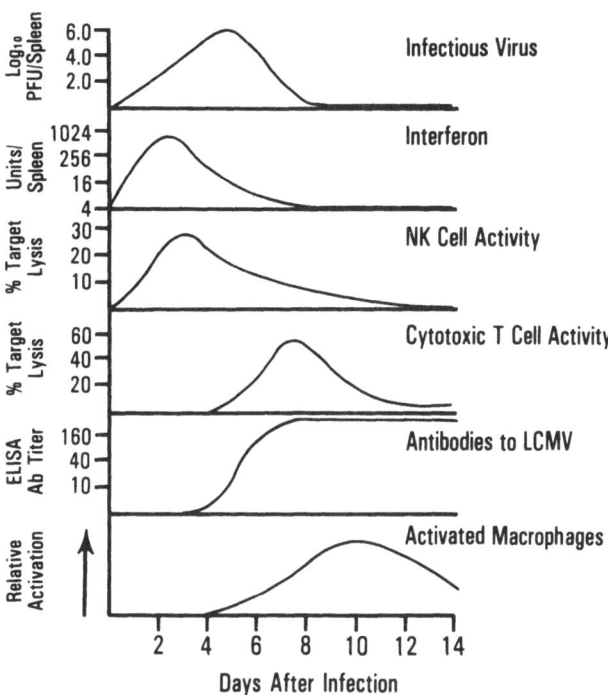

Fig. 1. Virus titer, nonspecific and specific host response following inoculation of lymphocytic choriomeningitis virus (virus) in adult mice. Figure modified from Buchmeier MJ, Welsh R, Dutko FJ, Oldstone MBA. (1980). Adv Immunol 30:275-331.

munoglobulin secretion by B cells, control melanoma tumor growth, and prevent the development of radiation-induced thymic leukemia [27]. Though it remains unclear whether the effects of the cloned cell lines are mediated directly or indirectly and what growth factors are responsible for maintaining the cells in an active state *in vivo*, such data point to a important regulatory role for NK cells *in vivo*. They also provide important models by which to explore the role of NK cells in viral infection.

Acknowledgments. This is publication no. 2995-IMM from the Department of Immunology, Scripps Clinic and Research Foundation. This research was supported by GCRC RR-00833, US PHS Grants AI-07007, NS-12428, CA-32898, CA-20833, CA-10815, and NATO Grant 069.81. P. Casali is recipient of an award from the Scripps Clinic Medical Group.

References

1. Trinchieri G, Santoli D, Dee RR, Knowles BB (1978) Antiviral activity induced by culturing lymphocytes with tumor-derived or virus-transformed cells. Identification of the antiviral activity as interferon and characterization of the human effector cells subpopulation. J Exp Med 147:1299-1313

2. Trinchieri G, Santoli D (1978) Antiviral activity induced by culturing lymphocytes with tumor-derived or virus-transformed cells. Enhancement of natural killer cell activity by interferon and antagonistic inhibition of susceptibility of target cell to lysis. J Exp Med 147:1314–1333
3. Welsh RM (1981) Natural cell-mediated immunity during viral infections. Curr Top Microbiol 92:83–106
4. Welsh RM (1978) Cytotoxic cells induced during lymphocytic choriomeningitis virus infection of mice. I. Characterization of the natural killer cell induction. J Exp Med 148:163–181
5. Bloom B (1982) Natural killer cells to rescue immunosurveillance? Nature 300:214–215
6. Herberman RB (ed) (1982) NK cells and other natural effector cells. Academic Press, New York
7. Herberman RB, Ortaldo JR (1981) Natural killer cells: Their role in defenses against disease. Science 214:24–30
8. Timonen T, Ortaldo JR, Herberman RB (1981) Characteristics of human large granular lymphocytes and relationship to natural killer cells. J Exp Med 153:569–582
9. Perussia B, Starr S, Abraham S, Fanning V, Trinchieri G (1983) Human natural killer cells analysed by B73.1, a monoclonal antibody blocking Fc-receptor function. I. Characterization of the lymphocyte subset reactive with B73.1. J Immunol 130:2133–2141
10. Abo T, Cooper MD, Balch CM (1982) Characterization of HNK-1(+) (Leu 7) human lymphocytes. I. Two distinct monotypes of human NK cells with different cytotoxic capability. J Immunol 129:1752–1757
11. Hercend T, Reinherz RL, Meuer S, Schlossman SF, Ritz J (1983) Phenotypic and functional heterogeneity of human cloned natural killer cell lines. Nature 301:158–160
12. Herberman RB, Ortaldo JR, Djeu SY, Holden HT, Sett S, Pestka S (1980) Role of interferon in regulation of cytotoxicity by natural killer cells and macrophages. Proc NY Acad Sci USA 350:63–71
13. Trinchieri G, Santoli D, Granato D, Perussia B (1981) Antagonistic effects of interferons on the cytotoxicity mediated by natural killer cells. Fed Proc 40:2705–2710
14. Djeu JY, Stocks N, Zoon K, Stanton GJ, Timonen T, Heberman RB (1982) Positive self regulation of cytotoxicity in human natural killer cells by production of interferon upon exposure to influenza and herpes viruses. J Exp Med 156:1222–1234
15. Timonen T, Ortaldo JR, Herberman RB (1982) Analysis by a single cell cytotoxicity assay of natural killer (NK) cell frequencies among human large granular lymphocytes and of the effects of IFN on their activity. J Immunol 128:2514–2521
16. Casali P, Sissons JGP, Buchmeier MJ, Oldstone MBA (1981) Generation of human cytotoxic lymphocytes by virus. Viral glycoproteins induce nonspecific cell mediated cytotoxicity without release of interferon in vitro. J Exp Med 154:840–855
17. Casali P, Oldstone MBA (1982) Mechanisms of killing of measles virus infected cells by human lymphocytes: Interferon associated and unassociated cell mediated cytotoxicity. Cell Immunol 70:330–344

18. Biron CA, Welsh RM (1982) Blastogenesis of natural killer cells during viral infection *in vivo*. J. Immunol 129:2788–2796
19. Quan P-C, Ishizaka T, Bloom BR (1982) Studies on the mechanism of NK cell lysis. J Immunol 128:1786–1791
20. Podack ER, Dennert G (1983) Cell mediated cytolysis: Assembly of two types of tubules with putative cytolytic function by cloned natural killer cells. Nature 302:662–666
21. Blanden RV, Gardner J (1976) The cell-mediated immune response to ectromelia virus infection. I. Kinetics and characteristic of the primary effector T cell response in vivo. Cell Immunol 22:271–282
22. Doherty PC, Zinkernagel RM (1974) Restriction of *in vitro* T cell mediated cytotoxicity in lymphocytic choriomeningitis within a syngeneic or semiallogeneic system. Nature 248:701–702
23. Kiessling R, Klein E, Wigzell H (1975) Natural killer cells in the mouse. I. Cytotoxic cells with specificity for mouse Maloney leukemia cells. Specificity and distribution according to genotype. Eur J Immunol 5:112–117
24. Welsh RM, Zinkernagel RM (1977) Hetero-specific cytotoxic cell activity induced during the first three days of acute lymphocytic choriomeningitis virus infections in mice. Nature 268:646–648
25. Quinnan GV, Kirmani N, Rock AH, Manischewitz J, Jackson L, Moreschi G, Santos GW, Saral R, Burns WH (1982) Cytotoxic T cells in cytomegalovirus infection. HLA-restricted T lymphocyte and non T lymphocyte cytotoxic responses correlate with recovery from cytomegalovirus infection in bone marrow-transplant recipients. N Engl J Med 307:7–13
26. Casali P, Rice GPA, Oldstone MBA (1984) *In vitro* alteration of human lymphocytes functions by virus. Differential effects of measles, influenza and cytomegalovirus on cytotoxic activity, proliferation and antibody production. J Exp Med, in press.
27. Warner JF, Dennert G (1982) Effects of a cloned cell line with NK activity on bone marrow transplants, tumour development and metastasis *in vivo*. Nature 300:31–34

CHAPTER 3

The Role of the Complement System in Host Defense Against Virus Diseases

NEIL R. COOPER*

Complement is an effector system capable of mediating a number of biological activities. Most familiar is the ability of the system to mediate the lytic destruction of many kinds of cells including bacteria and viruses surrounded by lipid membranes. Apart from such direct effects on a virus or other pathogens, the activated complement system also facilitates interactions with various effector cells including neutrophils, monocytes, basophils, mast cells, and lymphocytes. Depending on the effector cell involved, complement may trigger the release of histamine and other secondary mediators, stimulate oxidative metabolism, activate intracellular processes, initiate directed motion, facilitate phagocytosis, and modulate immunological responses and immune reactions (Fig. 1). Although complement can thus mediate numerous reactions *in vitro,* the most important *in vivo* role in diseases, including virus diseases, is probably related to the system's ability to produce an acute inflammatory response and to function as an opsonin in facilitating phagocytosis. Thus, activation of complement in a localized environment leads to changes in capillary permeability, edema, alterations in vessel contractility, and directed migration of leukocytes into the area. The phagocytic cells infiltrating the area of complement activation become fixed to specific opsonic sites on the complement molecules, which are in turn attached to the surface of the pathogens.

The complement system consists of 20 different plasma proteins termed components or factors that must be activated to produce biological effects. Activation triggers an orderly, sequential, and carefully regulated series of interactions between the complement proteins and the activator. There are two activation pathways, the classical and the alternative or properdin path-

---

* Scripps Clinic and Research Foundation, La Jolla, California.

Role of the Complement System in Host Defense

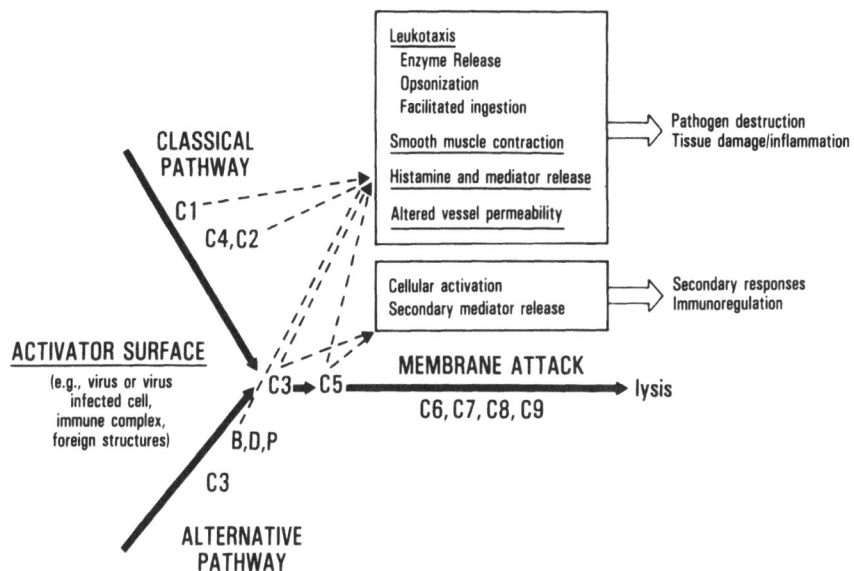

Fig. 1. Pathways and biology of the complement system.

ways, respectively, and a terminal membrane attack pathway (Fig. 1). Each activation pathway consists of, and is driven by, a series of enzymatic reactions. The enzymes involved are generally formed from inactive complement precursor molecules by limited proteolytic cleavage by other complement enzymes. Another feature characteristic of the activation pathways as well as the membrane attack pathway is the systematic formation of large multimolecular assemblies composed of several different complement proteins. The complement cleavage fragments and the protein–protein complexes possess various functional activities and are responsible for the biological activities of the complement system.

The classical complement pathway is triggered by IgG or IgM antibodies in complex with specific antigens including viruses as noted below. In addition, the classical pathway is directly triggered in the complete absence of antibody by a number of substances of extremely diverse chemical and physical composition [1]. These include certain viruses and bacteria. With activation, the first component, C1, is converted into an enzyme that cleaves C4 and C2, the next two reacting components (Fig. 1). These proteins form a protein–protein complex with *de novo* enzymatic activity directed against the next reacting factor, C3. With cleavage of C3 yet another enzyme is formed, which has the ability to cleave C5.

The alternative complement pathway, originally described as the properdin system, is directly activated by a number of different substances including certain viruses and bacteria [2]. Although activation generally occurs without antibody involvement, activation and biological reactivity in some cases are augmented by specific antibody. As with the classical pathway, activation triggers a series of interactions between the pathway proteins, factor B, factor D, properdin, and C3, which lead to the formation of protein–protein complexes with proteolytic activity. The pathway mediates the formation of an enzyme containing factor B, C3, and properdin, which has the ability to cleave C5.

The membrane-attack pathway is set in motion after C5 cleavage by the C5-cleaving enzymes of either the classical or the alternative pathways. The remaining proteins, C6, C7, C8, and C9 self-assemble with cleaved C5 into a large multimolecular complex called the membrane-attack complex. This complex, which is responsible for the cytolytic activity of the complement system, has the transient ability immediately after formation to lyse lipid-containing membranes.

The complement system can only play a role in controlling virus infection or dissemination after activation of the system. In this regard, it has been amply demonstrated that viruses and virus-infected cells coated with antibody behave as typical immune complexes and activate complement [3,4]. The antibody may be specific, elicited antibody, however, examples of "natural" or cross-reactive antiviral antibodies that activate complement have been described [5,6]. In addition to immune activation, however, it has become apparent that some viruses also directly activate complement in the complete absence of antibody [3]. Among these, retroviruses directly and very efficiently trigger the classical pathway [7]. The activator is a viral envelope protein, p15E, which avidly binds and activates C1 [8]. Sindbis virus activates either the classical or alternative pathways in the absence of antibody [9] while Newcastle disease virus can be neutralized via the classical pathway without antibody participation [10]. Epstein–Barr virus directly activates the alternative pathway [11]. Although relatively few viruses have been examined as yet, it is likely, based on the above examples and the many structurally different types of nonviral substances that activate the classical and alternative pathways in the absence of antibody, that further study will reveal that numerous viruses directly activate one or the other of the complement pathways.

All virus-infected cell models thus far examined, with one exception [12] interact with the alternative pathway directly [13,14] or in conjunction with specific antiviral antibody [4,15]. Because antibody- and complement-dependent effects on virus-infected cells are considered in a subsequent article, they will not be further analyzed here.

The activation of complement whether directly, or via antibody, may lead to virus neutralization. As a consequence of activation, complement proteins are deposited on the viral surface or envelope. These proteins may mask or cover structures required for the attachment of the virus to a potentially susceptible cell, and thereby reduce the infectivity of the virus. Neutralization in this manner by envelopment with complement proteins, primarily C3 and C4, has been demonstrated for a number of viruses including influenza, herpes simplex, Epstein–Barr, infectious bronchitis, vaccinia, and Newcastle disease viruses [3,5,6]. Complement also has the ability to lyse enveloped viruses. Most and probably all enveloped viruses are susceptible to complement-dependent lysis, and viral neutralization as a consequence of lysis has been observed for a large number of viruses [3]. Nevertheless, viral lysis is probably not a major mechanism of viral neutralization *in vivo*. In order to produce lysis, the complement-activation stimulus must be extremely potent. This can be accomplished only with large amounts of antibody [5], or a highly efficient, direct activator [7]. These conditions are not often met. In addition, the lipid bilayer of the viral envelope must be exposed to permit the C5 to C9 membrane-attack complex to insert into the bilayer, a requirement for lysis. The necessity for exposure of the bilayer may explain the peculiar requirement for specific antibody for lysis that has been observed in some systems [13]; the antibody may redistribute the surface viral glycoproteins and thereby expose the bilayer. Thus, complement-dependent neutralization of enveloped viruses frequently occurs in *in vitro* studies in the absence of lysis [3,5,6] and this situation is likely to pertain *in vivo*.

In addition to these mechanisms of viral neutralization, C3 cleavage, which characterizes complement activation, is likely to facilitate interactions with effector cells, which would also have consequences for viral infectivity. As C3b is an opsonin, its deposition on viral surfaces during activation should enable the C3b-coated virus particle to interact with specific C3b receptors located on the surface of polymorphonuclear leukocytes, monocytes and other phagocytic cells (Fig. 1). In the case of bacteria and other activator cells, such interactions either alone or together with other stimuli, such as traces of antibody, facilitate ingestion or extracellular destruction. This reaction sequence has not been investigated with either viruses or virus-infected cells.

Although not examined specifically in virus systems, the smaller activation cleavage products of C3 and C5 have numerous biological activities that produce an acute inflammatory reaction. The inflammatory reaction would limit dissemination of the virus infection. In addition, the probable opsonic activity of C3b attached to viruses, as noted above, and the complement-triggered release of lysosomal enzymes and other toxic constituents from leukocytes would facilitate virus destruction. Further, in a local environment the recently appreciated immunoregulatory functions of the C3 and C5 cleavage products [16] could modulate certain cellular reactions pertaining to antiviral responses.

The systemic symptoms characteristic of virus infections could be in part due to complement activation. Thus, histamine, pyrogens, and other mediators released from cells during complement activation as well as the products of the cyclooxygenase and lipoxygenase pathways triggered by complement-dependent stimulation of arachadonic acid metabolism could account for the myalgias, headaches, and fever characteristic of virus infection.

This review has selectively focused on the complement system and briefly described the effects of complement upon virus infectivity as analyzed in *in vitro* models of virus infection. It is important to appreciate, however, that the complement-dependent processes described here do not function alone but rather synergistically and cooperatively with other humoral and cellular antiviral defense mechanisms. The ultimate understanding of host defense against viral infection requires the detailed analysis of each constituent system and investigations of the interrelationships between them.

Acknowledgments. This is publication no. 2955-IMM from the Department of Immunology, Scripps Clinic and Research Foundation. Studies from the author's laboratory were supported by US PHS Grants CA-14692 and AI-17354. I wish to thank Bonnie Weier for assistance in the preparation of the manuscript.

References

1. Cooper NR (1983) Activation and regulation of the first complement component (C1). Fed Proc 42:134–138
2. Pangburn MK (1983) Activation of complement via the alternative pathway. Fed Proc 42:139–143
3. Cooper NR (1979) Humoral immunity to viruses. *In* Fraenkel-Conrat H, Wagner RR (eds) Comprehensive Virology. Plenum Press, New York p 123
4. Sissons JGP, Oldstone MBA (1980) Antibody-mediated destruction of virus-infected cells. Adv Immunol 29:209–260
5. Nemerow GR, Jensen FC, Cooper NR (1982) Neutralization of Epstein–Barr virus (EBV) by nonimmune human serum: Role of cross-reacting antibody to herpes simplex virus (HSV-1) and complement (C). J Clin Invest 70:1081–1091
6. Beebe DP, Schreiber RD, Cooper NR (1983) Neutralization of influenza virus by normal human sera: Mechanisms involving antibody and complement. J Immunol 130:1317–1322
7. Cooper NR, Jensen FC, Welsh Jr. RM, Oldstone MBA (1976) Lysis of RNA tumor viruses by human serum: Direct antibody independent triggering of the classical complement pathway. J Exp Med 144:970–984
8. Bartholomew RM, Esser AF, Müller-Eberhard HJ (1978) Lysis of oncornaviruses by human serum: Isolation of the viral complement (C1) receptor and identification as p15E. J Exp Med 147:844–853
9. Hirsch RL, Winkelstein JA, Griffin DE (1980) The role of complement in viral infections. J Immunol 124:2507–2510
10. Welsh Jr. RM (1977) Host cell modification of lymphocytic choriomeningitis virus and Newcastle disease virus altering viral inactivation by human complement. J Immunol 118:348–354
11. Mayes JT, Nemerow GR, Cooper NR (1983) Alternative complement (C) pathway (AP) activation by Epstein–Barr virus (EBV) infected normal B lymphocytes. Fed Proc 42:5530
12. Norley SG, Wardley RC (1982) Complement-mediated lysis of African swine fever virus-infected cells. Immunol 46:75–82
13. Sissons JGP, Oldstone MBA, Schreiber RD (1980) Antibody-independent activation of the alternative complement pathway by measles virus-infected cells. Proc Natl Acad Sci USA 77:559–562
14. McConnell I, Klein G, Lint TF, Lachmann PJ (1978) Activation of the alterna-

tive complement pathway by human B lymphoma lines is associated with Epstein–Barr virus transformation of the cells. Eur J Immunol 8:453–458
15. Perrin LH, Joseph BS, Cooper NR, Oldstone MBA (1976) Mechanism of injury of virus infected cells by antiviral antibody and complement: Participation of IgG, Fab'2 and the alternative complement pathway. J Exp Med 143: 1027–1041
16. Weigle WO, Morgan EL, Goodman MG, Chenoweth DE, Hugli TE (1982) Modulation of the immune response by anaphylatoxin in the microenvironment of the interacting cells. Fed Proc 41:3099–3103

CHAPTER 4
Proteolytic Cleavage and Virus Pathogenesis

ANDREAS SCHEID AND PURNELL W. CHOPPIN*

This chapter briefly summarizes instances in which proteolytic cleavage of viral structural proteins has been demonstrated to play a direct role in pathogenesis. It does not attempt to cover the extensive literature on the proteolytic processing of polyproteins, which is an essential step in virus replication, e.g., in picornaviruses or togaviruses, but has not been shown to be a determinant of pathogenesis other than in the broad sense that biosynthesis of viral components is required for viral replication. Emphasis is placed on those viruses for which cleavage by host proteases has been shown to be a major determinant of virulence, i.e., on paramyxoviruses and myxoviruses [1].

Paramyxoviruses

This large group of viruses includes the parainfluenza viruses types 1 to 5, mumps, measles, canine distemper, and Newcastle disease viruses. There are two glycoproteins on the surface of paramyxoviruses: HN, which has hemagglutinating (receptor binding) and neuraminidase activities, and F, which is involved in membrane fusion. The F protein is synthesized as a biologically inactive precursor, Fo, which may be cleaved by a host protease to yield two disulfide-bonded polypeptides, F1 and F2 [2–4]. The F2 polypeptide contains the original N-terminus of Fo, and the F1 polypeptide contains the C-terminus, which is embedded in the viral membrane. The cleavage of Fo generates a new N-terminus on F1 [4]. The F protein has been shown to be involved in penetration of the viral genome into the cell through

---

* The Rockefeller University, New York, New York.

fusion of the viral and cell membranes, and in virus-induced cell fusion and hemolysis [2–5], which also involve membrane fusion.

Several lines of evidence suggest that the new N-terminus generated on the F1 polypeptide by proteolytic cleavage is directly involved in membrane fusion. The amino acid sequence at this new N-terminus is highly conserved among different paramyxoviruses [6–8], suggesting the requirement for a specific amino acid sequence in this region. Furthermore, this region of F1 is highly hydrophobic, which suggested that a hydrophobic interaction could occur between the N-terminus and the target membrane [6–8]. This could result in the F1 polypeptide facilitating fusion by bringing the bilayers of the viral and cell membranes together, with its N-terminus inserted in the cell membrane and its C-terminus in the viral membrane. Further support for this hypothesis was obtained by the finding of a conformational change in the F protein upon cleavage, as shown by a change in the circular dichroism spectrum, and by an increase in the hydrophobicity, as measured by detergent binding [9]. Strong evidence for the role of F1 was obtained by the finding that oligopeptides with sequences that mimic that of the N-terminus of F1 specifically inhibit the membrane-fusing activity of F, as demonstrated by inhibition of virus penetration and virus-induced cell fusion and hemolysis [7,10]. Recent studies have shown that these inhibitors act on the cell membrane, presumably competing with the N-terminus of F1 [10].

The role of the F protein in pathogenesis was first suggested by the finding that some cells lack an appropriate protease to cleave the F protein and thus to activate viral penetration [2,3,5]. Viruses produced by such cells cannot undergo multiple cycle replication, spread in the host, and cause disease. Thus, the host range, tissue tropism, and virulence of the virus are dependent on the susceptibility of the F protein to cleavage by an available host protease. This concept was supported by the isolation of mutants of the parainfluenza virus Sendai virus that require different proteases to activate F and consequently have differing host ranges [5]. The results indicated that in cell culture and the chick embryo the pathogenesis of paramyxoviruses depends on the susceptibility of F to host proteases, and this led us to postulate that this would also be the case in the natural animal host [11]. Such a role for F-protein cleavage in determining viral virulence in the whole animal was promptly demonstrated in Newcastle disease virus in chickens [12].

The finding of proteolytic activation of the F protein has other important biological implications that are listed below. When a paramyxovirus is isolated in a laboratory host, the F protein of the isolated virus must be cleavable by an enzyme present in that host. This may result in the selection of a virus that is not representative of the majority of the virus population in the infected individual. Such a selected variant may not be efficiently cleaved by proteases available in the original natural host. This could provide an explanation for the often observed rapid attenuation for humans by a single or a few passages of the virus in the laboratory. The loss of a specific protease on serial passage of primary monkey kidney cells has explained the failure of established cell lines to support replication of certain paramyxoviruses [13]. In certain virus cell systems, cell death has been shown to be due to the membrane-damaging effects of the fusion protein [14,15]. Lack of antibodies to F has been implicated in the immunopathological response of atypical measles in patients receiving formalin-inactivated measles virus vaccine [16].

Myxoviruses

Myxoviruses (influenza viruses), like paramyxoviruses, possess two surface glycoproteins. Unlike paramyxovirus glycoproteins, the neuraminidase and hemagglutinin functions reside on separate proteins, HA and N. The HA protein serves two functions, receptor binding and penetration. Like the F protein of paramyxoviruses, the HA protein is cleaved by a host protease to yield two disulfide-bonded subunits [17]. This cleavage has no effect on hemagglutination, but it activates the infectivity of the virus [18,19] at the level of viral penetration. The specificity of this cleavage was demonstrated by the fact that trypsin could cleave and activate, and that other proteases like chymotrypsin could cleave at a nearby site without activation of activity [20,19], and after the initial cleavage, removal of one or few amino acids from the C-terminus by a carboxypeptidase also occurs [20]. The similarities between the HA and F proteins with respect to their function and activation by proteolytic cleavage was emphasized previously [8], as illustrated in Fig. 1. Unlike paramyxoviruses, influenza viruses do not induce cell fusion or hemolysis at neutral pH, however it has recently been shown that these activities are exhibited at acid pH [21–24]. These results suggested that both cleavage and acid pH are needed to activate the membrane-fusing activity of the HA protein, whereas cleavage alone of the F protein activates fusion at neutral pH. This was supported by the finding that cleavage and acid pH

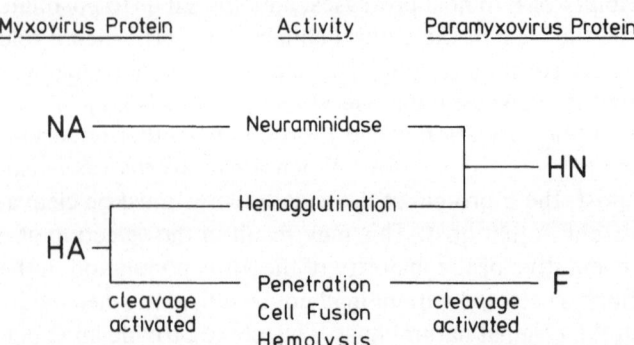

Fig. 1. The membrane glycoproteins of myxoviruses and paramyxoviruses and their biological activities. The functional homologies and their distributions among the proteins of the different viruses are illustrated. (From Scheid *et al.* [8]).

cause a conformational change in the HA protein with exposure of a hydrophobic region [25], a result analogous to the conformational change found previously on cleavage of the paramyxovirus F protein [9]. The analogy is further supported because the oligopeptides that mimic the N-terminus generated by the activating cleavage of HA inhibit the initiation of infection by influenza virus [7].

The above results indicate that, as is the case with the paramyxovirus F protein, cleavage of HA protein is necessary for the initiation of infection and spread in the host. It has also been demonstrated at the level of the whole animal with naturally occurring avian influenza strains, that cleavage of the HA protein is a major determinant of virulence of influenza virus in chickens [26].

Conclusion

The results summarized briefly above have shown that with two major groups of animal viruses, specific proteolytic cleavage of a membrane glycoprotein by an enzyme supplied by the host activates the ability of the virus to initiate infection through fusion of the viral and cell membranes, and consequently to undergo multiple cycle replication, spread in the host, and ultimately cause disease. The outcome of an encounter between the virus and the host is thus dependent on the susceptibility of the viral glycoprotein to cleavage by a protease and the availability of that protease in the host. In the case of paramyxoviruses, it has also been shown in certain virus cell systems that the proteolytic activation of the fusion protein is directly responsible for cell death as a result of membrane damage; the protein has also been implicated in a pathological immune reaction involving spread of infection from cell to cell by membrane fusion in the presence of antibodies to other viral proteins. In addition, selection of virus mutants whose proteins are resistant to cleavage by a host protease may be involved in the attenuation of viruses for their natural host by passage in the laboratory. These findings have clearly demonstrated that specific proteolytic cleavage of viral membrane proteins is a major determinant of the pathogenesis in important virus diseases.

References

1. Choppin PW, Scheid A (1980) The role of viral glycoproteins in adsorption, penetration and pathogenicity of viruses. Rev Inf Diseases 2:40–61
2. Homma M, Ohuchi M (1973) Trypsin action on the growth of Sendai virus in tissue culture cells. III. Structural difference of Sendai virus grown in eggs and in tissue culture cells. J Virol 12:1457–1465
3. Scheid A, Choppin PW (1974) Identification of biological activities of paramyxovirus glycoproteins. Activation of cell fusion, hemolysis and infectivity by proteolytic cleavage of an inactive precursor protein of Sendai virus. Virology 57:475–490

4. Scheid A, Choppin PW (1977) Two disulfide-linked polypeptide chains constitute the active F protein of paramyxoviruses. Virology 80:54–66
5. Scheid A, Choppin PW (1976) Protease activation mutants of Sendai virus: Activation of biological properties by specific proteases. Virology 69:265–277
6. Gething M-J, White JM, Waterfield MD (1978) Purification of the fusion protein of Sendai virus: Analysis of the NH_2-terminal sequence generated during precursor activation. Proc Natl Acad Sci USA 75:2737–2740
7. Richardson CD, Scheid A, Choppin PW (1980) Specific inhibition of paramyxovirus and myxovirus replication by oligopeptides with amino acid sequences similar to those at the N-termini of the F1 or HA2 viral polypeptides. Virology 105:205–222
8. Scheid A, Graves MC, Silver SM, Choppin PW (1978) Studies on the structure and functions of paramyxovirus glycoproteins. *In* Mahy BWJ, Barry RD (eds) Negative Strand Viruses and the Host Cell. Academic Press, London, p 181
9. Hsu M-C, Scheid A, Choppin PW (1981) Activation of the Sendai virus fusion protein (F) involves a conformational change with exposure of a new hydrophobic region. J Biol Chem 256:3357–3363
10. Richardson CD, Choppin PW (1983) Oligopeptides that specifically inhibit membrane fusion by paramyxoviruses: Studies on the site of action. Virology 131: in press
11. Scheid A, Choppin PW (1975) Activation of cell fusion and infectivity by proteolytic cleavage of a Sendai virus glycoprotein. *In* Reich E, Rifkin DB, Shaw E (eds) Proteases and Biological Control. Cold Spring Harbor Laboratory, New York, p 645
12. Nagai Y, Klenk H-D, Rott R (1976) Proteolytic cleavage of viral glycoproteins and its significance for the virulence of Newcastle disease virus. Virology 72:494–508
13. Silver SM, Scheid A, Choppin PW (1978) Loss on serial passage of rhesus monkey kidney cells of proteolytic activity required for Sendai virus activation. Infect Immun 20:235–241
14. Graves, MC, Silver SM, Choppin PW (1978) Measles virus polypeptide synthesis in infected cells. Virology 86:254–263
15. Holmes KV, Choppin PW (1966) On the role of the response of the cell membrane in determining virus virulence. Contrasting effects of the parainfluenza virus SV5 in two cell types. J Exp Med 124:501–520
16. Merz DC, Scheid A, Choppin PW (1980) Importance of antibody to the fusion glycoprotein of paramyxoviruses in the prevention of spread of infection. J Exp Med 151:275–288
17. Lazarowitz SG, Compans RW, Choppin PW (1971) Influenza virus structural and non-structural proteins in infected cells and their plasma membranes. Virology 46:830–843
18. Klenk H-D, Rott R, Orlich M, Bloedorn J (1975) Activation of influenza A viruses by trypsin treatment. Virology 68:426–439
19. Lazarowitz SG, Choppin PW (1975) Enhancement of the infectivity of influenza A and B viruses by proteolytic cleavage of the hemagglutinin polypeptide. Virology 68:440–454
20. Garten W, Bosch FX, Linder D, Rott R, Klenk H-D (1981). Proteolytic activation of the influenza virus hemagglutinin: The structure of the cleavage site and the enzyme involved in cleavage. Virology 115:361–374

21. Huang RTC, Rott R, Klenk H-D (1981) Influenza viruses cause hemolysis and fusion of cell. Virology 110:243–247
22. Lenard J, Miller DK (1981) pH-Dependent hemolysis by influenza, Semliki forest virus, and Sendai virus. Virology 110:479–482
23. Maeda T, Kawasaki K, Ohnishi S-I (1981) Interaction of influenza virus hemagglutinin with target membrane lipids is a key step in virus-induced hemolysis and cell fusion at pH 5.2. Proc Natl Acad Sci USA 78:4133–4137
24. White J, Matlin K, Helenius A (1981) Cell fusion by Semliki forest virus, influenza and vesicular stomatitis virus. J Cell Biol 89:674–679
25. Skehel JJ, Bayley PM, Brown EB, Martin SR, Waterfield MD, White JM, Wilson IA, Wiley DC (1982) Changes in the conformation of influenza virus hemagglutinin at the pH optimum of virus mediated membrane fusion. Proc Natl Acad Sci USA 79:968–972
26. Bosch, FX, Orlich M, Klenk H-D, Rott R (1979) The structure of the hemagglutinin, a determinant for the pathogenicity of influenza viruses. Virology 95:197–207

Immune Control—Humoral

CHAPTER 5
Mechanisms of Virus Neutralization

BENJAMIN MANDEL*

Viral infection elicits an immune response consisting of multiple populations of immunologically differentiated cells (immunocytes) and multiple populations of soluble immunoglobulins (antibodies). Subject to genetic restrictions, immunocytes recognize and destroy virus-infected cells that display unorthodox surface viral antigens, whereas antibodies react with extracellularly disseminated viruses and neutralize their infective capability. Although each of these immunological elements has its own *modus operandi,* they are also interdependent: e.g., cytotoxic "killer" cells depend on antibody as a mediator; antibody is synthesized by B-type immunocytes; B cells are "helped" by T-type immunocytes; under certain conditions B cells are suppressed by T cells. Subsequent discussions will unravel the above grossly oversimplified survey of the immune system.

This chapter focuses on virus–antibody interactions that result in neutralization of viral infectivity. In some instances these events are paradoxically and pathogenetically subverted with consequent manifestations of immune complex disease. This phenomenon is discussed separately.

The Antibody Response

Five main classes of antibody (Ab) are recognized that differ physicochemically and immunologically. Immunoglobulin (Ig) G, IgM, IgA are involved in neutralization, IgE is responsible for allergic manifestations, while no defi-

---

* The Public Health Research Institute of The City of New York, Inc., New York, New York and Long Island University, Brooklyn, New York.

nite role for IgD is known. Each class has been subdivided on the basis of immunologically detectable distinctions in its constituent polypeptide chains [1]. Antibodies are present in vascular fluids, gastrointestinal and urinary tracts, and glandular secretions such as tears, saliva, colostrum, etc. Significantly, only IgG molecules can traverse the placenta thereby endowing fetus and newborn with passive short-term specific immunity. IgA can supplement such passive immunity since it occurs in mother's milk.

The response to immunogenic stimulation varies. With small doses of viral antigen (Ag) administered for a short time, only IgM is produced. Soon after cessation of stimulation, the concentration of IgM in serum diminishes to an undetectable level. Intensive stimulation induces IgM synthesis and soon thereafter IgG and IgA. When stimulation is discontinued, IgM decays as above but IgG persists for variably long periods. Several ramifications are involved in these responses. (1) When IgM alone is produced, the response to a second stimulus is typical of a virgin response. (2) When IgG is produced, the response to restimulation is enhanced, indicating the establishment of "memory" (anamnesis). (3) The binding affinity of IgG molecules increases markedly with increased time after stimulation. The prolonged persistence of IgG synthesis may be due to several circumstances. (1) The initial priming and boosting of T and B lymphocytes results in large clones of these cells already sensitized to the homotypic antigen. (2) In an average environment there is a high probability that exposure to a specific antigen occurs repeatedly thereby rejuvenating the "memory" apparatus. (3) That a long-lasting built-in "memory" mechanism does exist has been shown by the persistence of antibody production in individuals first exposed to an antigen at an early age, then moved to a location where that antigen was not present. A peripheral message obtained from these observations is that an effective vaccine is one that can induce a strong prolonged response of high affinity IgG antibody.

IgA is the third antibody to appear. It has a tendency to form a dimer, and the acquisition of a small additional component (secretory piece) endows it with a predilection for regions that are rich in mucoid secretions, e.g., respiratory and gastrointestinal tracts. The increased proportion of IgA in these regions compared with that of serum suggests local production, possibly by plasma cells underlying the mucosal epithelium. On this basis it would seem reasonable to administer vaccines locally for viruses that multiply in such localized cells. This approach is under study with an influenza virus vaccine introduced by nasal spray. The orally administered poliovirus vaccine may be effective, at least in part, for the same reason, viz., IgA production in the gastrointestinal tract where poliovirus most frequently establishes the initial infection.

In some instances viral infection fails to evoke an immune response because the immune system *per se* (macrophages, lymphocytes, spleen, bone marrow) is the target of infection.

Neutralization of Viruses by Antibody—General Aspects

It was first shown in 1892 that the antibodies induced during a viral infection can neutralize the inciting virus. Subsequent studies of bacterial and animal virus neutralization have yielded much information as well as the realization of some problems [2–6]. Following are some salient aspects of viral neutralization. (1) The stage in the replication cycle that is affected by antibody is an early stage that precedes the start of virus-specific macromolecular synthesis. (2) As immunogens, some viruses possess one class of antigenic determinant, other viruses have several. In the latter case only one may be critical for neutralization. (3) The direct effect of antibody on the virion may be nontraumatic and reversible, or traumatic and therefore irreversible. (4) With virtually all viruses a small variable fraction of the population fails to be neutralized in spite of exposure to excess antibody. (5) In some instances neutralization requires, or is abetted by a third factor. These aspects are discussed below.

The replication of a virus is insensitive to antibody after the release of the viral genome (i.e., uncoating). Hence, destruction of the virus, or inhibition of any of the early stages—attachment, penetration, uncoating—may be the target of antibody action. Ordinarily, destruction is an uncommon event involving only viruses with lipoprotein envelopes that are sensitive to the combined effects of antibody and complement. Attachment to a cell may be prevented when a virion has been exposed to excess antibody causing all, or most, antigenic determinants to be obstructed by antibody. When the ratio of antibody to virion is low, the neutralized virion can attach, indicating that a subsequent stage is aborted. That attachment, even under normal conditions, does not preclude neutralization is shown by allowing virus to attach first, then adding antivirus antibody. With some viruses the probability that an attached virion will penetrate is reduced if antibody is bound to it; rather, there is an increased likelihood that it will elute from the cell. Finally, there are examples of inability of a cell to uncoat, or of aberrant uncoating, when the attached virion is coupled to antibody.

Some viruses contain two or more classes of surface antigens. Of these, one may be critical for neutralization, whereas antibody bound to the other(s) does not lead to neutralization. In the latter case, addition of a third component such as complement, rheumatoid factor, or antiglobulin antibody, by virtue of its cross-linking action can cause neutralization as a result of extensive accretion. Reduction of infectivity after noncritical Ag-Ab interaction is due to a pseudo-neutralizing effect since at equivalent concentrations of virus and antibody, the multivalency of each promotes the formation of latticelike aggregates of infectious virions, each aggregate registering as a single infectious unit. Complexes of virus and antibody that retain infectivity, and can be secondarily neutralized by intervention of a third factor are said to be "sensitized" [7].

The binding of antibody to viral antigen involves relatively weak chemical forces. Neutralized virus can therefore be reactivated by treatments (e.g., pH extremes, high salt concentration, protease, fluorocarbon) that dissociate or rearrange the complex. The ability to restore infectivity varies according to the binding affinity of antibody; the higher the affinity (i.e., late-appearing IgG) the more difficult is reactivation. Viruses with lipoprotein envelopes are reactivable after neutralization unless first exposed to complement, which inflicts irreparable lytic lesions in the envelope.

Sensitization of Viruses by Antibody

Almost without exception, the neutralization of a population of virus particles falls short of completion. Many studies have focused on the nature and reason(s) for the occurrence of this nonneutralized fraction. That nonneutralized virus had, in fact, combined with antibody was shown by its sensitivity to secondary neutralization by an accessory factor such as antiglobulin antibody. Of the three components involved—virus, cell, antibody—each has been implicated. The quality of the antibody is a factor as shown by the decreasing size of the nonneutralized fraction as the affinity of antibody increases. The size of the fraction varies according to cells used as assay host. Finally, under otherwise identical conditions, the same virus cultivated in different cells will yield different levels of nonneutralized virus. It is worthy of note that this laboratory curiosity has its counterpart in the "field"; it has been reported that circulating infectious virus–Ab complexes are characteristic of several infections of animals. The reason for the unsuccessful initial reaction, and the reason for the resultant refractory state have yet to be elucidated.

Recent Studies and Hypotheses

With refinements in the technology of studying viruses, the characteristics of neutralization of animal viruses were reexamined [8] using a nonenveloped and an enveloped virus as models. Kinetic analysis indicated that neutralization conforms to a single-hit phenomenon. Therefore, the implication is that one molecule of antibody is sufficient to neutralize one infectious virion although multiple antigenic sites may be present and accessible. This interpretation was challenged on the grounds of the questionable validity of data obtained by kinetic analysis, possible reversibility of the reaction during subsequent manipulations, and artifacts arising in the course of assay. Furthermore, it seemed unreasonable that a virion with many attachment and antigenic sites could be rendered neutral by one molecule of antibody.

Other studies have subsequently been undertaken to resolve the fundamental nature of viral neutralization. The salient problems are threefold: (1) Is the reaction single- or multi-hit? (2) What is the direct effect of antibody on the virion that deprives it of its infectivity? (3) What is the cause and the nature of the nonneutralized fraction, and what is its role, if any, in the clinical manifestations of viral infections?

Studies on a small nonenveloped virus confirmed that on the basis of kinetic data the reaction is single-hit [9]. These studies also showed that one, or very few molecules of antibody induced a conformational rearrangement in the outer protein shell as revealed by electrophoretic characterization. Although retaining the capability for attaching to a susceptible cell, the outer shell has become refractory to the uncoating mechanism of the cell. Hence, the replication cycle is aborted at the uncoating stage. Evidence for antibody-induced rearrangement of the protein shell was also observed in studies on the accessibility of the constituent polypeptides to external radioisotopic labeling [10]. An alternative hypothesis was proposed in support of a multi-hit mechanism [11]. A virus with a lipoprotein envelope containing multiple copies of critical and noncritical antigens served as the model. The viral functions that were examined were infectivity for different cell strains and hemagglutination. It was proposed that neutralization of a particular function requires a specific distribution of antibodies in accordance with the topological distribution of the respective antigens. Hence, infectivity and hemagglutination may be independently inhibited by the corresponding antibodies. Also, neutralization of infectivity may require different patterns of antibody distribution depending on the particular cell line used as host. For cells that fail to recognize the virus–Ab complex as neutralized, the higher level of stringency for neutralization requires the cooperation of an accessory factor such as anti-gammaglobulin Ab. Finally, the fundamental mechanism of neutralization, according to these findings, is the benign steric blockade of critical areas on the virus surface by correctly distributed molecules of the homologous antibodies. No other effect of antibody on the virion need be considered or invoked.

At the culmination of a series of studies on the neutralization of herpes virus, a hypothesis was proposed encompassing an explanation of both neutralization and sensitization [12]. The interactions of virus with antibodies of different classes and different affinities were examined with reference to the effect of accessory factors and environmental conditions on the results of interaction. Based on the findings it was proposed that neutralization is the final result of a series of alterations in the virion initiated by the binding of antibody. Whether the alterations go to completion or terminate before neutralization depends on the affinity of the antibody and, to some extent, on the environmental conditions. High-affinity IgG initiates and successfully drives the reaction to completion. Low-affinity IgG and early IgM initiate the reaction but it falters to a halt at a stage prior to completion. However, by addition of complement as an accessory substance, or by adjusting temperature and duration of the virus–Ab reaction, it can be prodded to completion. Herpes virus, which was used in these studies, is an enveloped virus. Therefore, antibody binds to antigens that are distributed over the lipoprotein surface of the virion. It was suggested [12] that the conformational changes induced in the antigens by antibody are transmitted to internal proteins by cooperative transitional modifications in the antigen molecules that traverse the envelope and associate with internal proteins, possibly those of the genomic shell.

Cross-Linking Laboratory Practice with Medical Practice

Control of viral infections can be implemented by three approaches: (1) eradication of the causative agent from the environment, (2) prevention and treatment via chemoprophylactic and chemotherapeutic agents, and (3) the universal application of effective vaccines. The first method, theoretically the best, is neither practical nor feasible, particularly in those situations where nonhuman reservoirs exist. The second is in its infant developmental stage and will undoubtedly occupy an important place in the physician's armamentarium of the future. On the other hand, the use of viral vaccines has already been successful in controlling (e.g., yellow fever, poliomyelitis, measles, mumps, rubella) and possibly eradicating (e.g., smallpox) some viral diseases. Problematically, some vaccines have been of dubious value, or associated with deleterious side effects. Many, if not the majority, of viral diseases are characteristically acute infections that are vulnerable to the action of humoral antibody. For these diseases, the present reasonable and applicable approach to control lies in the development of vaccines of high potency, purity, and safety. Achievement of this goal can be aided by: (1) a more complete understanding of the interrelationships of virus, antibody, and cell, (2) identification of the subviral component that is the crucial immunogen (The use of monoclonal antibody as a probe for this purpose has already been described.), and (3) the use of such subviral immunogens as the vaccine *per se*. It has already been shown that such an approach is feasible, particularly when used in conjunction with modern methods of protein extraction, purification, synthesis, and probably very soon, mass production through the application of recombinant DNA methods for dissecting out and cloning the essential genetic element(s).

References

1. Spiegelberg HL (1974) Biological activities of immuno-globulins of different classes and subclasses. *In* Advances in Immunology, Vol 19. Academic Press, New York, pp 259–294
2. Daniels CA (1975) Mechanisms of viral neutralization. *In* Notkins AL (ed) Viral Immunology and Immunopathology. Academic Press, New York, p 79
3. Fazekas de St. Groth S (1962) The neutralization of viruses. *In* Advances in Virus Research, Vol 9. Academic Press, New York, p 1
4. Mandel B (1978) Neutralization of animal viruses. *In* Advances in Virus Research, Vol 23. Academic Press, New York, p 205

5. Mandel B (1979) Interactions of viruses with neutralizing antibodies. *In* Fraenkel-Conrat H, Wagner RR (eds) Virus–Host Interactions—Immunity to Viruses (Comprehensive Virology, Vol 15). Plenum Press, New York, p 37
6. Svehag S-E (1968) Formation and dissociation of virus–antibody complexes with special reference to the neutralization process. *In* Progress in Medical Virology, Vol 10. Karger, New York, p 1
7. Majer M (1972) Virus sensitization. *In* Current Topics in Microbiology and Immunology, Vol 58. Springer-Verlag, New York, p 69
8. Dulbecco R, Vogt M, Strickland AGR (1956) A study of the basic aspects of neutralization of two animal viruses, Western equine encephalitis virus and poliomyelitis virus. Virology 2:162–205
9. Mandel B (1976) Neutralization of poliovirus: A hypothesis to explain the mechanism and the one-hit character of the neutralization reaction. Virology 69:500–510
10. Carthew P (1976) The surface nature of a bovine enterovirus, before and after neutralization. J Gen Virol 32:17–23
11. Della-Porta AJ, Westaway EG (1978) A multi-hit model for the neutralization of animal viruses. J Gen Virol 38:1–19
12. Yoshino K, Isono N (1978) Studies on the neutralization of herpes simplex virus. IX. Variance in complement requirement among IgG and IgM from early and late sera under different sensitization conditions. Microbiol Immunol 22:403–414

CHAPTER 6

Antibody- and Complement-Dependent Lysis of Virus-Infected Cells

J. G. P. Sissons*

The individual roles of antibody and complement in neutralizing virus, and their synergism in this function, which are described in Chapter 3 and 5 of this volume, are usually thought to be the main importance of these humoral systems in host defense against viruses. However, once infection is established its resolution depends on elimination of virus from infected cells. Complement has the capacity for cytolysis by virtue of its terminal components, C5–C9, and should thus have the potential for lysing virus-infected cells. This chapter discusses the mechanisms by which antibody and complement lyse virus-infected cells.

Complement

The mechanism of activation of the alternative complement pathway is particularly relevant to the ensuing discussion and is, therefore, worth emphasizing. The important points to grasp are (1) C3, the major protein of the complement system undergoes slow spontaneous hydrolysis in plasma with formation of C3b, (2) C3b is the major biologically active complement fragment; it can form a bimolecular complex with factor B, C3bB. Upon cleavage of factor B by factor D, the C3-cleaving enzyme of the alternative pathway, C3bBb, is generated. (3) Properdin can bind to multiple C3b molecules in the $C3b_{(n)}Bb$ complex, and retard the decay of the labile active site on the Bb fragment. (4) This powerful positive feedback loop is normally regulated by factors H and I, which inactivate C3b in the fluid phase or when it is deposited on the surface of nonactivating particles. (5) The reason why

---

* Royal Postgraduate Medical School, London, England.

certain particulate agents "activate" the alternative pathway is because C3b deposited on their surface is relatively protected from factors H and I. Assembly of C3bBb can thus proceed on their surface removed from the regulatory effects of the inactivators of C3b.

Hence what distinguishes an activator of the alternative complement pathway is some particular intrinsic surface property it possesses, rather than any specific recognition of it by the complement proteins themselves. However, the precise structures on activators of the alternative pathway that afford deposited C3b protection from its inactivators in plasma have not been identified with any certainty. In view of the above mechanism of activation, the alternative complement pathway is now viewed as a nonspecific mechanism of host defense, which is capable of discriminating against foreign and microbial surfaces in the absence of specific antibody [1,2].

Lysis of Virus-infected Cells by Serum

It has been shown convincingly that cells of human origin infected with virus can be lysed by specific antiviral antibody and fresh human serum as a source of complement. This applies to cells infected with a variety of viruses including herpes simplex virus 1 and 2, influenza, parainfluenza, mumps, and measles viruses. It was initially rather surprising when an examination of which pathway of complement activation was required showed that virus-infected cells were lysed by specific antibody in C2- and C4-depleted serum, but not in factor B-depleted serum. Consistent with the lack of requirement for the classic pathway for lysis of virus-infected cells was the finding that F(ab)$_2$ fragments of antiviral antibody were as effective as the whole IgG in inducing lysis. F(ab)$_2$ fragments cannot bind C1q and thus cannot activate the classic pathway [3,4].

This apparent dependence on antibody and the alternative pathway of complement activation (as assessed by the requirement for factor B but not C2 or C4) for lysis of virus-infected cells seemed paradoxical, when antibody is known to activate the *classic* pathway of complement activation by binding C1q and activating C1. However, as the phenomenon appeared to be a fairly general one, applying to cells infected with a number of different RNA and DNA viruses, it seemed worthy of further study.

Lysis of Virus-infected Cells by Antibody and Purified Complement Components

Once the constituent proteins of the alternative complement pathway had been isolated and their mechanism of interaction defined, it became possible to assemble the alternative pathway of complement activation *in vitro*. A mixture composed of C3, P, and factors B, D, I, and H, all at physiologic concentration, could reproduce the functions of the pathway in serum. When the five proteins of the membrane-attack complex were added (C5,C6,C7,C8, and C9) to the six proteins of the alternative pathway, the system was capable of producing cytolysis of conventional alternative pathway activators such as rabbit erythrocytes [5,6].

Taking measles-virus-infected Hela cells as a model, it was shown that they could be lysed by this "isolated cytolytic alternative pathway" with an efficiency comparable to whole serum, but again lysis only occurred if IgG antiviral antibody was present. Additionally if P was omitted from the purified proteins, no lysis occurred. The amount of IgG required on the cell surface to produce lysis was high, approximately 5×10^6 molecules bound per cell to produce 50% lysis. An equivalent amount of F(ab)$_2$ bound per cell was comparably effective whereas Fab' fragments were incapable of producing lysis [7]. These experiments conclusively excluded a necessary role for the classic pathway in the antibody- and complement-dependent lysis of virus-infected cells, and established that the alternative pathway alone was sufficient. However, the role of antibody in the system remained unclear.

Antibody-independent Activation of the Alternative Pathway by Virus-Infected Cells

IgG was not known to be required for activation of the alternative pathway in other systems, and altered cell surfaces were themselves capable of activating the pathway as described above. Hence, the possibility arose that virus-infected cells might themselves activate the alternative pathway directly. Again using measles virus as a model, it was shown that Hela cells infected with measles virus, in the absence of antibody would activate the alternative pathway, as assessed by uptake of ^{125}I-C3b from the isolated alternative pathway composed of its six purified proteins. Despite progressive time-dependent deposition of C3b on the surface of the infected cells, they were not lysed in the cytolytic alternative pathway unless IgG was present. An examination of the kinetics of C3b deposition on the surface of measles-virus-infected cells showed an obvious difference in the absence and presence of IgG. In the presence of IgG there was accelerated uptake of C3b, although the final amount of radiolabel deposited on the cell was similar at 90 minutes in the presence and absence of antibody [8] (see Fig. 1).

In what may be a parallel phenomenon it has been shown that EBV-transformed lymphoblastoid cell lines activate the alternative pathway in serum, but are only lysed very slowly (over 12–24 hours); this is an antibody-independent process. The ability of the various lymphoblastoid cell lines to activate the alternative pathway correlated with the presence of the EBV nuclear antigen (EBNA), suggesting that a virus-induced protein or other transformation marker may be responsible for alternative pathway activation by these transformed cells [9]. This may well be a rather different

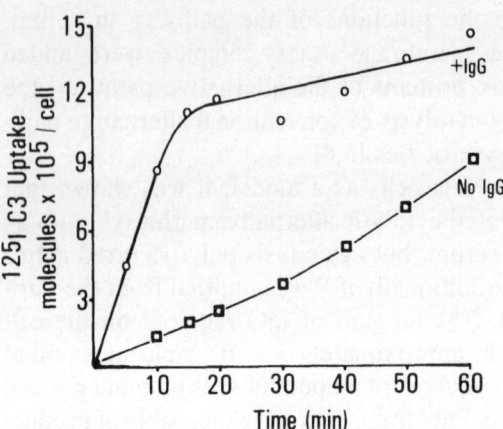

Fig. 1. Kinetics of radiolabeled C3b uptake from the isolated alternative pathway of complement activation (C3, factors B, D, I, H, and properdin) onto measles-virus-infected Hela cells. Virus-infected cells alone (□) activate the pathway, but there is enhanced early uptake of C3b in the presence of cellbound antiviral IgG (○).

system in that this is a transforming infection rather than a lytic one as in the other systems.

There are also various reports of virus-infected cells activating the alternative pathway, or being lysed by the classic pathway (reviewed in [10]) in heterologous serum. It is difficult to know how to interpret such observations because heterologous sera may contain natural antibody to cell membrane determinants, which can result in classic pathway-mediated lysis. There are also some species differences in the ability of the alternative complement pathway to recognize a given activating particle. However, there may be real species differences. For instance pig cells infected with African swine fever virus are lysed by pig antibody and complement by the classic pathway in a homologous system [11].

Mechanisms Involved

The above evidence shows quite clearly that at least for homologous human systems, virus-infected cells can be lysed by antiviral antibody and the alternative pathway of complement activation in serum. What are the likely mechanisms involved in this process?

The experiments with measles virus show that the virus-infected cell can activate the alternative pathway independently of antibody; this suggests that the main role of antibody must, therefore, be to facilitate actual lysis of the cell. It should be made clear that the classic pathway of complement is *activated* by antibody-coated virus-infected cells as shown by the fact that C4b deposition occurs on them in whole or factor B-depleted serum [4]. However, in the absence of an intact alternative pathway no lysis ensues, indicating that the classic pathway is by itself incapable of producing lysis of the virus-infected cell. The structure on the virus-infected cell membrane that "protects" deposited C3b from factor H may be the viral glycoproteins themselves. For instance, there is some evidence from the measles system that C3b deposition on virus-infected cells occurs in very localized sites on the cell membrane adjacent to the viral glycoprotein spikes, as judged by immunoelectron microscopy [12,13]. However a definite answer awaits definition of the structures involved in protecting C3b.

As it is clear that IgG is not required for activation of the alternative pathway by virus-infected cells in human serum, the role of IgG in inducing lysis needs additional explanation. Some experiments on the mechanism of lysis of one EBNA positive lymphoblastoid line may be relevant here. It was mentioned that these cell lines activate the alternative pathway in serum, but only lyse very slowly. In fact it has been shown that deposition of C9 and leakage of intracellular radiolabeled rubidium (used as an indicator of membrane damage sufficient to cause leakage of intracellular cations) both occur simultaneous with C3b deposition, although the cell is not lysed. However, if cell protein synthesis is inhibited with cycloheximide, then cell death occurs concurrent with C9 deposition [14]. This suggests that alternative pathway activation by a relatively weak activator (such as the EBNA positive or virus-infected cell) produces insufficient complement membrane lesions to overwhelm the membrane-repair mechanism of the nucleated cell. It seems likely that the role of IgG in producing lysis of the virus-infected cell may be related to its ability to enhance C3b deposition on the cell surface, which in turn generates more sites for the cleavage of C5 and subsequent assembly and insertion of the membrane-attack complex, C5b-9. It should be noted that IgG antibody can also enhance C3b deposition on other alternative pathway activators (such as zymosan or rabbit erythrocytes), but the mechanism of this effect is again unclear [15,16]. Two possible explanations for the phenomenon are (1) that IgG can directly bind C3b in some way by virtue of a binding site on the IgG molecule, or (2) that the action of IgG in increasing C3b deposition results from redistribution or patching of viral glycoproteins on the cell surface. Such redistribution might also facilitate the binding of P, which is dependent on close spatial association of multiple C3b molecules for its solid-phase binding. At present no hard evidence exists to distinguish clearly between these two or other possibilities. A schematic summary of the possible events is shown in Fig. 2.

Significance *in vivo*

Experiments in mice have shown that depletion of C3 *in vivo* using cobra venom results in more severe influenza or Sindbis virus infection. Congenitally C5-deficient mice develop more severe influenza virus infection [17,18]. However there is, as yet, no very obvious way of determining whether any protective effect of complement results from its effect on virus in the fluid phase, or on virus-infected cells. On the other hand, there is evidence to

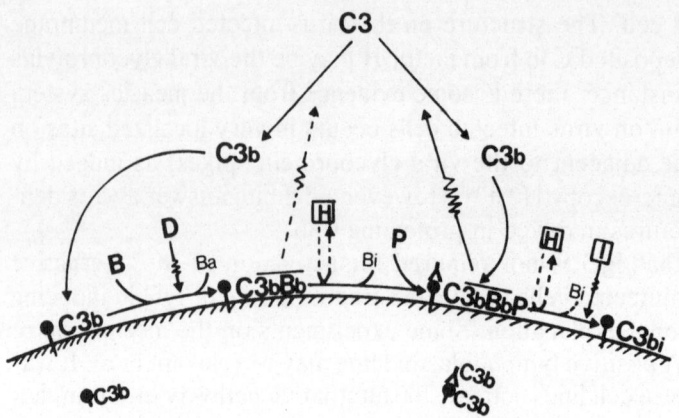

Fig. 2. A schematic summary of the events that may be involved in the activation of the alternative complement pathway by virus-infected cells. C3b deposited from the fluid phase onto the plasma membrane adjacent to viral glycoprotein (♀) is relatively protected from factor H. This allows assembly of C3bBb. Binding of IgG (λ) to the glycoprotein facilitates further C3b deposition and binding of P. ⟶⟶ indicates the production of an enzymatic cleavage.

suggest that complement may contribute to the immunopathology of virus infections by damaging virus-infected cells [19].

The genetic complement-deficiency states in humans have been reviewed elsewhere [2,20]. They provide no conclusive evidence for an undue susceptibility to virus infection in the absence of any one of the classic pathway components C1q–C9. Most patients with C3 deficiency exhibit recurrent bacterial infection, while patients with deficiency of a component of the membrane-attack pathway (C5–C9) are prone to chronic or recurrent Neisserial infections. Patients with C1q, C4, or C2 deficiency are prone to immune complex disease, which probably reflects the physiological role of the classic pathway in the normal handling and disposal of immune complexes. The only deficiency of an alternative pathway protein (other than C3 itself) so far described is of properdin—one family was prone to Neisserial infections.

Thus the role of complement in virus infections in humans remains an open question. However, the activation of the alternative pathway by cells infected with viruses certainly accords with it being an important humoral system of natural immunity, capable of activation by a range of different microorganisms.

References
1. Müller-Eberhard HJ, Schreiber RD (1980) The alternative pathway of complement activation. Adv Immunol 29:285–310
2. Lachman PJ, Peters DK (1982) Complement. *In* Lachman PJ, Peters DJ (eds) Clinical Aspects of Immunology. Blackwell Scientific, Boston

3. Joseph BS, Cooper NR, Oldstone MBA (1975) Immunologic injury of cultured cells infected with measles virus. I. Role of IgG antibody and the alternative complement pathway J Exp Med 141:761–774
4. Perrin LH, Joseph BS, Cooper NR, Oldstone MBA (1976) Mechanism of injury of virus infected cells by antiviral antibody and complement: Participation of IgG,F(ab)$_2$ and the alternative complement pathway. J Exp Med 143:1027–1038
5. Schreiber RD, Pangburn MK, Lesavre PH, Müller-Eberhard HJ (1978) Initiation of the alternative pathway of complement: Recognition of activators by bound C3b and assembly of the entire pathway from six isolated proteins. Proc Natl Acad Sci USA 75:3948
6. Schreiber RD, Müller-Eberhard HJ (1978) Assembly of the cytolytic alternative pathway of complement from eleven isolated plasma proteins. J Exp Med 148:1722
7. Sissons JGP, Schreiber RD, Perrin LH, Cooper NR, Oldstone MBA, Müller-Eberhard HJ (1977) Lysis of measles virus infected cells by the purified cytolytic alternative complement pathway and antibody. J Exp Med 150:445–454
8. Sissons JGP, Oldstone MBA, Schreiber RD (1980) Antibody independent activation of the alternative complement pathway by measles virus infected cells. Proc Natl Acad Sci USA 77:559–562
9. McConnell I, Klein, G, Lint TF, Lachman PJ (1980) Activation of the alternative complement pathway by human B cell lymphoma lines is associated with Epstein–Barr virus transformation of the cells. Eur J Immunol 8:453–458
10. Sissons JGP, Oldstone MBA (1980) The antibody mediated destruction of virus infected cells. Adv Immunol 29:209–260
11. Norley SG, Wardley RC (1982) Complement mediated lysis of African swine fever virus infected cells. Immunology 46:75
12. Oldstone MBA, Lampert PW (1979) Antibody mediated complement dependent lysis of virus infected cells. Springer Semin Immunopathol 2:261
13. Cooper NR, Oldstone MBA (1983) Virus infected cells, IgG, and the alternative complement pathway. Immunol Today 4:107
14. Schreiber RD, Pangburn MK, Medicus RG, Müller-Eberhard HJ Raji cell injury and subsequent lysis by the purified cytolytic alternative complement pathway. Clin Immunol Immunopathol 15:384–396
15. Schenkein HA, Ruddy S (1981) The role of immunoglobulins in alternative complement pathway activation by zymosan. II. J Immunol 126:11
16. Moore FD, Fearon DT, Austen KF (1981) IgG on mouse erythrocytes augments activation of the human alternative complement pathway by enhancing deposition of C3b. J Immunol 126:1805
17. Hicks JT, Ennis FA, Kim E, Verbonitz M (1978) The importance of an intact complement pathway in recovery from a primary viral infection. Influenze in decomplemented and in C5-deficient mice. J Immunol 121:1437–1445
18. Hirsch RL, The complement system—its importance in the host response to viral infection. Microbiol Rev 46:71
19. Oldstone MBA, Dixon FJ (1971) Acute viral infection: Tissue injury mediated by antiviral antibody through a complement effector system. J Immunol 107:1274–1280
20. Lachmann PJ, Rosen FS (1979) Genetic defects of complement in man. Springer Semin Immunopathol 1:339–353

Immune Control—Cellular

CHAPTER 7

The Inflammatory Response to Acute Viral Infections

DIANE E. GRIFFIN*

The inflammatory response to viral infections is characteristically composed of mononuclear cells. The full development of this response is an immunologically specific event that is dependent on the presence of sensitized T lymphocytes [1-4]. These mononuclear inflammatory cells represent a variety of functional cell types indistinguishable, for the most part, on the basis of conventional morphology. Critical studies of the inflammatory response have, to date, been limited almost exclusively to examination of mice with infections of the central nervous system, lung, or liver with enveloped viruses. Under these circumstances, inflammatory infiltrates at various stages of development are known to consist of T lymphocytes of the helper/delayed-type hypersensitivity (Lyt-1) and cytotoxic/suppressor (Lyt-2) phenotypes, B lymphocytes at various stages of differentiation, natural killer (NK) cells, and monocyte/macrophages. Infections of other organs and/or of other species with different, particularly nonenveloped, viruses may generate inflammatory patterns different from those reviewed here.

T Lymphocytes

T lymphocytes are generally considered to be the cells providing the immunologic specificity for the subsequent full development of the inflammatory response. These conclusions are based on studies of virus-induced inflammation in athymic mice [5,6] and reconstitution of the inflammatory response in immunosuppressed mice [1-4]. The earliest morphologically detectable cellular infiltration usually occurs coincident with the onset of the

---

* The Johns Hopkins University School of Medicine, Baltimore, Maryland.

immune response [1,3,4]. T cells are, therefore, likely to be among the earliest inflammatory cells from the peripheral blood to arrive at the site of virus replication. The exact mechanisms by which circulating antigen-reactive T cells recognize the tissue site of viral replication and "know" to stop at that site is unclear.

Two possible mechanisms seem likely to be relevant: (1) Recirculating T lymphocytes constantly monitor various tissues and, when an antigen-reactive cell comes in contact with a relevant antigen, it remains in that location and initiates the inflammatory process. (2) Factors "attractive" to inflammatory cells are produced by the infected tissue and encourage entry of circulating lymphocytes from the periphery. Some of the cells attracted will be antigenically relevant and capable of responding specifically to the presence of viral antigen. Both mechanisms may contribute to the initiation of inflammation.

Some leukocyte chemotactic factors are generated by lytically infected cells either directly or through complement activation [7,8]. Tissue macrophages that have engulfed locally produced virus also produce chemotactic factors [9]. These factors are chemotactic for polymorphonuclear as well as mononuclear leukocytes [7-9] and do not account for the mononuclear character of the fully developed response. Such factors may, however, help to attract the first cells to a site of virus replication. In the earliest stages of inflammation, lymphocytes attach to capillary endothelial cells prior to entering the infected tissue [10] suggesting that this vascular lining cell may play a crucial role in identifying a site of virus replication to circulating cells. Some viruses replicate in capillary endothelial cells, but it is now recognized that endothelial cells, as well as tissue macrophages, may function as antigen-presenting cells [11] and may, therefore, serve as a site of immunologic recognition in the absence of virus replication.

It is known that the interaction of T lymphocytes with virus-infected tissue to initiate the inflammatory response requires compatibility at the major histocompatibility (MHC) region between the lymphocyte and the infected cell. This fact suggests that the T cell must recognize not only viral antigen, but viral antigen on infected cells [12,13]. Virus-specific T cells may be of either the helper/delayed-type hypersensitivity (Lyt-1) or suppressor/cytotoxic (Lyt-2) phenotype, both of which can serve as effector cells in viral infections. Both of these cell types have been identified either functionally [14-16] or by surface markers [17] in inflammatory infiltrates (Fig. 1). Cytotoxic Lyt-2 (Tc) cells require compatibility of the effector cell with the transplantation (class I, H-2) region antigens of the MHC for lysis of the virus-infected cell to occur [18] and for accumulation at or near the site of tissue infection [12,13]. These cells lyse infected cells expressing viral antigen on the surface and may serve *in vivo* to reduce the number of cells producing virus since in some infections, the presence of these cells correlates with virus clearance [19,20].

Quantitative studies of togavirus encephalitis indicate that Lyt-1 cells pre-

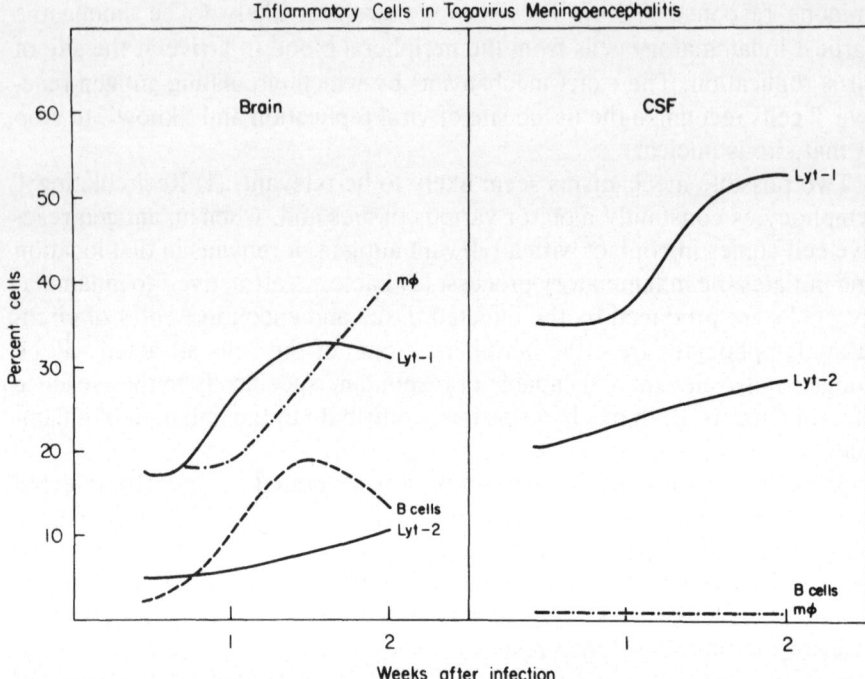

Fig. 1. Schematic diagram of the entry of the various mononuclear cell types into the central nervous system inflammatory lesions induced by infection with Sindbis virus.

dominate at all stages of the inflammatory response, both in the brain and the cerebrospinal fluid ([17], Fig. 1). These delayed-type hypersensitivity Lyt-1 (T_D) cells respond to the stimulation of cell-associated viral antigen by producing a variety of lymphokines: interferon, mononuclear chemotactic factor, lymphotoxin, macrophage inhibitory factor, macrophage activating factor, etc. [21]. Many of these lymphokines are focused on marshaling and directing the functions and activities of monocytes. In most infections, interaction with virus-infected tissue appears to require compatibility of the T_D cell with the immune response (class II, Ia) region antigens of the MHC [16,22] although some investigators have reported that T_D cells are also restricted at the transplantation (H-2) region [23,24]. Difficulty with measurement of the delayed-type hypersensitivity response [25] may account for some of this confusion. The requirement for recognition of viral antigen in the context of self antigen (either transplantation or immune response regions of the MHC) implies that these effector T cells cannot respond to free infectious virus in tissue or to virus circulating in the blood stream, but rather respond to virus infection in the tissue, at the site of virus replication. The amount of transplantation antigen present on the surface of a normal cell varies among cell types and among tissues [26]. Many cells do not express immune response region antigens. Therefore, the cells in which a virus replicates may determine both the quality and quantity of the inflammatory response elicited. In fact, during the same infection, different inflammatory compartments may show different patterns of cellular infiltration. For instance, the CSF of mice with togavirus-induced encephalitis contains almost exclusively T cells, while the perivascular cuffs in the brain parenchyma contain significant numbers of B cells and monocytes in addition to T cells ([17], Fig. 1).

It is not known whether a majority or a minority of the T lymphocytes present in an inflammatory focus are specific for the virus in question. However, when this question was addressed for vaccinia meningitis, only 5%–10% of the Lyt-2 cells were estimated to be vaccinia specific [27]. In the same infection, some inflammatory cells in previously primed mice could be shown to have specificity for influenza. It therefore seems likely that T lymphocytes are also recruited nonspecifically into the developing lesions.

Monocytes

Parenchymal inflammatory infiltrates usually contain a large number of phagocytic mononuclear cells which come from a rapidly dividing peripheral pool [28–30]. These cells are assumed to be peripheral blood monocytes that have been recruited to the local area of virus replication by chemotactic factors produced by the locally stimulated virus-sensitized T lymphocytes. Among the lymphokines produced by Lyt-1 cells in response to specific antigenic stimulation is a factor preferentially chemotactic for mononuclear cells [9,21]. In some tissues, such as the brain and skin, the entry of monocytes into the tissues through capillary endothelia may be impeded by the presence of tight junctions. Locally produced lymphokines also stimulate mast cells to release monoamines, which can open endothelial tight junctions and facilitate the full development of the inflammatory response [31,32]. Monocytes attracted to the site are probably induced to stay and serve important effector functions by the presence of other locally produced lymphokines and/or immunoglobulins [21].

B Lymphocytes

B lymphocytes represent a third cell population, which comprises only a very small proportion of the mononuclear cells early in the viral inflammatory process but increases in later phases [17,33,34]. The factors governing the entry of these cells and the stage of functional differentiation at which they enter have been exposed to only limited investigation in viral infections. It is, however, likely that B cells, like monocytes, also enter local inflammatory reactions in response to lymphokines produced by T cells [21]. This mechanism would be expected to recruit irrelevant as well as relevant B cells. B lymphocytes appear first in the perivascular cuff and then migrate into the tissue [17,34]. Differentiation appears to occur *in situ* resulting in the presence of plasma cells secreting IgG and IgA in the recovery phase of acute infections [34] and in chronic infections [35]. Virus-specific antibody of local origin can be detected in animals recovering from acute infections [36]. Memory B cells also persist in previously infected tissues of recovered animals and can be demonstrated by transfer to naive animals [37].

Natural Killer Cells

Other mononuclear cell subtypes, most notably natural killer cells, are known to participate in the response to virus infections [38]. The location of these cells in tissue and mechanisms of recruitment are largely unknown. In vaccinia meningitis they were most easily demonstrated, using functional assays, in the CSF of young mice early in the inflammatory process [39]. Since it is known that NK cells are induced and activated by interferon [40] it is logical that they may be involved in the very earliest stages of the local inflammatory response.

Summary

The generation of the inflammatory response to viral infections involves a complex series of events that is, as yet, incompletely understood. T lymphocytes appear to be central, both as primary effector cells and for their ability to recruit and facilitate the entry of other mononuclear cells, primarily monocytes and B lymphocytes, into the inflammatory focus.

Acknowledgments. Work from this laboratory was supported in part by a research grant from the National Multiple Sclerosis Society and Grant NS-18596 from the National Institutes of Health.

References

1. McFarland HF, Griffin DE, Johnson RT (1972) Specificity of the inflammatory response in viral encephalitis. I. Adoptive immunization of immunosuppressed mice infected with Sindbis virus. J Exp Med 136:216–226
2. Doherty PC, Zinkernagel RM (1974) T-cell-mediated immunopathology in viral infections. Transplant Rev 19:89–120
3. Blanden RV (1974) T cell response to viral and bacterial infection. Transplant Rev 19:56–88
4. Berger ML (1982) Immunologic requirements for the adoptive transfer of ectromelia virus meningitis. J Neuropath Exp Neurol 41:18–33
5. Doherty PC (1973) Quantitative studies of the inflammatory process in fatal viral meningoencephalitis. Am J Pathol 73:607–621
6. Hirsch RL, Griffin DE (1979) The pathogenesis of Sindbis virus infection in athymic nude mice. J Immunol 123:1215–1218

7. Brier AM, Snyderman R, Mergenhagen SE, Notkins AL (1970) Inflammation and herpes simplex virus: Release of a chemotaxis generating factor from infected cells. Science 170:1104–1106
8. Ward PA, Cohen S, Flanagan TD (1972) Leukotactic factors elaborated by virus-infected tissues. J Exp Med 135:1095–1103
9. Mokhtarian F, Griffin DE, Hirsch RL (1982) Production of mononuclear chemotactic factors during Sindbis virus infection of mice. Infect Immun 35:965–973
10. Baringer JR, Griffith JF (1970) Experimental herpes simplex encephalitis: Early neuropathologic changes. J Neuropath Exp Neurol 29:89–104
11. Hirschberg H, Hirschberg T, Jaffe E, Thorsby E (1981) Antigen-presenting properties of human vascular endothelial cells: Inhibition by anti-HLA-DR antisera. Scand J Immunol 14:545–553
12. Doherty PC, Dunlop MBC, Parish CR, Zinkernagel R (1976) Inflammatory process in murine lymphocytic choriomeningitis is maximal in H-2K or H-2D compatible interactions. J Immunol 117:187–189
13. Berger ML (1982) The role of the major histocompatibility complex in the adoptive transfer of ectromelia virus meningitis. J Neuropathol Exp Neurol 41:34–44
14. Yap KL, Ada GL (1978) Cytotoxic T cells in the lungs of mice infected with an influenza virus. Scand J Immunol 7:73–80
15. Ennis FA, Wells MA, Butchko GM, Albrecht P (1978) Evidence that cytotoxic T cells are part of the host's response to influenza pneumonia. J Exp Med 148:1241–1250
16. Leung KN, Ada GL (1980) Cells mediating delayed-type hypersensitivity in the lungs of mice infected with an influenza A virus. Scand J Immunol 12:393–400
17. Moench TR, Griffin DE (1984) Immunocytochemical identification and quantitation of mononuclear cells in cerebrospinal fluid, meninges, and brain during acute viral encephalitis. J Exp Med 159:77–88
18. Zinkernagel RM, Doherty PC (1979) MHC-restricted cytotoxic T cells: Studies on the biological role of polymorphic major transplantation antigens determining T-cell restriction-specificity, function, and responsiveness. Adv Immunol 27:51–171
19. Zinkernagel RM, Welsh RM (1976) H-2 compatibility requirement for virus-specific T cell-mediated effector functions in vivo. I. Specificity of T cells conferring antiviral protection against lymphocytic choriomeningitis virus is associated with H-2K and H-2D. J Immunol 117:1495–1502
20. Yap KL, Ada GL, McKenzie IFC (1978) Transfer of specific cytotoxic T lymphocytes protects mice inoculated with influenza virus. Nature 273:238
21. Oppenheim JJ (1981) Lymphokines. In Oppenheim JJ, Rosenstreich DL, Potter M (eds), Cellular Immunity and Inflammation. Elsevier, New York, p 259
22. Nash AA, Phelan J, Wildy P (1981) Cell-mediated immunity in herpes simplex virus-infected mice: H-2 mapping of the delayed-type hypersensitivity response and the antiviral T cell response. J Immunol 126:1260–1262
23. Zinkernagel RM (1976) H-2 restriction of virus-specific T cell mediated effector function in vivo. II. Adoptive transfer of delayed-type hypersensitivity to murine lymphocytic choriomeningitis virus is restricted by the K and D region of H-2. J Exp Med 144:766–787
24. Leung KN, Ada GL (1980) Two T-cell populations mediate delayed-type hypersensitivity to murine influenza virus infection. Scand J Immunol 12:481–487

25. Liew FY (1982) Regulation of delayed-type hypersensitivity to pathogens and alloantigens. Immunol Today 3:18-23
26. Edidin M (1972) The tissue distribution and cellular location of transplantation antigens. In Kahan BD, Reisfeld RA (eds) Transplantation Antigens. Academic Press, New York, p 125
27. Hurwitz JL, Korngold R, Doherty PC (1983) Specific and nonspecific T cell recruitment in viral meningitis: Possible implications for autoimmunity. Cell Immunol 76:397-401
28. Johnson RT (1971) Inflammatory response to viral infection. In Immunological Disorders of the Nervous System, Vol 69. Res Pub Assoc Res Nerv Ment Dis, p 305
29. Kitamura T (1975) Hematogenous cells in experimental Japanese encephalitis. Acta Neuropathol 32:341-353
30. Blinzinger K, Herrlinger H, Luh S, Anzil AP (1978) Ultrastructural cytochemical demonstration of peroxidase-positive monocyte granules: An additional method for studying the origin of mononuclear cells in encephalitic lesions. Acta Neuropathol 43:55-61
31. Askenase PW, Bursztajn S, Gershon MD, Gershon RK (1980) T cell-dependent mast cell degranulation and release of serotonin in murine delayed-type hypersensitivity. J Exp Med 152:1358-1374
32. Mokhtarian F, Griffin DE (1984) Role of mast cells in virus induced CNS inflammation. Cell. Immunol. (in press)
33. Owens SL, Osebold JW, Zee YC (1981) Dynamics of B-lymphocytes in the lungs of mice exposed to aerosolized virus. Infect Immun 23:231-238
34. Esiri MM (1980) Poliomyelitis: Immunoglobulin-containing cells in the central nervous system in acute and convalescent phases of the human disease. Clin Exp Immunol 40:42-48
35. Esiri MM, Oppenheimer DR, Brownell B, Haire M (1982) Distribution of measles antigen and immunoglobulin-containing cells in the CNS in subacute sclerosing panencephalitis (SSPE) and atypical measles encephalitis. J Neurol Sci 53:29-43
36. Griffin DE (1981) Immunoglobulin in the cerebrospinal fluid: Changes during acute viral encephalitis in mice. J Immunol 126:27-31
37. Gerhard W, Koprowski H (1977) Persistence of virus-specific memory B cells in mice CNS. Nature 266:360-361
38. Leung KN, Ada GL (1981) Induction of natural killer cells during murine influenza virus infection. Immunobiology 160:352-366
39. Doherty PC, Korngold R (1983) Characteristics of poxvirus-induced meningitis: Virus-specific and nonspecific cytotoxic effectors in the inflammatory exudate. Scand J Immunol 17:1-7
40. Herberman RB, Djeu JY, Kay D, Ortaldo JR, Riccardi C, Bonnard GD, Holden HT, Fagnani R, Santoni A, Puccetti P (1979) Natural killer cells: Characteristics and regulation of activity. Immunol Rev 44:43-70

CHAPTER 8
MHC Restriction and Cytotoxic T Lymphocytes

P. C. Doherty* and R. M. Zinkernagel†

Cell-mediated immunity (CMI) is concerned with the elimination of foci of virus growth from tissues [1]. The basic need is to prevent the progressive spread of the infectious process, which would eventually result in compromised organ function and death. Effector thymus-derived lymphocytes (T cells), the operators of CMI, interact with target cells that are supporting virus replication and are thus expressing surface changes seen as nonself. This focusing of virus-immune lymphocytes onto virus-infected cells reflects that T-cell function can only be mediated via recognition of neoantigen (virus) in the context of self major histocompatibility (MHC) determinants [2].

The operational significance of this MHC restriction phenomenon is thus that effector lymphocytes are recruited into sites of virus-induced pathology and not blocked, or enhanced, by recognizing free virus particles in blood or other body fluids [2]. Isolated virions are obviously handled much better by neutralizing antibody, which can be rapidly produced in considerable excess, whereas the number of T cells that are involved in any immune response is a function of the precursor frequency and the doubling time of the lymphocyte effectors. The realization of the existence of MHC restriction thus provides a basis for proposing a "law of conservation of T-cell function."

Any specific host response is likely to involve at least two classes of virus-immune T cells, the cytotoxic effectors and the helper-inducers. Both have been shown to contribute to the development of inflammatory process, but the best evidence to date indicates that the cytotoxic set is principally responsible for eliminating viruses [2–4]. The collaboration of activated mac-

---

* The John Curtin School of Medical Research, Canberra, Australia.
† University of Zurich, Zurich, Switzerland.

rophages may also be essential for the termination of the infectious process because, unless the T cells kill the virus "factories" before new progeny are produced, lysis of the target could simply act to enhance virus release and spread.

Cytotoxic T cells are, at least in the *in vitro* situation, known to lyse virus-infected fibroblasts, tumor cells, or macrophages subsequent to direct cell-to-cell contact. Some cloned, virus-immune cytotoxic T-cell lines also produce γ interferon [5]. The absolute constraint governing such functions is that the T cell must see, both at the priming and effector phases, virus-induced changes presented in the context of a particular class I MHC glycoprotein. The relevant entities are encoded at the *H-2K*, *D*, and *L* loci in the mouse, and *HLA-A* and *B* in humans [2]. They consist of an integral membrane protein, approximately 45,000 molecular weight, which is noncovalently associated with the 12,000 molecular weight polypeptide $\beta 2$ microglobulin. A number of other class I genes have now been identified in the region of the MHC, but no CMI-related function has yet been ascribed to them [6].

The basic role of the MHC glycoprotein is to be recognized in a highly specific way by the T cell, a conclusion that may be reached from a variety of experimental approaches including (most recently) transfection of class I MHC genes into MHC-different target cells [7]. Obviously, the T cell is normally constrained to see virus in association with self-MHC glycoproteins because it is stimulated in the context of self, though T-cell responses to nonself MHC and virus can sometimes be generated subsequent to removal of lymphocyte precursors (alloreactive) specific for the foreign MHC glycoprotein alone [8]. In general, however, it seems that most T-cell precursors are committed to seeing virus presented with self-MHC components, the spectrum of self being learned by exposure to MHC antigens expressed on radiation-resistant thymus cells during ontogeny [2,9].

The need for recognition of class I MHC glycoproteins means that virus-immune cytotoxic T cells can be potentially targeted to, and function in, any tissue site throughout the body. In some organs, such as the brain, the level of expression of class I MHC gene products is generally very low [10]. However, this can be rapidly enhanced by exposure by γ interferon [10,11], which is probably produced by at least a subset of the invading virus-immune T cells [5]. The class II MHC glycoproteins, which target the helper-inducer set, are principally expressed on lymphocytes and macrophages. These cell types are found mainly in lymphoid tissue, where factors produced by the helpers may promote the differentiation and proliferation of both T cells and B cells. However, it is also worth asking how important such events are in inflammatory sites, a question that has received little attention as yet.

The molecular nature of the T-cell recognition event is still not resolved. Debate continues as to whether the receptors on T cells are seeing virus determinants and MHC glycoproteins as distinct entities, or the T cell is specific for some neoantigen reflecting, perhaps, a complex of virus and

MHC glycoproteins or virus-induced changes of the MHC structures themselves. A compromise position is to argue that a single receptor with two "chains" interacts with viral and MHC antigens that are in close proximity. Obviously, this will only be resolved when the molecular nature of the T-cell recognition structure(s) is defined [12]. Recent experiments with T-cell hybridomas from a number of laboratories have led to the identification of a two-chain structure, which is precipitated by monoclonal anti-idiotype reagents [13]. Each chain has a molecular weight of 40,000–44,000, and is probably not encoded by the immunoglobulin genes [14]. We already have a good operational definition of the significance and nature of the MHC-restriction phenomenon, and it seems possible that we may soon also understand its basic chemistry [2,12].

Much of the work on T-cell specificity has, because of the lack of information to date about the actual receptors, been directed at defining antigenicity. As yet, this has failed to provide any definitive answers, and cannot be expected to do so until the nature of the receptor is clear. The precision of the MHC-restriction phenomenon was shown early, using mice expressing mutations in the H-$2K^b$ gene [2,15]. However, when individual clones of virus-immune T cells from the H-$2K^b$ mutants are analyzed, some surprising cross-reacting patterns are found between the various mutant mice [16]. The overall conclusion from this, and a variety of other approaches with nonviral antigens, is that the T cell is probably recognizing conformational determinants on the MHC glycoprotein, rather than just linear sequences of amino acids.

The nature of specificity for virus has also received a great deal of attention. The first major surprise was that most influenza-immune T cells are type specific, discriminating between influenza A and B rather than being restricted to a particular hemagglutinin (HA) or neuraminidase (N) subtype [2]. Similarly unexpected (from serological criteria) cross-reactivity patterns have been shown for vesicular stomatitis virus and the alphaviruses [17]. However a minority of influenza immune T cells do seem to be subtype specific. The latest surprise is that recognition by this latter category of T-cell clones cannot, using recombinant virus technology, be shown to depend on the presence of either the HA or N glycoproteins of the particular virus subtype [18].

Another imperfectly understood phenomenon is the great polymorphism of class I MHC genes encoded at H-$2K$ and H-$2D$, in the mouse, and HLA-A and B in humans. A convenient, but facile, explanation for this is that the polymorphism reflects a need on the part of the species to avoid total nonresponsiveness to any potentially pandemic, pathogenic virus [19]. Clear evidence has been found for virus-specific hypo-, or nonresponsiveness associated with individual class I MHC genes: It now seems abundantly substantiated that the class I MHC glycoproteins are, indeed, the immune-response gene products concerned with regulating the level of virus-immune cytotoxic T-cell responses. Does this mean, as suggested originally [19], that lack of responsiveness reflects failure to form an appropriate association between MHC and viral components in the target cell membrane? Alternatively, is nonresponsiveness determined at the level of the T-cell repertoire, perhaps as a result of deletion of cross-reactive clones that are also specific for self components? Both may be important. It seems obvious that, when considering a situation where potent effector T cells are focused (at least in part) on self components, there must be a close interaction between mechanisms for maintaining self tolerance on the one hand and the development of the T-cell repertoire on the other [20]. The restriction of functional T-cell recognition to a very limited range of self components is, presumably, central to the balance between the avoidance of autoimmunity and the need for immune responsiveness.

The fact that virus-immune T cells may be involved in the induction of immunopathological processes cannot be ignored. Obviously, in any situation where large numbers of virus-infected cells (some of which may still be functioning) are acutely destroyed there is a possibility of severe physiological distress. Evidence that this does occur has, in particular, been documented for lymphocytic choriomeningitis (LCM), influenza, and coxsackieviruses in mice [2]. Even so, the trade-off between tissue destruction and virus clearance generally benefits the host. An exception in humans might be hepatitis B infection, as is LCM in the mouse.

References

1. Blanden RV (1974) T cell response to viral and bacterial antigens. Transplant Rev 19:56–88
2. Zinkernagel RM, Doherty PC (1979) MHC-restricted cytotoxic T cells: Studies on the biological role of polymorphic major transplantation antigens determining T cell restriction-specificity, function and responsiveness. Adv Immunol 27:51–177
3. Ada GL, Leung KN, Ertl H (1981) An analysis of effector T cell generation and function in mice exposed to influenza A or Sendai viruses. Immunol Rev 58:5–24
4. Lin YL, Askonas BA (1981) Biological properties of an influenza A virus specific killer T cell clone: Inhibition of virus replication *in vivo* and induction of delayed-type hypersensitivity reactions. J Exp Med 54:225–234
5. Morris AG, Lin YL, Askonas BA. Immune interferon release when a cloned cytotoxic T-cell line meets its correct influenza-infected target cell. Nature 295:150–152
6. Goodenow RS, McMillon M, Nicholson M, Taylor-Sher B, Eakle K, Hood L (1982) Identification of class I genes of the mouse major histocompatibility complex by DNA-mediated gene transfer. Nature 300:231–237
7. Mellor AL, Golden L, Weiss E, Bullman H, Hurst J, Simpson E, James RFL, Townsend ARM, Taylor PM, Schmidt W, Ferluga J, Leben L, Santamaria M, Atfield G, Festenstein H, Flavell RA (1982) Expression of $H\text{-}2K^b$ histocompatibility antigen in cells transformed with cloned *H-2* genes. Nature 298:529–533
8. Doherty PC, Korngold R, Schwartz DH, Bennink JR (1981) Development and loss of virus-specific thymic competence in bone marrow radiation chimeras and normal mice. Immunol Rev 58:37–72

9. Zinkernagel RM (1982) Selection of restriction specificities of virus-specific cytotoxic T cells in the thymus: No evidence for a crucial role of antigen-presenting cells. J Exp Med 156:1842–1847
10. Wong GHW, Bartlett PJ, Clark-Lewis I, McKimm-Breschkin JL, Schrader JW (1983) Induction of H-2 and Ia antigens on cultured brain cells by an interferon-gamma like molecule. Neurosci Let [supp 11] p 84
11. Wallach D, Fellow M, Ravel M (1982) Preferential effect of gamma-interferon on the synthesis of HLA antigens and their mRNA's in human cells. Nature 299:833–836
12. Jensenius JC, Williams AF (1982) The T lymphocyte antigen receptor—paradigm lost. Nature 300:583–588
13. Mackenzie IFC, Pang T, Blanden RV (1977) The use of *H-2* mutants as models for the study of T cell activation. Immunol Rev 35:181–230
14. Haskins K, Kubo R, White J, Pigeon M, Kappler J, Marrack P (1983) The MHC-restricted antigen receptor T cells. I. Isolation with monoclonal antibody. J Exp Med, in press
15. Marrack P, Kappler J (1982) Use of somatic cell genetics to study chromosomes contributing to antigen plus I recognition by T cell hybridomas. J Exp Med 157:404–418
16. Hurwitz JL, Pan S, Wettstein P, Doherty PC (1983) Cross-reactivity patterns of vaccinia-specific T lymphocytes from H-$2K^b$ mutants. Immunogenetics 17:79–88
17. Zinkernagel RM, Rosenthal S (1981) Experiments and speculation of antiviral specificity of T cells and B cells. Immunol Rev 58:131–155
18. Townsend ARM, Skehel JJ (1982) Influenza A specific cytotoxic T cell clones that do not recognize viral glycoproteins. Nature 300:655–657
19. Doherty PC, Zinkernagel RM (1975) A biological role for the major histocompatibility antigens. Lancet i:1406–1409
20. Doherty PC, Bennink JR (1980) An examination of MHC restriction in the context of a minimal clonal abortion model for self-tolerance. Scand J Immunol 12:271–280

CHAPTER 9

Antibody-Dependent Cellular Cytotoxicity (ADCC) Mediated by Human Killer Lymphocytes (K-Cells)

PETER PERLMANN*

During the past two decades a variety of *in vitro* procedures has been designed to elucidate the mechanisms of cellular immunity. The most thoroughly studied systems deal with the extracellular, nonphagocytic destruction of various types of "target" cells. Three major models of this cell-mediated cytotoxicity can be distinguished. (1) Certain cytolytic T lymphocytes (CTL) may destroy target cells after contact interaction mediated by lymphocyte receptors specific for surface antigens of the targets. These receptors are not conventional immunoglobulins and are synthesized by the cells that carry them. A further important feature of CTL-mediated cytotoxicity is its (major histocompatibility complex) MHC-restriction, i.e., CTL recognize foreign antigens only in association with "self"-MHC antigens [1]. (2) A variety of leukocytes, including polymorphs (PMN), mononuclear phagocytes, and certain lymphocytes may destroy target cells after similar cell-to-cell interactions but with humoral antibodies serving as the major target recognition factors (*antibody-mediated cellular cytotoxicity, ADCC*). Commonly, the effector cells exhibiting ADCC have surface receptors for the Fc-part of immunoglobulin (FcR), enabling them to interact with target cellbound antibody molecules. Since antibody-producing lymphocytes (B cells) are not active in ADCC, the effector cells are dependent on recognition factors (antibodies) produced by other cells. Lymphocytic cells display-

---

* The University of Stockholm, Stockholm, Sweden.

ing ADCC are operationally designated as *K cells* (K = killer), and this review focuses on ADCC mediated by human K cells. (3) Certain lymphoid cells may also kill a large variety of targets spontaneously *in vitro* without previous immunization of the effector cell donor against target antigens. This spontaneous cytotoxicity or *natural killing (NK)* is nonspecific in an immunological sense. Nevertheless, the effector cells (NK cells) frequently exhibit a certain selectivity with regard to which target cells they are able to kill. The molecular basis of the underlying recognition is unknown [2].

In principle, the specificity requirements and the nature of the effector cells involved make it possible to decide whether or not a given cytotoxic reaction is of the ADCC type. In practice this is not always easy. Thus, T-cell activation leads to the generation of both specific CTL and FcR-bearing lymphocytes with K-cell potential [3]. Moreover, K cells and NK cells are largely overlapping cell populations, capable of exerting both antibody-dependent and antibody-independent cytotoxicity [2]. Target cell-recognizing antibodies are often present in an adsorbed form in the lymphocyte samples isolated from a donor or may be released during the *in vitro* assay, thereby giving rise to a cellular cytotoxicity that may be partly or entirely antibody dependent [4,5].

ADCC Assays and Target Cells

K-cell mediated ADCC is most commonly assayed by incubating lymphocytes from nonimmune donors with target cells for several hours in the presence of antitarget cell antibodies. Target cell death is assessed by measuring release of an isotopic marker (usually ^{51}Cr) from the prelabeled targets. These procedures permit precise quantitative analysis of the cytotoxic reaction and its kinetics at the level of the effector cell population [6]. Direct information regarding the number and nature of the effector cells is obtained by analyzing ADCC at the level of the individual effector cells, e.g., in a conjugate assay or an ADCC plaque assay [6,7]. In these, one physically counts the number of K cells present in a lymphocyte preparation. The morphology, surface characteristics and mode of interaction of the cells with the targets are assessed by light or electron microscopy [8].

All target cells expressing surface antigens giving rise to humoral immune responses may be expected to be susceptible to ADCC. This has been shown to be the case for erythrocytes, normal, tumor-transformed, or virus-infected nucleated cells, bacteria, and parasites. However, different types of targets vary profoundly in their susceptibility to lymphocytic as compared with nonlymphocytic effector cells, probably reflecting differences in susceptibility of different targets to the various cytolytic mechanisms utilized by individual effector cell types. The factors determining target cell susceptibility to ADCC also include cell surface density and stability of antigen, cell size, and capacity to repair surface lesions [6].

Effector Cells

The K-cell activity of lymphocyte preparations derived from human spleen or peripheral blood (PBL) is high, whereas that of lymph nodes, tonsils, or thymuses is low. In the peripheral blood of healthy adults, from 5%–10% of the lymphocytes have K-cell potential, but their numbers vary considerably for different donors and are also influenced by factors such as disease, hormone levels, drugs, etc. [6]. Their numbers and activities are also regulated by interferons, lymphokines, and cells producing such factors [2]. The K cells are heterogeneous in regard to surface characteristics and morphology. B cells and the majority of the T cells in PBL are devoid of K-cell activity. Nevertheless, a significant fraction of PBL-derived K cells appear to be FcR-bearing cells of T-cell lineage, called Tγ cells, as revealed by surface marker studies [3] and supported by cloning experiments [9]. Another major K-cell fraction consists of medium-sized lymphocytes with azurophilic granules in their cytoplasm (LGL) [2]. These cells exhibit antibody-independent natural cytotoxicity and carry a monoclonal antibody-defined antigen, designated HNK-1. Dependent on their state of maturation, these cells also carry either T-cell-associated or myeloid-cell-associated surface antigens [10], but the question of the lineage of these cells is unresolved. K cells of LGL type also appear to have a greater cytotoxic capacity than those of Tγ-type [11]. Whether the latter may mature in K cells having the LGL phenotype or represent a different cell lineage is unknown. In addition to these two major K-cell categories in human PBL, several minor ones also exist. It is important to note that the type of K cells most prominent in different systems varies with the target cells used, indicating a process of effector cell selection governed by the fine structure of the target cell surface and the presence of accessory recognition systems on the effector cells [12] as further described below.

Recognition and Fc-Receptor Function

The major recognition factors mediating effector-to-target cell contacts in ADCC are antibodies of the IgG-isotype. Under optimal conditions, as few as approximately 50 antibody molecules/target cell suffice to bring about lysis, indicating that ADCC constitutes a highly efficient cellular defense system. All four human IgG subclasses appear to be capable of inducing K-cell-mediated ADCC, but IgG1 or 3 is more efficient than IgG2 or 4.

The FcRs involved in K-cell ADCC have low avidity for monomeric IgG. Immune complex formation and FcR-clustering are important for cytotoxicity induction. Hence, efficient K-cell interactions with antibody-carrying target cells are not inhibited by physiological concentrations of nonantibody IgG. On the other hand, ADCC is highly susceptible to inhibition by aggregated IgG, soluble immune complexes, cellbound IgG or anti-immunoglobulin reagents. In blood and tissues, the balance between antibodies and these inhibitory factors is decisive in determining the outcome of a K-cell reaction [6]. Sera from patients with cancer, inflammatory diseases, or chronic infections may, therefore, often be inactive in an ADCC assay even when the sera contain relevant antitarget antibodies.

Although lymphocytes have been reported to have FcR for immunoglobulin isotypes other than IgG [13], such antibodies appear not to be capable of inducing K-cell-mediated ADCC. IgM antibodies may either inhibit or enhance IgG-dependent ADCC, depending on their antigen specificity and the balance of isotype concentrations. Enhancement reflects improved contact interaction between the targets and K cells possessing FcR for IgM as well as for IgG. This IgM-dependent amplification significantly decreases the concentration of IgG antibodies required for ADCC induction and may be important in early phases of an immune response when the amount of available IgG antibodies is low [14,15]. Similarly, target cellbound C3 fragments (particularly C3bi) do not induce, but strongly amplify K-cell-mediated ADCC by recruiting effector cells with the appropriate complement receptors [16]. Thus, complement activation in the course of an inflammatory reaction may efficiently contribute to the production of K-cell-mediated tissue lesions. Available evidence suggests that several similar accessory recognition systems have an important regulatory function in ADCC and may be responsible for the phenotypic diversity of K cells, a diversity that greatly enhances the versatility of this cellular defense system.

The existence of such nontriggering, accessory recognition systems in ADCC indicates that the FcR for IgG on K cells is an important triggering device, necessary for initiation of the lytic reaction. This is supported by experiments in which nonantibody IgG can be shown to induce ADCC, provided effector cell–target cell contacts are brought about by accessory recognition [6,12]. However, the nature of the putative triggering reactions is unknown. IgG-binding to certain macrophage-derived FcR has been shown to generate phospholipase A2 activity [17]. In addition, such receptors may also mediate triggering signals by functioning as ligand-dependent ionophores [18]. Similar FcR activities may be implicated in K-cell-mediated ADCC.

Mechanisms

K-cell-mediated target cell lysis proceeds linearly with time but ceases after approximately 4 to 6 hours, apparently owing to saturation and loss of FcR. While the effector cells are not killed in the course of the reaction, one K cell usually kills more than one target cell (recycling), implying that contact interaction between the cells, although necessary for induction, is a temporary event. K cells are considered to kill their targets by energy-dependent mechanisms similar to those involved in CTL- or NK-cell-mediated cytotoxicity. These can be divided into at least three different phases (recognition, lethal hit, and target cell disintegration), differing in their dependency on bivalent cations, cytoskeleton involvement, and direct participation of the effector cells [6,8,19]. It has been suggested that target cell disintegration involves colloid osmotic lysis following membrane damage inflicted by the effector lymphocytes. In contrast, others have reported that lysis is initiated intracellularly without previous damage to the outer target cell membrane [8]. ADCC mediated by monocytes or PMN appears to be linked to a respiratory burst, involving formation of reactive (toxic?) oxygen intermediates [20], but available evidence speaks against such mechanisms in lymphocyte cytotoxicity, including K-cell-mediated ADCC. Studies with various inhibitors and antimetabolites suggest the involvement of stimulation-dependent secretory events that could give rise to local release of hydrolytic enzymes or other mediators from secretory granules [6,8,19]. However, no effector molecules mediating unidirectional target cell lysis in K-cell ADCC have thus far been identified.

ADCC and Virus Infection

Antibodies inducing cell-mediated destruction of virus-infected target cells *in vitro* are common in virus infection. Although antiviral ADCC can be mediated by PMN as well as macrophages, most reports indicate that K cells appear to have an important function in this respect since they require low antibody concentrations and kill their targets rapidly. K-cell-mediated ADCC *in vitro* has been shown to parallel the appearance of antiviral IgG *in vivo* following vaccination with vaccinia or measles virus. PBL isolated from vaccinated or infected patients exhibit a direct virus-specific cytotoxicity reflecting the presence or release of antiviral antibodies within the assay system [21,22].

In spite of much indirect evidence, there is presently little *direct* proof establishing the relative role of ADCC or other cellular defense systems in viral infection. In nonimmune individuals, ADCC- and CTL-mediated cytotoxicity are believed to influence and modify the course of an acute infection by that virus after the onset of specific antiviral immune responses. These pathways may also contribute to the appearance of tissue lesions in persistent infection. In contrast, antibody-independent cellular reactions of the NK type probably constitute an important first line of defense by limiting the spread of an infection soon after onset [21–23]. This is supported by the fact that NK cells are activated by viral infection concomitantly with the induction of interferon production [19]. In addition, certain viral proteins activate NK cells in an interferon-independent manner [24]. However, as indicated above, a major fraction of the NK cells have FcR for IgG and ADCC-effector function. Moreover, both interferons and viral proteins have also been shown to amplify ADCC by substantially reducing the antibody concentrations necessary for induction [12]. Since small amounts of antiviral IgG may be present or appear rapidly after onset, the cellular responses observed in early phases of an infection may frequently reflect both antibody-dependent and antibody-independent reactions [19,21,22]. The relative roles of these different mechanisms in inhibiting virus spread remain to be determined.

References
 1. Zinkernagel RM, Doherty PC (1979) MHC-restricted cytotoxic T-cells: Studies on the biological role of polymorphic major transplantation antigens determining T-restriction, specificity, function and responsiveness. Adv Immunol 27:1–50
 2. Herberman RB (ed) (1982) NK cells and other natural effector cells. Academic Press, New York
 3. Moretta L, Moretta A, Canonica GW, Bacigalupo A, Mingari MC, Cerottini JC (1981) Receptors for immunoglobulins on resting and activated human T cells. Immunol Rev 56:141–161
 4. Perrin LH, Zinkernagel RM, Oldstone MB (1977) Immune response in humans after vaccination with vaccinia virus: Generation of a virus specific cytotoxic activity by human peripheral lymphocytes. J Exp Med 146:949–969
 5. Pape GR, Troye M, Axelsson B, Perlmann P (1979) Simultaneous occurrence of immunoglobulin-dependent and immunoglobulin-independent mechanisms in natural cytotoxicity of human lymphocytes. J Immunol 122:2251–2260
 6. Perlmann P, Cerottini JC (1979) Cytotoxic lymphocytes. In Sela M (ed) The Antigens, Vol 5. Academic Press, New York, p 173
 7. Neville ME, Grimm E, Bonavida B (1980) Frequency determination of K-cells by a single cell cytotoxic assay. J Immunol Methods 36:255–268
 8. Sanderson CJ (1981) The mechanism of lymphocyte-mediated cytotoxicity. Biol Rev 56:153–197
 9. Moretta L, Mingari MC, Sekaly PR, Moretta A, Chapuis B, Cerottini JC (1981) Surface markers of cloned human T cells with various cytolytic activities. J Exp Med 154:569–580
10. Abo T, Miller CA, Gartland GL, Balch CM (1983) Differentiation stages of human natural killer cells in lymphoid tissues from fetal to adult life. J Exp Med 157:273–282
11. Wåhlin B, Perlmann P (1983) Characterization of human K cells at the single cell level by monoclonal antibodies and cell morphology. J Immunol 131:2340–2347.
12. Perlmann P (1982) Associative recognition in ADCC. In Clark WR, Golstein P (eds) Mechanisms of Cell Mediated Cytotoxicity. Plenum Press, New York, p 249
13. Unkeless JC, Fleit H, Mellman IS (1981) Structural aspects and heterogeneity of immunoglobulin Fc receptors. Adv Immunol 31:247–270
14. Perlmann H, Perlmann P, Moretta L, Rönnholm M (1981) Regulation of IgG antibody dependent cellular cytotoxicity in vitro by IgM antibodies. Scand J Immunol 14:47–60
15. Öhlander C, Perlmann H, Perlmann P (1982) Regulation of IgG-IgM interplay by antibody specificity in human K cell mediated cytotoxicity. Scand J Immunol 15:409–417
16. Wåhlin B, Perlmann H, Perlmann P, Schreiber RD, Müller-Eberhard HJ (1983)

C3 receptors on human lymphocyte subsets and recruitment of ADCC effector cells by C3 fragments. J Immunol 130:2831–2836
17. Suzuki T, Saito-Taki T, Sadasivan R, Nitta T (1982) Biochemical signal transmitted by Fcγ receptors: Phospholipase A2 activity of Fcγ2b receptor of murine macrophage cell line P388D1. Proc Natl Acad Sci USA 79:591–595
18. Young JDE, Unkeless JC, Kaback HR, Cohn ZA (1983) Mouse macrophage Fc receptor for IgGγ2b/γ1 in artificial and plasma membrane vesicles functions as a ligand dependent ionophore. Proc Natl Acad Sci USA 80:1636–1640
19. Welsh RM (1983) Natural killer cells and interferon CRC Critical Reviews in Immunology (in press)
20. Conkling P, Klassen DK, Sagone L Jr (1982) Comparison of antibody dependent cytotoxicity mediated by human polymorphonuclear cells, monocytes and alveolar macrophages. Blood 60:1290–1297
21. Sissons JGP, Oldstone MBA (1980) Antibody-mediated destruction of virus-infected cells. Adv Immunol 29:209–260
22. Welsh RM (1981) Natural killer cell mediated immunity during viral infection. *In* Haller O (ed) Natural Resistance to Tumors and Viruses. Springer-Verlag, Berlin, p 83
23. Rager-Zisman B, Bloom BR (1982) Natural killer cells. *In* Resistance to Virus Infected Cells. Springer Semin Immunpathol 4:397–414
24. Casali P, Oldstone MBA (1982) Mechanisms of killing of measles virus-infected cells by human lymphocytes: Interferon associated and unassociated cell mediated cytotoxicity. Cell Immunol 70:330–334

CHAPTER 10
Cloning of Functional T Lymphocytes

JEAN-CHARLES CEROTTINI AND H. ROBSON MACDONALD*

While there is general agreement that T lymphocytes play a central role in the immunological effector mechanisms operative in viral diseases, the diversity of T lymphocytes has made it difficult to unravel the molecular and cellular basis of these mechanisms. This diversity is accounted for by the existence of multiple, distinct lymphocyte subsets that are responsible for a variety of immunological functions. Thus, in addition to cytolytic, helper, and suppressor activities, effector T lymphocytes release soluble, biologically active mediators known as lymphokines. While it is generally accepted that a single effector T cell cannot perform all these functions, there is still much uncertainty as to the number of functionally distinct subsets. A direct approach to this question would be to examine the function(s) of individual effector T cells. Unfortunately, with the exception of cytolytic T lymphocytes (CTL), there are no single cell assays available for functional T cells. Therefore, it has been necessary to turn to the more practical approach of determining the function(s) of clonal progeny of individual effector T cells. Such an approach has been suggested by the recognition that effector T cells, unlike their B cell counterparts (i.e., plasma cells), are not necessarily end cells, but can undergo extensive proliferation *in vitro* in the presence of an adequate source of T-cell growth factor (TCGF), also designated Interleukin 2 (IL-2) (reviewed in [1]).

IL-2, a T-cell product, has been purified from murine and human sources. Human IL-2 is a glycosylated polypeptide of 15,000 molecular weight and is active on both human and murine T cells, whereas murine IL-2, which has a molecular weight of 25,000–30,000, is only active on murine T cells. Binding studies using biosynthetically labeled IL-2 have shown that surface recep-

---

* Ludwig Institute for Cancer Research, Lausanne, Switzerland.

tors for IL-2, although they are not detectable on virgin, resting T cells, are expressed in appreciable amount ($\geq 5{,}000$ sites/cell) upon activation of T lymphocytes with mitogen or antigen [2]. From the limited data available, it appears that IL-2 exerts its effect upon the control of proliferation of activated T cells in a manner similar to polypeptide growth factors such as epidermal growth factor and fibroblast growth factor, namely it allows the passage of dependent cells from the G_1 phase to the S phase of the cell cycle [3].

With the discovery of IL-2, two developments have been made possible: (1) the establishment of long-term cloned lines derived from effector T cells, and (2) the quantitation of effector T cells and their immediate precursors in limiting dilution assays. These topics have recently been the subject of several reviews [4–6]. Although application of these developments to the analysis of T-cell-mediated responses in viral diseases is still limited, studies at the clonal level in a variety of antigenic model systems have already contributed information pertinent to antivirus effector T cells.

Two major protocols for the derivation of T-cell clones have been employed [7]. The first method involves cloning T cells from long-term cultures that have been first enriched for antigen-specific cells by repeated stimulation and then adapted to continuous growth in IL-2-containing medium in the absence of antigen and "filler" cells. There is suggestive evidence that T-cell clones obtained under these conditions are derived from rare variants that are selected for during the culture period in medium containing only IL-2. As all the clones that have been analyzed carry at least some and often many gross karyotypic abnormalities, it is evident that such clones are not necessarily representative of the effector T-cell populations from which they are derived.

An alternative method involves cloning T cells directly after stimulation with antigen either *in vivo* or *in vitro*. The highest efficiency is obtained when cells are cloned in the presence of the original stimulating antigen, filler cells, and IL-2-containing medium. Similar culture conditions are used to maintain these clones in continuous growth. Such clones possess normal karyotype and are functionally stable for long periods of culture. However, for reasons that are not yet clear, some clones cannot be maintained for more than a few months.

The majority of clones reported have been derived by techniques of limiting dilution, in which cells are diluted to low density and plated in microtiter wells. Since the probability of cloning is calculated according to Poisson statistics, it is necessary to prove by dose-response analysis that such statistics apply to the limiting dilution culture system employed. Most often, recloning is necessary to ensure monoclonality. Under optimal conditions, the plating efficiency obtained by limiting dilution in liquid culture may approach 100%. While this high plating efficiency has first been demonstrated for activated T cells, recent reports indicate that virtually every resting T cell of mouse or human origin can undergo extensive clonal expansion in liquid microculture systems containing a suitable T-cell mitogen, filler cells, and IL-2-containing medium [8,9]. Another approach to deriving T-cell clones in liquid culture is micromanipulation of single cells. The advantage of this method is that it allows the derivation of clonal populations in a single step. In contradistinction to cloning in liquid culture, the use of semisolid agar to derive T-cell clones appears to be much less satisfactory.

T-cell clones derived by either of these methods are dependent on repeated exposure to antigen, filler cells, and/or IL-2 for continuous growth. Although some clones have been maintained in culture for several years, there is no report on the spontaneous generation/selection of variants able to proliferate in an autonomous fashion. For this reason, "immortalization" of effector T lymphocytes through cell fusion with a malignant lymphoma line has been developed as an alternate approach for obtaining homogeneous cell populations [10]. Hybrids so obtained grow constitutively and exhibit at least some of the functions attributed to effector T cells.

There are several areas where T-cell clones already have contributed valuable information. First, functional analysis of T-cell clones *in vitro* has provided direct evidence that soluble mediators such as IL-2, gamma interferon (IFNγ), granulocyte-macrophage colony-stimulating factor (GM-CSF), macrophage-activating factor (MAF), B-cell helper factor, and mast-cell growth factor are T-cell products that are secreted upon interaction of effector T cells with the appropriate antigen [11]. Moreover, it is now evident that lymphokine production is not restricted to the noncytolytic subset of effector T cells. Most, if not all, CTL clones can produce IFN, MAF, and GM-CSF. In addition, some of these CTL clones produce detectable amounts of IL-2. Furthermore, while it is evident that several lymphokine activities can be produced by individual noncytolytic and cytolytic T-cell clones, there is considerable heterogeneity among clones in the array of lymphokines they secrete. Such heterogeneity has been used to confirm that some (but not all) lymphokine activities are mediated by different molecules [12]. Thus, it is evident that IL-2, GM-CSF, and MAF activities are mediated by different molecules by the fact that supernatants from some clones were found to contain only one of them. In this context, it is of interest that IFN and MAF activities have not been dissociated in an extensive study of clonal supernatants obtained from a variety of murine T-cell clones with a wide range of antigenic specificity [12]. Based on these findings, it has been suggested that MAF and IFNγ activities are mediated by the same molecules. The recent demonstration [13] that murine IFNγ obtained by recombinant DNA technology is able to render macrophages nonspecifically cytolytic is certainly in line with this hypothesis.

Another example of the usefulness of T-cell clones is the unambiguous demonstration that recognition of alloantigens of the major histocompatibility complex (MHC) can overlap with recognition of nominal antigen in the context of self-MHC molecules. Although such a possibility was inferred from studies at the population level, definitive proof for dual specificity exhibited by single T cells has been provided by the finding that some CTL clones that specifically lysed virus-infected syngeneic target cells also could lyse allogeneic, noninfected target cells. Similar findings have been obtained in studies of the antigen-induced proliferative response of noncytolytic T-cell clones [14]. Although it is not yet proved that the same receptors are involved in such a dual reactivity, these results strongly support the "modified self" hypothesis of T-cell recognition.

T-cell clones have also been used to analyze the diversity of the specificity repertoire of T cells. While these studies have confirmed the phenomenon of MHC restriction at the clonal level, they have also revealed the diversity of MHC products that can serve as restriction elements. Individual cell clones, however, can recognize only one of these restriction elements. Moreover, there is evidence of some heterogeneity among T-cell clones that recognize the same nominal antigen in association with the same restriction molecules. Thus, it is clear that the effector T cells generated in the immune response to a given antigen are highly heterogeneous, both in terms of functional activity and specificity.

Surprisingly, the availability of T-cell clones and hybridomas has not yet resulted in significant progress in our understanding of the nature of T-cell receptors. Attempts in many laboratories to produce antireceptor monoclonal antibodies using T-cell clones have met with limited success. Very recently, however, several promising results concerning the production of idiotypically specific antireceptor monoclonal antibodies have been reported [15,16], and it is likely that such developments will lead to rapid progress in the elucidation of the structure of the receptors on the various functional effector T cells. Moreover, now there is evidence for surface structures which, although they do not function as antigen receptors, are nevertheless required for efficient interactions of effector T cells with the corresponding antigen-bearing cells [17]. The identification of such structures has been done mainly by studying the effect of monoclonal antibodies directed against surface membrane constituents on the functional activity of T-cell clones. The most studied system has been inhibition of CTL-mediated lysis. Thus, in the mouse both the Lyt-2/3 molecular complex and the so-called lymphocyte-function-associated antigen (LFA-1) play an important role in the initial (binding) step between CTL and their target cells. Similar studies in man have indicated further complexity [15]. CTL clones that recognize class I (HLA-A/B/C) MHC antigens express surface antigens homologous to the mouse Lyt-2/3 complex, which are referred to as OKT8, Leu-2, or B9 according to the monoclonal antibodies with which they are defined. While the cytolytic activity of such clones is inhibited by these antibodies, CTL clones that are directed against class II MHC antigens do not express OKT8 and hence are not inhibited. However, the latter clones bear the OKT4/Leu-3 surface antigens and are inhibited by antibodies directed against these structures. The mechanism by which these surface structures contribute to antigen recognition remains to be defined. It is now evident that anti-class I CTL clones are heterogeneous in terms of their susceptibility to inhibition by anti-Lyt 2/3 (or anti-B9) antibodies. As resistance of CTL clones to inhibition by these antibodies appears to correlate quantitatively (but not absolutely) with *in vivo* immunization, it has been suggested that the role of the Lyt-2/3 molecular complex is to stabilize the interaction of the antigen receptors with the corresponding antigenic molecules when the receptors are of low affinity or density [18]. In this context, it is of interest that another surface structure involved in CTL-mediated lysis, the OKT3 glycoprotein, appears to be closely associated with human CTL receptors. Monoclonal antibodies against OKT3, however, do not inhibit the initial binding step between CTL and the target cells but appear to prevent the subsequent step leading to the delivery of "lethal hits."

Finally, the use of T-cell clones is providing insight into the effector functions of T cells *in vivo*. From the limited data available, it appears that murine T-cell clones that exhibit helper function *in vitro* also are able to function effectively *in vivo* [19]. Some of these helper T-cell clones have been shown to mediate delayed-type hypersensitivity, thus suggesting that at least some effector T cells can perform several functions *in vivo*. While there have been only few reports describing the activities *in vivo* of cloned CTL, recent work using T-cell clones directed against allogeneic or virus-induced syngeneic tumor cells has shown rapid destruction of these tumor cells given intraperitoneally following intravenous injection of cloned CTL [20]. It is thus likely that further analysis of the *in vivo* activities that can be displayed by functionally defined T-cell clones will provide valuable information as to the role of distinct effector T cells in viral immunity and pathogenesis. Moreover, the availability of cloned populations of distinct effector T lymphocytes makes it now possible to assess the possible use of such clones for cellular immunotherapy.

References

1. Möller G (ed) (1982) Interleukins and lymphocyte activation. *In* Immunological Reviews, Vol 23. Munksgaard, Copenhagen
2. Robb RJ, Munck A, Smith KA (1981) T cell growth factor receptors. Quantitation, specificity and biological relevance. J Exp Med 154:1455–1474
3. Sekaly RP, MacDonald HR, Zaech P, Nabholz M (1982) Cell cycle regulation of cloned cytolytic T cells by T cell growth factor: Analysis by flow microfluorometry. J Immunol 129:1407–1415
4. Möller G (ed) (1981) T cell clones. *In* Immunological Reviews, Vol 54. Munksgaard, Copenhagen
5. Paul WE, Sredin B, Schwartz RH (1981) Long-term growth and cloning of non-transformed lymphocytes. Nature 294:697–699
6. Fathman CG, Fitch FW (eds) (1982) Isolation, characterization, and utilization of T lymphocyte clones. Academic Press, New York
7. Nabholz M (1982) The somatic cell genetic analysis of cytolytic T lymphocyte functions. *In* Fathman CG, Fitch FW (eds) Isolation, Characterization, and Utilization of T Lymphocyte Clones. Academic Press, New York, p 165

8. Chen WR, Wilson A, Scollay R, Shortman K (1982) Limit-dilution assay and clonal expansion of all T cells capable of proliferation. J Immunol Methods 52:307–322
9. Moretta A, Pantaleo G, Moretta L, Cerottini JC, Mingari MC (1983) Direct demonstration of the clonogenic potential of every human peripheral blood T cell. Clonal analysis of HLA-DR expression and cytolytic activity. J Exp Med 157:743–754
10. von Boehmer H, Haas W, Köhler G, Melchers F, Zeuthen J (eds) (1982) T cell hybridomas. In Current Topics in Microbiology and Immunology, Vol 100. Springer-Verlag, Berlin
11. Prystowsky MB, Ely JM, Beller DI, Eisenberg L, Goldman J, Goldman M, Goldwasser E, Ihle J, Quintans J, Remold H, Vogel SN, Fitch FW (1982) Alloreactive cloned T cell lines. VI. Multiple lymphokine activities secreted by helper and cytolytic cloned T lymphocytes. J Immunol 129:2337–2344
12. Kelso A, Glasebrook AL, Kanagawa O, Brunner KT (1982) Production of macrophage-activating factor by T lymphocyte clones and correlation with other lymphokine activities. J Immunol 129:550–556
13. Pace JL, Russel SW, Schreiber RD, Altman A, Katz DH (1983) Macrophage activation: Priming activity from a T-cell hybridoma is attributable to interferon-γ Proc Natl Acad Sci USA 80:3782–3786
14. Schwartz RH, Sredin B (1982) Alloreactivity of antigen-specific T cell clones. In Fathman CG, Fitch FW (eds) Isolation, Characterization, and Utilization of T Lymphocyte Clones. Academic Press, New York, p 375
15. Reinherz EL, Meuer SC, Schlossman SF (1983) The delineation of antigen receptors on human T lymphocytes. Immunol Today 4:5–8
16. Lancki DW, Lorber MI, Loken MR, Fitch FW (1983) A clone-specific monoclonal antibody that inhibits cytolysis of a cytolytic T cell clone. J Exp Med 157:921–935
17. Möller G (ed) (1982) Effects of anti-membrane antibodies on killer T cells. In Immunological Reviews, Vol 68. Munksgaard, Copenhagen
18. MacDonald HR, Glasebrook AL, Cerottini JC (1982) Clonal heterogeneity in the functional requirement for Lyt-2/3 molecules on cytolytic T lymphocytes: Analysis by antibody blocking and selective trypsinization. J Exp Med 156:1711–1722
19. Schreier MH, Tees R, Nordin AA, Benner R, Bianchi ATJ, van Zwieten MJ (1982) Functional aspects of helper T cell clones. Immunobiology 161:107–138
20. Engers HD, Glasebrook AL, Sorenson GD (1982) Allogeneic tumor rejection induced by the intravenous injection of Lyt-2^+ cytolytic T lymphocyte clones. J Exp Med 156:1280–1285

Genetic Control

CHAPTER 11
Host Genes that Influence Susceptibility to Viral Diseases

MARGO A. BRINTON,* KENNETH J. BLANK,
AND NEAL NATHANSON†

Virus infections that induce permanent impairment or death in their hosts have undoubtedly exerted a selective pressure in the evolution of both plants and animals. Therefore, it is not surprising that those host alleles that fortuitously provided a reduced susceptibility to viral diseases would be maintained in the host population. A classic example is the introduction of myxomatosis virus into populations of wild European rabbits in Australia. The European rabbit had become an agricultural pest by 1950 and myxomatosis virus was introduced in an attempt to reduce the rabbit population. Initially, rabbit mortality rates were observed to be in excess of 99% [1], but a small number of rabbits recovered from infection even during the spread of the original highly virulent strain of virus. These survivors bred and, in areas where the rabbit population was reexposed to virus annually, the mortality rate had decreased to 25% by the end of 7 years [2]. In some areas attenuated virus strains arose and this provided additional opportunity for the selection of resistant animals. Thus, natural selection operated to produce an ecological balance, that ensured survival of the virus in the presence of a disease resistant host population.

The majority of the experimental studies of genetically controlled resistance in animals have been carried out in inbred mice. Mice are an excellent species for such studies because of the ease with which they can be bred and because of the wealth of genetic information already available. To date, murine genes that can specifically modulate the outcome of infections with three families of DNA viruses and five families of RNA viruses have been identified. A representative list of these genes is given in Table 1.

---

* The Wistar Institute, Philadelphia, Pennsylvania.
† University of Pennsylvania, Philadelphia, Pennsylvania.

Table 1. Mouse genetic loci which influence virus susceptibility[a]

Virus	Disease or effect	No. of genes or designation	Dominant trait	Maps to H-2	Reference
DNA viruses					
Herpes simplex	Encephalitis	2+	R	−	[3]
Cytomegalo	Encephalitis	1+	S	+	[3]
Polyoma	Tumors	1+	S	?	[3]
	Runting	1+	R	−	[3]
Ectromelia	Mouse pox	1+	?	?	[3]
RNA viruses					
Flavi	Encephalitis	1	R	−	[4]
LDV	Polioencephalitis	1+	R	−	[3]
Influenza A	Pneumonia	1	R	−	[5]
Measles	Encephalitis	1+	R	−	[3]
EMC	Diabetes	1	R	−	[3]
Rabies	Encephalitis	1	R	−	[7]
Corona	Hepatitis	1	S	−	[3]
	Persistence	2	?	−	[3]
	Encephalitis	1	S	−	[6]
	Demyelination	*Rhv-1*	R/S	−	[3]
		Rhv-2	S	−	[3]
Murine leukemia viruses					
Ecotropic FLV	Replication	*Fv-1*	R	−	[8]
Ecotropic	Replication	*CV*	S	−	[9]
Ecotropic	Replication	*Akv-1*	S	−	[9]
Xenotropic(?)	Replication	*Akv-2*	S	−	[9]
Xenotropic	Replication	*hr*	S	−	[10]
Xenotropic	Replication	*BXV*	S	−	[9]
Xenotropic	High replication	*Nzv-1*	S	−	[9]
Xenotropic	Low replication	*Nzv-2*	S	−	[9]
FLV	Focus formation	*Fv-2*	S	−	[9]
FLV	Suppressed lymphocyte proliferation	*Fv-3*	−	−	[11]
FLV	Replication helper	*Fv-4*	R	−	[9]
FLV	Anemia or polycythemia	*Fv-5*	−	−	[9]
FLV	Generation recombinants	*Fv-6*	R	−	[9]
FLV	Recovery splenomegaly	*Rfv-1*	R/S	+	[9]
FLV	Recovery splenomegaly	*Rfv-2*	R/S	+	[9]
FLV	Recovery splenomegaly	*Rfv-3*	R	−	[9]
GLV	Recovery leukemia	*Rgv-1*	R	+	[9]
GLV	Recovery leukemia	*Rgv-2*	R	−	[9]
WM 1504 E	Slow disease of CNS	2	R	−	[12,13]

[a] R, resistance; S, susceptibility; LDV, lactate dehydrogenase-elevating virus; FLV, Friend leukemia virus; GLV, Gross leukemia virus.

Some important generalizations can be made about the murine systems of genetically controlled resistance. In many instances, a single genetic locus is responsible for the difference observed between susceptible and resistant strains, but in other examples, two and, occasionally, more than two loci are involved. Those systems in which a single locus accounts for a change in virus-induced susceptibility offer the best chance to identify the host gene product involved. The opportunity to elucidate the mechanism of resistance is further improved if the resistant and susceptible alleles can be bred into mice of the same genetic background by the development of congenic strains. If an effect of the resistance gene can be detected in tissue culture cells obtained from disease-resistant animals, an even better chance of pinpointing the mechanism exists. It is important to recognize that the phenotypic expression of genetic resistance in animals can be modified by host variables such as experimental enhancement or suppression of the immune response or the age of the animal at the time of infection, and by virus variables such as strain virulence and infecting route and dose. The contribution of virus genes is discussed in Chapter 15 of this volume.

Each of the described murine virus resistance loci is probably a unique gene, since in all cases tested each locus has been found to segregate independently [14]. The dominant allele may code for either resistance or susceptibility and in a few instances co-dominance is observed. Only a few of the murine resistance genes have been found to map within the histocompatibility locus, a fact that is sometimes overlooked.

Each resistance gene appears to influence infections with only a particular family of viruses or, sometimes, as in the case of several of the coronaviruses, one species of virus. This virus-specific feature distinguishes genetically controlled resistance from other types of host-defense mechanisms that are virus nonspecific. Specificity is not surprising, since the various families of viruses differ significantly in their modes and sites of replication. Furthermore, specificity implies that each resistance gene product interacts with a unique event characteristic of only one type of virus. Such an interaction could occur at the level of virus receptor binding, transcription of viral nucleic acids, translation of viral proteins, or assembly of viral progeny.

Three of the murine resistance genes are discussed in detail and serve as specific examples of the different types of interactions that can occur between resistance gene products and viral molecular events. Further information on other genetic loci can be obtained from several reviews [3,14–16].

Orthomyxoviruses

Resistance to orthomyxoviruses, primarily influenza A viruses, is controlled by an autosomal dominant allele that has been designated *Mx* [5]. This allele has been identified only in A2G mice. Although comparable amounts of influenza A virus are required to initiate an infection in resistant A2G and susceptible A/J mice, virus titers produced in A2G tissues were always found to be at least 100-fold lower than those produced in A/J tissues [5]. Experimental immunosuppression of resistant A2G animals does not enhance their ability to replicate influenza viruses nor does this type of treatment alter the resistance of A2G mice to influenza virus-induced disease. However, anti-interferon antibody (AIF) rendered A2G mice susceptible to a lethal infection with influenza viruses [5], and the virus titers in the tissues of the A2G mice given AIF serum were equal to those observed in susceptible A/J mice.

A2G peritoneal macrophage cultures produce low yields of virus when infected during the first two weeks in culture, but can replicate influenza virus as efficiently as A/J cultures if infected after the second week in culture. This change in susceptibility correlates with the disappearance of interferon from the A2G cultures [5]. Influenza virus-specific resistance in A2G cells is elicited by Type I murine interferon (a mixture of α and β interferons), but not by γ interferon and this resistance is greater than that elicited in susceptible A/J cells [17].

Recent studies by Horisberger et al. [18] have shown that treatment of A2G cells with Type I interferon results in the production of a 72,500 dalton cellular protein that was identified in cytoplasmic extracts subjected to two-dimensional gel electrophoresis. When the *Mx* allele was back-crossed onto a number of different murine genetic backgrounds, the inducibility of the 72,500 dalton protein by interferon correlated directly with the presence of the *Mx* allele in the cells tested. These data suggest that the 72,500 dalton protein is directly associated with the activity of the *Mx* allele. How this protein acts to inhibit influenza virus replication is as yet unknown.

Flavivirus

Resistance to flavivirus-induced encephalitis is controlled by an autosomal dominant allele. Princeton Rockefeller Institute (PRI) mice and the strains (BRVR and BSVR) are the only commonly available inbred mouse strains that carry the resistance allele. A study of wild mice obtained from California and Maryland demonstrated that the flavivirus resistance allele was present in wild mouse populations from both locations [3]. The resistance allele from PRI has been back-crossed onto the C3H/HE background yielding the congenic resistant strain, C3H/RV. The development of these congenic strains, by reducing the background of unrelated variables, has facilitated studies of genetically controlled resistance to flaviviruses.

Flaviviruses replicate in mice that possess the resistance allele. However, virus titers in tissues are lower (by as much as 10,000-fold) and the spread of the infection is slower and usually self limiting. Although C3H/RV mice can be killed by large doses of flaviviruses administered by the intracerebral route or by infection after experimental immunosuppression, a reduced titer of virus is always observed as compared with that produced by similarly treated susceptible mice. Even though resistant animals produce less virus after infection than susceptible animals, a functioning immune system is necessary for the phenotypic expression of resistance. Resistant animals are not rendered susceptible to flavivirus-induced disease by injection of anti-interferon antibody. It appears that although both the immune system and the antiviral state may interact with the flavivirus resistance gene in a synergistic manner, neither of these host responses is specifically involved in the action of the gene product of the resistance allele.

Cell cultures obtained from resistant mice produce lower yields of flaviviruses than do comparable cultures of cells from susceptible mice [4]. This differential ability to produce flaviviruses is maintained in established lines of SV40-transformed embryofibroblasts prepared from C3H/RV and C3H/HE animals.

Defective interfering (DI) particles, which are viral deletion mutants, have been found to occur in cells infected with virtually any animal virus. The generation and amplification of DI particles is suspected to be, in part, regulated by host factors. In experiments using C3H/RV and C3H/HE embryo fibroblasts infected with the flavivirus, West Nile virus (WNV) strain E101, it was found that the host cell significantly influenced the composition of the viral progeny populations. WNV replication in C3H/RV cells yields progeny with a high proportion of DI particles, whereas progeny virus produced by C3H/HE cells consists primarily of standard virus. The ratio of DI to standard virus among progeny can be shifted by passaging virus from one cell type to the other [19]. When C3H/HE and C3H/RV mice are injected intracerebrally with purified WNV genomic RNA, the virus subsequently produced in the brains of the C3H/RV mice contains a high proportion of DI particles (Brinton, unpublished data). These data are consistent with the conclusion that if DI particles are already present in the WNV inoculum, they are amplified more efficiently and interfere more effectively with standard virus replication in cells containing the resistance allele. In addition, DI RNAs are generated more rapidly during replication of WNV genome RNA in C3H/RV cells.

It is tempting to speculate that the protein specified by the flavivirus resistance allele interacts specifically at the level of the flavivirus RNA replication complex, affecting template-polymerase interactions. If a host factor does interact with the flavivirus polymerase to aid in template recognition or binding, then an alteration in this host factor could influence the production of and/or extent of interference by DI RNAs. A WNV mutant has now been isolated [20] from a C3H/RV persistently infected culture that is not affected by the resistance allele.

Murine Leukemia Viruses

The induction of erythroleukemia by Friend murine leukemia virus, spontaneous thymic lymphoma in AKR and C58 mice and leukemia induced by Gross virus are controlled by multiple host genes whose expression affects various aspects of the infectious process [3,8,9]. It is now possible to identify both the individual genes and their phenotypic effects so that their distinct roles in the complex series of events leading to resistance can be determined. Many of these genes are listed in Table 1. This discussion will be limited to the *Fv-1* gene, the most intensely studied, and to two genes linked to the murine major histocompatibility complex (*H-2*), *Rfv-1* and *Rfv-2*, which appear to control immune responsiveness to virus-induced antigens. These genes demonstrate the diverse phenotypic effects that can mediate resistance to virus-induced disease.

Fv-1

Friend leukemia virus (FLV) causes a polycythemia as a result of malignant transformation of erythroid stem cells. This virus actually consists of two components: a replication-competent helper virus (F-MuLV) and a defective, transforming, spleen-focus-forming virus (SFFV). The *Fv-1* gene confers relative resistance to an FLV-infected host [8]. Mice homozygous with respect to $Fv-1^n$ (N-type) or $Fv-1^b$ (B-type) alleles are permissive (susceptible) for infection by N- and B-tropic viruses respectively. Heterozygous $Fv-1^{n/b}$ (NB-type) mice are nonpermissive (resistant) for both N- and B-tropic viruses indicating that the resistance allele is dominant. Resistance is not absolute, however, although focus formation is reduced 1000-fold. Viruses produced by repeated forced passage in susceptible hosts are NB-tropic, that is they infect N-, B-, and NB-type mice with equal efficiency. *Fv-1*-mediated resistance is observed in mouse embryo fibroblast cultures derived from N- and B-type mice, which suggests that *Fv-1* is expressed at the cellular level and not at the level of the immune system.

Molecular studies have demonstrated that significantly less proviral DNA is present in the genome of the infected *Fv-1*-resistant cells than in susceptible cells, although the levels of unintegrated DNA were the same in both cell types. These studies suggest that both the viral reverse transcriptase used to transcribe viral RNA and the cellular machinery used to produce double-stranded viral DNA are normal in resistant cells. In addition, no differences in degradation of vDNA has been observed. The only major difference between resistant and susceptible cells detected so far has been a low level of circular (Form I) DNA in *Fv-1*-resistant cells. Thus, *Fv-1* may inhibit proviral integration by blocking the circularization of linear double-stranded DNA. However, the lack of circular DNA may only reflect defective linear DNA production.

The genome of replication-competent retroviruses can typically be divided into three regions: *gag*, coding for the virus core proteins; *pol*, encoding the reverse transcriptase; and *env*, encoding the envelope glycoprotein. The viral sequences associated with virus tropisms have been found to be associated with the *gag* region by oligonucleotide mapping. Electrophoretic studies have detected differences in the p30, the major core protein, in viruses expressing different tropisms. Exactly how p30 interacts with the *Fv-1*-encoded gene product to produce the observed patterns of viral tropism is unclear.

Rfv-1 and *Rfv-2*

The *Rfv-1* and *Rfv-2* genes [9] have been found to affect the recovery from Friend virus-induced splenomegaly, i.e., mice inoculated with very low doses of FLV begin to develop splenomegaly, which then regresses in resistant mice. This effect was first ascribed to the *Rfv-1* gene, which was mapped to the *H-2D* region. Although no mechanism has been described for the effect of this gene, it is possible that *Rfv-1* and the gene encoding the *H-2D* molecule may be identical. The *H-2D* locus could affect regression of virus-induced splenomegaly since it has previously been demonstrated that a specific T lymphocyte-mediated immune response to FLV-infected cells is only generated when virus-induced antigen is seen in association with an *H-2D* molecule encoded by a resistant *H-2* haplotype.

On the other hand, *Rfv-2* has been mapped to the *H-2I* region and appears to exert direct control over antibody responses to certain viral antigens. In this respect, this gene appears to function in the same way as a classic immune response gene. Thus, *Rfv-2* may confer resistance by allowing the production of antibodies that are either cytotoxic for FLV-induced tumor cells or neutralize FL virions.

Comment

Mouse models establish the existence of a large number of host genetic loci that influence susceptibility to virus-induced disease. The great majority of these loci map distant from the *H-2* loci and do not act through general host defenses such as the immune response. Instead, most genes influence replication of one virus group only, and appear to involve different virus-specific steps in the intracellular replication cycle. It seems reasonable to assume that analogous loci exist in the human genome, although they are technically difficult to detect.

References

1. Myers KM, Marshall ID, Fenner F (1954) Studies in the epidemiology of infectious myxomatosis of rabbits. III. Observations on two succeeding epizootics in Australian wild rabbits on Riverine Plain of southeastern Australia, 1951–1953. J Hyg 52:337–360
2. Sobey WR (1969) Selection of resistance to myxomatosis in domestic rabbits (*Oryctolagus cuniculus*). J Hyg 67:743–754
3. Brinton MA, Nathanson N (1981) Genetic determinants of virus susceptibility: Epidemiologic implications of murine models. Epidemiol Rev 3:115–139

4. Darnell MB, Koprowski H (1974) Genetically determined resistance to infection with group B arboviruses. II. Increased production of interfering particles in cell cultures from resistant mice. J Infect Dis 129:248–256
5. Haller, O (1981) Inborn resistance of mice to orthomyxoviruses. *In* Haller O (ed) Natural Resistance to Tumors and Viruses, Springer-Verlag, Berlin, p 25
6. Knobler RL, Haspel MV, Oldstone MBA (1981) Mouse hepatitis virus type 4 (JHM strain)-induced fatal central nervous system disease. I. Genetic control and the murine neuron as the susceptible site of disease. J Exp Med 153:832–843
7. Lodmell DL (1983) Genetic control of resistance to street rabies virus in mice. J Exp Med 157:451–460
8. Jolicoeur P (1979) The *Fv-1* gene of the mouse and its control of murine leukemia virus replication. *In* Current Topics in Microbiology and Immunology, Vol 86. Springer-Verlag, New York, p 67
9. Teich N, Wyke J, Mak T, Bernstein A, Hardy W (1982) Pathogenesis of retrovirus-induced disease. *In* Weiss R, Teich N, Varmus H, Coffin J (eds) Molecular Biology of Tumor Viruses: RNA Tumor Viruses (2nd ed). Cold Spring Harbor Laboratory, New York, p 785
10. Greene N, Hiai H, Elder JH, Schwartz RS, Khiroya RH, Thomas CY, Tsichlis PN, Coffin JM (1980) Expression of leukemogenic recombinant viruses associated with a recessive gene in HRS/J mice. J Exp Med 152:249–256
11. Kumar V, Bennett M (1976) Mechanisms of genetic resistance to Friend virus leukemia in mice. II. Resistance of mitogen-responsive lymphocytes mediated by marrow-dependent cells. J Exp Med 143:713–727
12. Oldstone MBA, Lampert PW, Lee S, Dixon FJ (1977) Pathogenesis of the slow disease of the central nervous system associated with WM 1504 E virus. I. Relationship of strain susceptibility and replication to disease. Am J Pathol 88:193–212
13. Oldstone MBA, Jensen F, Dixon FJ, Lampert PW (1980) Pathogenesis of the slow disease of the central nervous system associated with wild mouse virus. II. Role of virus and host gene products. Virology 107:180–193
14. Bang FB (1978) Genetics of resistance of animals to viruses. I. Introduction and studies in mice. *In* Advances in Virus Research, Vol 23. Academic Press, New York, p 269
15. Pincus T, Snyder HW (1975) Genetic control of resistance to viral infection in mice. *In* Viral Immunology and Immunopathology. Academic Press, New York, p 167
16. Blumberg BS (1961) Inherited susceptibility to disease. Arch Environ Health 3:612–636
17. Staeheli P, Horisberger, MA, Haller, O (1983) Mx-Dependent resistance to influenza viruses is induced by mouse interferons α and β but not γ. Virology 132:456–461.
18. Horisberger MA, Staeheli P, Haller O (1983) Interferon induces a unique protein in mouse cells bearing a gene for resistance to influenza virus. Proc Natl Acad Sci USA 80:1910–1914
19. Brinton MA (1983) Analysis of extracellular West Nile virus (WNV) particles produced by cell cultures from genetically resistant and susceptible mice indicates enhanced amplification of DI particles by resistant cultures. J Virol 46:860–870
20. Brinton MA, Fernandez AV (1983) A replication efficient mutant of West Nile virus in insensitive to DI particle interference. Virology 129, 107–115

CHAPTER 12
Host Genes Controlling the Immune Response

HUGH O. MCDEVITT*

As with any complex organ system, the immune system is controlled by a large number of genes regulating many discrete functions in the several different cell types that make up the system. Deliberate efforts to select animals for high or low antibody response to several different antigens have demonstrated quite clearly that the resultant high responder and low responder strains differ at least by ten or more genes (Biozzi, personal communication). These genes determine proliferation rates for cellular components of the immune system, and other immunologically *nonspecific* cellular functions.

However, there are at least two genetic systems that regulate not only the *magnitude* of the immune response, but its *specificity*. The two genetic systems that regulate the specificity of the immune response are: (1) the structural genes for the immunoglobulin heavy chains, which determine the ability to produce particular antibody-combining sites against a foreign antigenic determinant (epitope); and (2) the genes of the major histocompatibility system, which regulate both the *specificity* and the *magnitude* of the immune response to a wide variety of foreign antigens, by mechanisms not yet completely understood.

Studies from a number of laboratories have demonstrated that the ability to produce a particular antibody-combining site (idiotype) complementary to a foreign epitope is linked to the Ig heavy chain locus on chromosome 12 in the mouse. Detailed studies have demonstrated recombination between the constant region genes of the heavy chain linkage group and the ability to produce a particular idiotype. This indicates that the responsible genes are,

---

* Stanford University, Stanford, California.

in all likelihood, germline heavy chain variable region (V_H) genes, which place some limit on the ability to produce particular idiotypes.

While it seems reasonable that the structural genes for the polypeptides that determine the structure of the antibody-combining site would regulate the specificity of the immune response, it is less clear why genes of the major histocompatibility system should affect not only the magnitude, but also the specificity of the immune response.

Fig. 1 is a schematic diagram of the murine major histocompatibility system and Fig. 2 is a schematic diagram of the human major histocompatibility system. As the diagrams indicate, both systems are complex, multigene systems with at least three classes of gene products. The class I MHC antigens are 45,000 molecular weight glycoproteins expressed on the surface of all or nearly all cells (except in the early embryo, and in the trophoblast) and found on the cell surface in close noncovalent association with beta two microglobulin (β_{2m}). Class II MHC antigens on the other hand, are two-chain dimers of a 34,000 and a 29,000 molecular weight glycoprotein and are expressed on the surface of B lymphocytes, antigen-presenting macrophages, and activated T lymphocytes. Class III MHC molecules include several components in the early steps in activation of either the classic or the alternate pathway of complement activation.

Class I MHC molecules, class II MHC molecules and β_2 microglobulin all share a number of characteristics with the immunoglobulins. They are all organized into domain structures based on 110 amino acid domains (with an intrachain disulfide loop in which the two cysteines are approximately 60 residues apart), each with quite definite sequence homology with the constant regions of immunoglobulin heavy chain constant region domains. However, the MHC molecules do not show the molecular heterogeneity that is the most striking structural aspect of the immunoglobulins.

Class I and class II MHC molecules regulate the specificity of the immune response by interacting in some manner with the antigen so that in the course of *induction* of an immune response, the responding T cells "see" the antigen only in the context of the class I or the class II MHC molecule with which it is associated.

Class I MHC molecules regulate the specificity of cytotoxic T lymphocytes (T killer cells) and can function, along with the foreign epitope, as an immunogenic component on the surface of any cell type in the body. Class II MHC molecules, in contrast, regulate the specificity of helper T cells (T_h) and suppressor T cells (T_s) and appear to interact with foreign epitopes primarily on the surface of the *antigen-presenting macrophage,* or *accessory cell,* and perhaps on the surface of B lymphocytes and activated T cells.

Cytotoxic T cells are "restricted" by class I molecules while helper T cells and suppressor T cells are "restricted" by class II MHC molecules. The T cell will recognize a foreign antigen on the surface of a target cell (for cytotoxic T cells) or an antigen-presenting cell (for helper and suppressor T cells) *only* if it occurs on the cell in association with the same MHC mole-

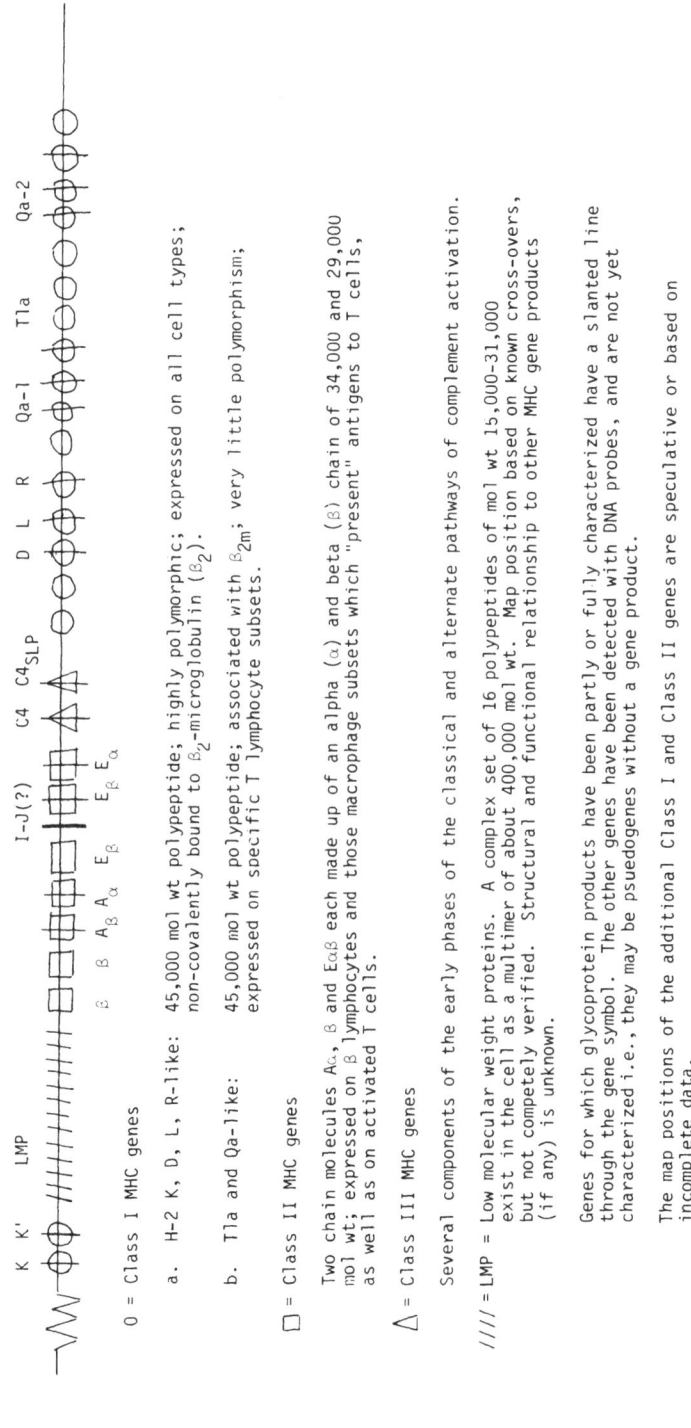

Fig. 1. Schematic diagram of the genetic organization of the murine (H-2) major histocompatibility complex.

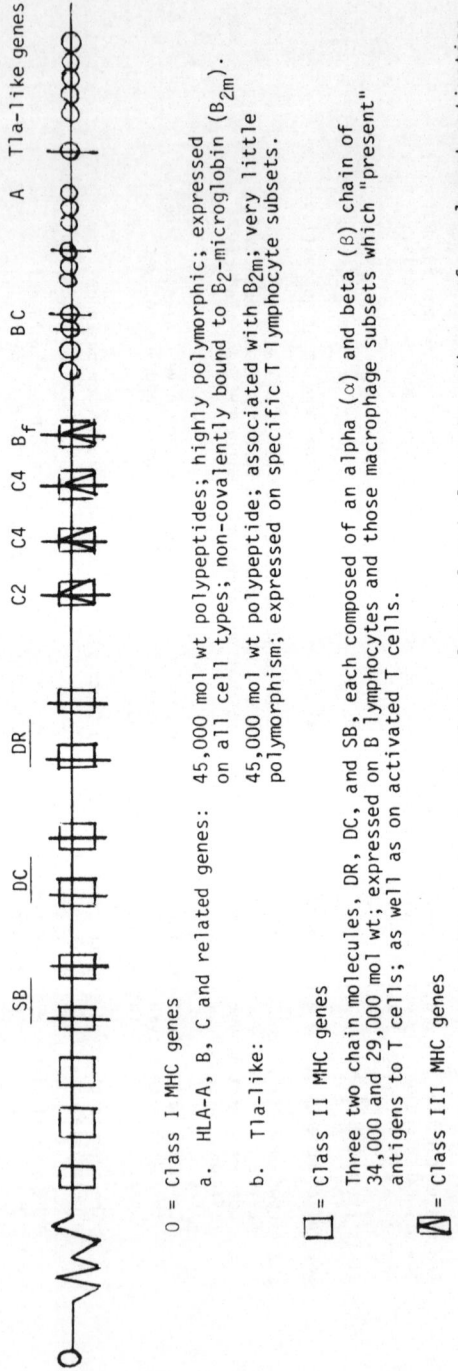

Fig. 2. Schematic diagram of the genetic organization of the *HLA* system.

cules (the same alleles) as were present on the original immunizing cells during the induction of cytotoxic or helper or suppressor T cells. Thus, the immunized T cell has specificity not only for the foreign antigen, but also for the class I or class II MHC molecule.

A number of lines of evidence suggest quite strongly that this MHC "restriction" is due to some type of molecular association between the foreign antigen and a class I or class II MHC molecule, which may occur only at the crucial moment of interaction between the target cell and the cytotoxic T lymphocyte, or the antigen-presenting cell and the helper or suppressor T cell. The nature of this presumed molecular association, and whether it involves some type of antigen internalization and processing in the case of the antigen-presenting macrophage (APC), is not yet clear. However, it is clear that antibodies to class II MHC molecules completely inhibit the ability of APC to present antigen to T cells, either during the course of primary immunization, or in the course of interaction of APC with previously immunized T helper cells.

At the present time, our understanding of the precise nature of these molecular interactions is seriously hampered by our lack of knowledge of the genes encoding the T-cell receptor, and the structural nature of the T-cell receptor. Further detailed studies of the interaction of foreign antigens with class I and class II MHC molecules, as well as an understanding of the structure of the T-cell receptor, will be required before the precise mechanism by which MHC molecules determine the *specificity* of the immune response becomes clear. Suffice it to say that, at the phenomenologic level, it is clear that MHC genotype determines which *epitopes* on a foreign antigen elicit a response. Thus, there are numerous examples in which inbred animals of two different MHC genotypes respond to different epitopes on the same antigen. Both strains may make immune responses of equal *magnitude* to the antigen, but the response is specific for two completely different epitopes on the antigen. For example, one inbred strain of guinea pigs responds only to the A chain of bovine insulin, while another inbred strain responds only to the B chain of bovine insulin.

While this might seem to be a trivial distinction, it is clear that this genetic control of the *specificity* of the antibody response has profound effects of considerable biological importance. Susceptibility to a wide variety of diseases, both in animal models and in humans, is strongly associated with the genotype of the major histocompatibility complex. This is particularly true for the class II MHC molecules. One major category of diseases, the spondylo-arthropothies, show a very strong association with a class I MHC molecule in humans—the *HLA-B27* allele of the *HLA-B* locus of the MHC. On the other hand, there are some 40 diseases in which susceptibility is strongly influenced by genotype of the class II MHC molecules—the *HLA-DR* locus and its associated *HLA-DC* and *SB* loci. Thus, individuals with juvenile-onset diabetes (insulin-dependent diabetes mellitus) are almost exclusively *HLA-DR3* or *DR4* and the preponderance of individuals with rheumatoid arthritis possess *HLA-DR4*. There is a marked increase of *HLA-DR2* in patients with multiple sclerosis and similar observations involving *HLA-DR2, 3,* or *4* haplotypes have been found in hyperthyroidism, myasthenia gravis, celiac disease, chronic active hepatitis, Addison's disease, systemic lupus erythromatosus, pemphigus, and psoriasis. It is not yet clear why these three *DR* alleles, out of a total of at least ten *HLA-DR* alleles, are the ones primarily associated with autoimmune disease, or why they are associated with such a wide variety of autoimmune diseases, many of which involve the production of quite specific autoantibodies. For example, the pathogenesis of hyperthyroidism clearly involves the production of antibody against the TSH receptor, which results in stimulation of that receptor, while the pathogenesis of myasthenia gravis clearly involves antibody production to the acetylcholine receptor.

The latter disease is particularly informative because both in experimental autoimmune myasthenia gravis and in myasthenia gravis in humans, predisposition is associated with a particular class II MHC genotype *and* the immunoglobulin heavy chain linkage group, and the correct allele of *both* predisposed to development of disease. Both genetic systems regulate antibody *specificity* as well as quantity.

It is quite possible that humans, mammals, and vertebrates in general, have been selected during the course of evolution for MHC and Ig genotypes that mediate an optimal immune response to pathogens in the environment. (Despite a number of studies, there are relatively few examples in which MHC or Ig genotypes play a major role in differential susceptibility to infectious agents. This presumably reflects the effect of stringent selection during the course of evolution, so that the particularly susceptible genotypes have been eliminated.) The surviving MHC genotypes, while being optimal for an immune response to pathogens in the environment, such as the common and frequent bacterial and viral infections, may mediate hypernormal immune responses to environmental antigens that cross-react with self antigens, or with self antigens themselves, leading to development of autoimmunity and the appearance of clinical disease entities such as juvenile-onset diabetes mellitus (in which antibodies that react with islet cell surface antigens ultimately lead to destruction of the islets of Langerhans and insulin deficiency). Similar pathways can be suggested for the development of autoantibodies leading to rheumatoid arthritis, systemic lupus erythromatosus, hyperthyroidism, and myasthenia gravis.

Viewed in this sense, the MHC genotypes associated with susceptibility to autoimmune diseases have been selected for rapid production of antibodies of particular specificities, some of which may have the unfortunate side effect of reacting with self components and leading to autoimmune disease. In this respect, it is useful to note that most of the diseases associated with the HLA system occur after the child-bearing years and would not result in elimination of these genotypes from the gene pool.

In addition to the immunoglobulin structural genes, and the genes of the major histocompatibility system, it is quite likely that a number of genes controlling the immune response remain yet to be discovered. These include the influence of other cell surface molecules, including minor histocompatibility antigens, on the *specificity* and *magnitude* of the immune response as well as genes determining the structure of the T-cell receptor, and genes determining the structure of macrophage cell surface molecules, which may influence the ability of macrophages and *antigen-presenting cells* to induce an immune response. There is already evidence suggesting that minor histocompatibility antigens affect the *magnitude* of some immune responses, and it is likely that further discoveries of this nature will be forthcoming. Availability of cloned copies of the genes of the major histocompatibility system make it likely that our understanding of the molecular basis by which these genes influence the *specificity* of the immune response will be significantly advanced in the near future. Knowledge of the biochemical nature of the T-cell receptor for antigen, and of the postulated molecular interactions between MHC molecules and foreign antigens, would serve to complete our understanding of how host genes effect the *magnitude* as well as the *specificity* of the immune response to foreign antigens and self antigens.

Acknowledgment. The work reported herein was supported by NIH Grant AI-07757.

CHAPTER 13

On the Role of Recombinant Retroviruses in Murine Leukemia

JOHN H. ELDER*

Under normal circumstances, a compatible relationship exists between the mouse and the extensive family of endogenous retroviruses present in the mouse genome. These retroviruses are present as proviruses in all somatic and germline cells and are inherited according to Mendelian expectations [1]. The expression of either infectious virus, or more commonly expression of certain viral proteins, is regulated to various degrees in different mouse strains. Whether the presence of these retroviruses is simply tolerated or whether they serve some selective advantage to the host is yet to be determined. Suffice it to say, however, that in certain circumstances, the compatible relationship breaks down, resulting in leukemia. Evidence now suggests that leukemia results from recombination between otherwise innocuous, endogenous viruses, to produce a virus with oncogenic potential. The following is a synopsis of our understanding of this phenomenon as well as an attempt to outline possible mechanisms by which virus-induced transformation can occur.

Definition of a Leukemia Virus

For those not versed in the jargon of the field, it may be helpful to define some of the terms commonly used to describe the murine retroviruses. First of all, there are three types defined by morphology of their particles, types A, B, and C. All share the common features that they contain reverse transcriptase and are RNA viruses. The type A particles are noninfectious intracellular particles that have not been well characterized, although some

---

* Research Institute of Scripps Clinic, La Jolla, California.

evidence suggests that they may be precursors to the second group, the type B viruses [2]. Type B viruses include the mouse mammary tumor viruses, which have been linked to the development of spontaneous mammary tumors in susceptible strains [3]. The third group, the type C viruses, are the leukemia viruses, and will be the only group examined below.

To begin, we now know that all the C-type retroviruses do not cause leukemia, and thus the term leukemia virus is more convenient than functionally significant. All the retroviruses have the basic genomic structure 5'-LTR-GAG-POL-ENV-LTR-3'. LTR refers to long terminal repeat, the region that facilitates integration of the viral genome into the host chromosome; GAG stands for the group antigen region which encodes the core proteins of the virus; POL is the polymerase gene which encodes reverse transcriptase; and ENV is the envelope gene which encodes the surface glycoprotein gp70 on the 5' side of the gene and a small protein, p15E on the 3' side. Our attention below will focus on the envelope gene and the 3' LTR, as this is the primary region of change in the recombinant viruses. The C-type retroviruses can be divided into three groups, defined by host range: (1) the ecotropic [4] viruses, which grow in and can infect only mouse cells; (2) the xenotropic [4] viruses, which are endogenous to mouse cells, but can only infect cells of other species; and (3) the dualtropic viruses, which can grow in and infect both mouse and heterologous species cells. Each of these categories can be further subdivided by other criteria. The ecotropic viruses are primarily separated by differential mouse tropism controlled by a locus termed *Fv-1* [5] i.e., some ecotropic viruses grow preferentially on NIH 3T3 fibroblasts (*Fv-1 N*), others grow better on BALB 3T3 cells (*Fv-1 B*), and still others show no preference (*Fv-1 NB*). The xenotropic viruses are primarily separated based on differential inducibility with drugs, such as BUDr [6], and more recently radioimmunoassays [7] or peptide fingerprinting [8]. The dualtropic viruses are comprised of certain naturally occurring wild mouse viruses, termed amphotropic [9] and a host of recombinant viruses, termed dualtropic, polytropic, or MCF-type viruses. The term MCF stands for mink cell focus-forming, used by Hartley *et al.* [10] to describe the unique effect these latter viruses have on mink cells in culture. This latter group of dualtropic viruses are the most recently recognized group and are the primary topic of the remainder of this review.

Historical Background

Extensive work has been done in several mouse systems, but for our purposes, we will concentrate on AKR mice, which has been the classic system for study of retrovirus-induced leukemia. AKR mice spontaneously develop thymic leukemia starting at 4 to 6 months of age. From the early work of Gross [11], it was recognized that cell-free extracts from these tumors could accelerate the disease in newborns. The oncogenic activity of these extracts was associated with a virus preparation from these tumors, termed Gross Passage A. However, subsequent work demonstrated that virus could also be isolated from young, normal mice, which could not cause acceleration of the disease, as did Gross Passage A. Thus, it was obvious that the viral repertoire of tumor tissue possessed a greater complexity than could be expected from simple release of the predominant endogenous virus. In fact, AKR mice were noted to be viremic from birth with an ecotropic virus, termed Akv, which was produced from two distinct loci [12]. Intensive effort was directed at determining what factors were involved in creating an oncogenic virus from what appeared to be an innocuous group of retroviruses found in the AKR genome. Finally, it was noted by Kawashima and colleagues [13] that although ecotropic virus production remained constant with age, xenotropic virus production increased dramatically in AKR mice at around 4 to 6 months, coinciding with the onset of spontaneous tumor induction. However, from other studies it was clear that at least the characterized xenotropic viruses could not, in themselves, cause disease. Then, in 1977, Hartley and colleagues at the National Institutes of Health, in collaboration with Old and colleagues at Sloan-Kettering, reported the isolation of a unique class of retroviruses with tropism for both mouse and heterologous species cells [10]. These viruses were similar to a unique virus, termed HIX, isolated from Moloney virus stocks by Peter Fischinger 2 years earlier [14]. These isolates, termed MCF viruses (see above) were only isolated from preleukemic and leukemic animals and could not be recovered from newborn AKR mice. Subsequent studies revealed that the MCF viruses could accelerate leukemia in newborn AKR mice and also cause leukemia in certain other mouse strains upon intravenous injection. The expanded host range of the MCF viruses coupled with extensive relationship to Akv led the authors to suggest that this new class of viruses resulted from recombination between Akv and an unidentified endogenous xenotropic-like virus. Still to be determined was the mechanism by which these viruses could induce transformation.

Characterization of Recombinant Viruses

The host range of the dualtropic recombinant viruses suggested that the primary changes occurred in the envelope genes of this new family of recombinant viruses. The envelope gene resides at the 3' end of the retrovirus genome and encodes the surface glycoprotein gp70, which is responsible for binding of virus to the cell and is the primary target of neutralizing antibodies. Subsequent analyses of the AKR recombinant viruses by peptide map [15], T1 oligonucleotide [16] and heteroduplex [17] analyses revealed that indeed, almost all the detectable changes in the recombinants relative to their respective ecotropic virus parents were in the proteins encoded by the 3' portion of the viral genome. The size of the substituted region varied from recombinant to recombinant, but each virus had very similar substitutions in the gene encoding gp70. Both protein and nucleic acid studies revealed striking homologies with known xenotropic viruses. However, close examination, particularly by T1 oligonucleotide analyses, failed to show identity between the substituted xenotropic virus-like information and any known xenotropic virus isolate. Further studies using a recombinant-specific probe made from the envelope gene of the virus to examine the DNA of AKR mice revealed, however, that at least 10–15 copies of this xenotropic virus-like information was present in the genome [18]. The authors proposed that the dualtropic nature of the recombinant viruses was inherited totally from the unidentified endogenous parental virus, rather than as a result of a combination of the host ranges of the two parental viruses. It now seems probable that the expanded host range was, in fact, a fortuitous observation for a number of reasons. First, direct sequence analyses of a Moloney-derived recombinant virus [19] as well as AKR 247 recombinant virus [20] has revealed that all but the C-terminal one-third of the gp70 genes of these two viruses has been substituted with xenotropic virus-like information. Thus, if the expanded host range of the dualtropic viruses was a combination of the host ranges of the parental viruses, then one would have to construct a model whereby cellular binding was accommodated on different species cells by different regions of the molecule. Furthermore, examination of a broad spectrum of recombinants by T1 oligonucleotide analysis [21] had shown that certain of these viruses appear to be totally substituted in gp70 and still retain dualtropic properties. Interestingly, the majority of this latter category have been shown to be nononcogenic. Regardless, the above analyses show that the recombinants are substituted to various degrees in their envelope genes with viral-related sequences found in the DNA of normal cells. As will be outlined below, additional substitutions in the extreme 3' portion of the recombinant may be responsible for imparting oncogenicity to certain of these viruses.

What Comprises the Leuk Gene

The obvious question regarding the mechanism by which a recombinant virus could induce leukemia remains unanswered. However, considerable progress has been made. There have been two approaches to these studies, one virological, the other cellular. Below, the basic tenets of both approaches are outlined and the findings to date are summarized.

The virological approach starts with the premise that the causitive agent in murine leukemia is an infectious virus and uses as the criteria for leukemogenesis the ability of certain viruses to accelerate or induce leukemia when injected into newborn animals. As will be pointed out below, this may or may not be true. However, the system has the advantage of being directly testable and thus considerable progress has been made. The most extensive analyses have been carried out by Hopkins and her colleagues at MIT in collaboration with Rowe and his colleagues at the National Institutes of Health. They have performed T1 oligonucleotide analyses of an extensive group of recombinant viruses and have tried to relate specific changes in the nucleotide sequence to the oncogenicity of the virus [21]. The authors were primarily able to separate these viruses into two groups, termed class I and class II. The class I isolates were oncogenic and of thymic origin, while the class II isolates were of splenic origin and failed to induce leukemia. The class I isolates exhibited thymotropic behavior; i.e., they preferentially grew in the thymus of injected animals, an observation first described by Kaplan and colleagues for the recombinant radiation-induced leukemia virus of C57Bl/6 mice [22]. The class II isolates appear not to replicate in the thymus. Comparative analysis of the T1 oligonucleotides of these isolates revealed that both the class I and class II isolates were substituted in gp70, but that class II isolates were more extensively substituted 3' to the gp70 gene in the region which encodes *p15E*. The 3' long terminal repeat (LTR) of the class II isolates appeared to be identical to the parental Akv LTR. Class I isolates, on the other hand, retained a small portion of the C-terminus of gp70 and the N-terminal half of *p15E* identical to the parental Akv virus, but also contained small substitutions in the C-terminal portion of *p15E* and the 5' portion (U3 region) of the LTR. Direct sequence analysis has now been performed on one class I AKR isolate, AKR 247 (Holland and Hopkins, personal communication), which confirms these findings. Studies are now in progress using *in vitro* constructs of viruses made from various combinations of these three regions of substitution to elucidate which regions are critical for leukemogenicity. Preliminary data indicate that simple replacement in Akv by any one of the three regions is insufficient to transform Akv into a leukemogenic virus. Rather, it seems that more than one substitution is necessary for efficient transformation (Holland and Hopkins, personal communication). Within the next year, these studies should lead to answers regarding what regions of the virion are responsible for the ability of certain exogenously applied viruses to cause leukemia. Clearly, the growth potential of these viruses is critical to the experimental design, since regardless of whether a virus carries a "leuk" gene, it must first survive and replicate in order for expression and subsequent transformation. This is particularly important in regard to the nonleukemogenic class II isolates, which do not replicate in the thymus when exogenously administered and thus may have the means but not the opportunity to induce a thymic disease. As will be seen below, the production of infectious virus may not be a prerequisite for spontaneous transformation in AKR.

There are several reasons why the murine leukemia viruses are associated with an etiology distinct from other viral systems. Most importantly, the murine viruses are only transmitted vertically, not horizontally as is true of most totally unrelated viruses described elsewhere in this book. Furthermore, the thymic tumors arising in AKR are clonal in origin and thus the disease does not require the horizontal recruitment one might expect if infectious virus facilitates transformation. Certainly, infectious virus is not required for maintenance of the transformed state, since many lymphoma lines do not produce virus, although viral antigens are expressed. Though we may not preclude infectious virus as an agent in all cases, a scenario can be constructed whereby it is not a prerequisite. With this in mind, it is worthwhile to examine retrovirus-induced leukemia from a cellular, rather than virological standpoint. The basic premise of this approach is that the viral genes are inherited in Mendelian fashion and for the most part are under host regulatory controls. Therefore, for the sake of argument, we may consider these genes as essentially host. Whether viral genes play any positive role in murine processes is unknown, though their presence is clearly tolerated. This inconvenience makes the cellular approach infinitely less testable than the above approach, where virus is treated as an exogenous agent. Nonetheless, it is worthwhile to pursue viral genes as cellular entities because of the unique expression of some of the viral constituents, particularly gp70. Early studies revealed that different mouse strains regulated the expression of viral components to different extents. Furthermore, only certain strains carried the AKV ecotropic virus endogenously [5,6,12] a trait shared by all high leukemia strains. It was noted that dramatic differences existed in the amount of gp70 detectable in the serum [23]. Further studies reported the presence of gp70 on lymphoid and epithelial tissues of the mouse [24], even in strains such as 129, where no infectious virus has ever been isolated. Both immunological [7] and peptide fingerprint [8] analyses of both retroviral and endogenous gp70 molecules revealed that the surface glycoproteins comprised a polymorphic family. The question was whether distinct gp70s were expressed at different sites in the mouse. From the studies of Boyse and colleagues [1], it was learned that a serologically distinct gp70 (termed G_{IX},) was present on the thymus of certain mouse strains, but not others. Analyses by peptide fingerprinting of gp70s from various tissues revealed that each tissue examined possessed highly related, but distinct gp70s that had peptide profiles very similar to xenotropic virus gp70s [25]. The implication from these findings was that the expression of distinct gp70s was regulated differentially in various tissues and thus, gp70 could be considered a differentiation antigen. This definition is artificial in that no role has been assigned to gp70 in differentiation. We can, however, for the sake of argument propose that the constitutive expression of a differentiation antigen might lead to arrested differentiation and subsequent tumor development. Herein is a possible tie between the appearance of recombinant viruses and a role of altered gp70 expression in tumorigenesis. The deregulation of endogenous gp70 sequences via recombination with an unregulated (or loosely regulated) virus genome could result in constitutive expression of inappropriate "host" sequences on the surface of the cell. How much of this scenario is fact and how much is fantasy remains to be determined. However, we do have preliminary evidence in our laboratory that a gp70-bearing recombinant virus determinant is specifically expressed on normal thymus cells of all mouse strains tested. This gp70 is not expressed on mature T cells in the spleen, even though multiple gp70s can be seen using a broadly reactive anti-gp70 serum [26]. This molecule thus fulfills the criteria for a differentiation antigen and its relationship to the MCF viruses offers an intriguing link between recombinant viruses and the specific expression of viral sequences in normal murine development.

In summary, although we do not know the mechanism by which recombinant retroviruses can induce leukemia, the evidence is compelling that they are intimately involved. Dramatic strides have been made in the structural characterization of oncogenic and nononcogenic recombinants, which should soon indicate what changes in the genomic structure are just and sufficient to create an oncogenic virus. Whether constitutive expression of recombinant viral molecules or induction of truly cellular *onc* genes is responsible for transformation is still to be determined.

Acknowledgment. This is publication no. 3038-IMM from the Research Institute of Scripps Clinic.

References

1. Stockert E, Old LJ, Boyse EA (1971) The GIX system: A cell surface alloantigen associated with murine leukemia virus implications regarding chromosomal integration of the viral genome. J Exp Med 133:1334–1355
2. Tanaka H, Tamura A, Tsuyimura D (1972) Properties of the intracytoplasmic A particles purified from mouse tumors. Virology 49:61–78
3. Bernhard W (1958) Electron microscopy of tumor cells and tumor viruses. A review. Cancer Res 18:491–509
4. Levy J (1973) Xenotropic viruses: Murine leukemia viruses associated with NIH Swiss, NZB, and other mouse strains. Science 182:1151–1153
5. Lilly R, Pincus T (1973) Genetic control of murine viral leukemogenesis. Adv Cancer Res 17:231–277
6. Aaronson SA, Stephenson JR (1973) Independent segregation of loci for activation of biologically distinguishable RNA C-type viruses in mouse cells. Proc Natl Acad Sci USA 70:2055–2058
7. Hino S, Stephenson JR, Aaronson SA (1976) Radioimmunoassays for the 70,000 molecular-weight glycoproteins of endogenous mouse type C viruses: Viral antigen expression in normal mouse tissues and sera. J Virol 18:933–941
8. Elder JH, Jensen FC, Bryant ML, Lerner RA (1977) Polymorphism of the major envelope glycoprotein (gp70) of murine C-type retroviruses: Virion associated and differentiation antigens encoded by a multi-gene family. Nature 267:23–28
9. Gardner MB (1978) Type C viruses of wild mice: Characterization and natural history of amphotropic, ecotropic and xenotropic MuLV. Curr Top Microbiol Immunol 79:215–259
10. Hartley JW, Wolford NK, Old LJ, Rowe WP (1977) A new class of murine leukemia virus associated with development of spontaneous lymphomas. Proc Natl Acad Sci USA 74:789–792
11. Gross L (1951) "Spontaneous" leukemia developing in C3H mice following

inoculation, in infancy, with Ak-leukemic extracts, or Ak-embryos. Proc Soc Exp Biol Med 76:27–32
12. Rowe WP (1977) Leukemia virus genomes in the chromosomal DNA of the mouse. Harvey Lect Ser 71:173–192
13. Kawashima K, Ikeda H, Hartley JW, Stockert E, Rowe WP, Old LJ (1976) Changes in expression of murine leukemia virus antigens and production of xenotropic virus in the late preleukemic period in AKR mice. Proc Natl Acad Sci USA 73:4680–4684
14. Fischinger PJ, Nomura S, Bolognesi DP (1975) A novel murine oncornavirus with dual eco- and xenotropic properties. Proc Natl Acad Aci USA 72:5150–5155
15. Elder JH, Gautsch JW, Jensen FC, Lerner RA, Hartley JW, Rowe WP (1977) Biochemical evidence that MCF murine leukemia viruses are envelope (env) gene recombinants. Proc Natl Acad Sci USA 74:4676–4680
16. Rommelaere J, Faller DV, Hopkins N (1978) Characterization and mapping of RNase T1-resistant oligonucleotides derived from the genomes of AKV and MCF murine leukemia viruses. Proc Natl Acad Sci USA 75:495–499
17. Chien Y-H, Verma IM, Shih TY, Scolnick EM, Davidson N (1978) Heteroduplex analysis of the sequence relations between RNAs of mink cell focus-inducing and murine leukemia viruses. J Virol 28:352–360
18. Chattopadhyay SK, Cloyd MW, Linemeyer DL, Lander MR, Rands E, Lowry DR (1982) Cellular origin and role of mink cell focus-forming viruses in murine thymic lymphomas. Nature 295:25–31
19. Bosselman RA, Van Straaten F, Van Beveren C, Verma IM, Vogt M (1982) Analysis of the *env* gene of a molecularly cloned and biologically active Moloney mink cell focus-forming (MCF) proviral DNA. J Virol 44:19–31
20. Holland C, *et al.* (1983) J Virol 47:413–420
21. Lung ML, Hartley JW, Rowe WP, Hopkins NH (1983) Large RNase T1-resistant oligonucleotides encoding p15E and the U3 region of the long terminal repeat distinguish two biological classes of mink cell focus-forming type C viruses of inbred mice. J Virol 45:275–290
22. Decleve A, Lieberman M, Kaplan HS (1977) In vivo interaction between RNA viruses isolated from the C57BL/ka strain of mice. Virology 81:270–283
23. Strand M, Lilly F, August JT (1974) Host control of endogenous murine leukemia virus expression: Concentrations of viral proteins in high and low leukemia mouse strains. Proc Natl Acad Sci USA 71:3682–3686
24. Lerner RA, Wilson CB, DelVillano BC, McConahey PJ, Dixon RJ (1976) Endogenous oncornaviral gene expression in adult and fetal mice: Quantitative, histologic and physiologic studies of the major viral glycoprotein, gp70. J Exp Med 143:151–166
25. Elder JH, Gautsch JW, Jensen JC, Lerner RA, Chused TM, Morse HC, Hartley JW, Rowe WP (1980) Differential expression of two distinct xenotropic viruses in NZB mice. Clin Immunol Immunopathol 15:493–501
26. Johnson D, Elder J (1983) J Exp Med 159:1751–1756

Host Range and Tissue Tropism

14. Viral Receptors: Expression, Regulation and Relationship to Infectivity
    PATRICK R. MCCLINTOCK AND ABNER LOUIS NOTKINS......... 97

15. Viral Genes and Tissue Tropism
    BERNARD N. FIELDS...................................... 102

16. Utilization of Host Proteins as Virus Receptors
    GREGORY H. TIGNOR, ABIGAIL L. SMITH, AND
    ROBERT E. SHOPE 109

17. Host Range and Tissue Tropisms: Antibody-Dependent Mechanisms
    JAMES S. PORTERFIELD AND M. JANE CARDOSA 117

18. Enveloped Virus Maturation at Restricted Membrane Domains
    RICHARD W. COMPANS 123

19. Viruses and Differentiation: The Molecular Basis of Viral Tissue Tropisms
    ARNOLD J. LEVINE 130

CHAPTER 14
Viral Receptors: Expression, Regulation and Relationship to Infectivity

PATRICK R. MCCLINTOCK AND ABNER LOUIS NOTKINS*

The presence or absence of specific cell surface receptors can influence the host range and tissue tropism of viruses. For example, receptors for poliovirus are known to confer susceptibility on human and some primate cells, while cells of other species lack receptors and are resistant to infection [1]. Although postattachment restrictions are known, virtually all viruses appear to be restricted initially by appropriate receptor expression on the target cells. The most intensively studied viral receptors have been those for the picornaviruses, myxoviruses, and retroviruses. For these, detailed information on the mechanisms of attachment and penetration have been reported [2]. Even the chromosomal locations of the structural genes for some of these receptors are known and have been assigned to human chromosome 19 [3].

Very little is known about the nature or identity of viral receptors; however, a few viruses appear to bind to receptors associated with known cellular functions. The rabies virus receptor appears to be closely associated with the acetylcholine receptor [4]. Epstein–Barr virus has a strong tropism for B lymphocytes and was thought to use the complement C3 receptor [5]; however, later reports have shown that the C3 receptor is distinct from the EBV receptor [6,7]. The human T-cell leukemia virus preferentially infects T cells of the helper/inducer class [8], but the receptor has not been identified. The identification of the major histocompatibility antigens of murine (H-2) and human (HLA) cells as receptors for Semliki forest virus [9] has been challenged [10]. The cardioviruses and influenza A and B viruses bind to the glycophorin A molecule of human erythrocytes [11], and this glycoprotein also carries the M and N blood group antigens. It is possible that many viral

---
* National Institutes of Health, Bethesda, Maryland.

receptors are, in fact, receptors that serve other purposes (e.g., binding of hormones or other biologically active molecules). Of the receptors that have been characterized chemically or enzymatically, most are glycoproteins of which sialic acids appear to be important in the virus–receptor interaction, but no other identification has been made.

For most viruses, the identity of the receptor is unknown and the receptors are widely distributed among different cell types. Viruses such as influenza virus, encephalomyocarditis (EMC) virus, and coxsackie B viruses bind to a wide variety of cell types from different species. In fact, it has been shown that two or more different viruses can compete for attachment sites [12], suggesting that they share the same receptor. Among these receptor-sharing viruses are influenza virus with EMC virus, coxsackievirus B_3 (CB_3) with adenovirus 2 (Ad_2), and coxsackievirus A_{21} (CA_{21}) with human rhinovirus 14 (HR14). For the latter two pairs, it is interesting that the sharing of receptors may also be correlated with their tropism (CB_3 and Ad_2 are enteropathic and CA_{21} and HR14 are upper respiratory viruses). Whether these correlations are merely coincidences or are biologically relevant remains to be studied.

Although it has long been known that there are viral variants, it is now becoming clear that there are also receptor variants. For example, some antigenically indistinguishable, but biologically different, cardiovirus variants appear to utilize different receptors on the same cells [13]. Moreover, receptors on murine and rat mammary cells can differentiate between different isolates of mouse mammary tumor viruses [14]. In fact, selection of viral variants with altered tropisms may occur by repeated passage *in vitro* or *in vivo* in specific tissues through selection of virions with preference for the receptor expressed on these tissues.

Evidence that viral receptors are in a dynamic state on the plasma membrane comes from recent studies using EMC virus [15]. In these experiments, receptors were shown to be induced or modulated by a variety of stimuli and to be regulated by cell growth and differentiation. Receptors on macrophages were increased and receptors on lymphocytes induced by *in vivo* stimulation. Resident splenic or thymic lymphocytes (after removal of adherent cells), as well as purified B and T lymphocytes, did not bind detectable amounts of EMC virus. However, if the lymphocytes were cultured in the presence of mitogens such as phytohemagglutinin, concanavalin A or *E. coli* lipopolysaccharide, receptors appeared within 24 hours after the addition of the mitogen. The induction was dependent upon DNA synthesis and was greatest 48 to 66 hours after stimulation.

Cells undergoing differentiation may also alter the expression of viral receptors, and receptors may be modulated at different stages of cell growth [15]. Cells taken from high density stationary cultures were compared with cells taken from exponentially growing cultures. The rapidly growing cells bound virus at a rate ten times greater than the nondividing cells. If, however, dimethylsulfoxide was added to the medium to induce erythroid differentiation in FLC cells [16], the rate of virus binding did not increase during exponential growth, but decreased slowly over the course of the experiment. In contrast to erythrocytes from other species [17], murine erythrocytes are known to lack EMC virus receptors [18] and thus the inhibition seen may be a programmed feature of murine erythropoiesis. The induction and modulation of EMC virus receptors is probably typical of many virus receptors and follows a pattern shown for hormone receptors. Receptors for insulin and epidermal growth factor have been reported to increase on cells [19,20] as the cells differentiate *in vitro*. Insulin receptors can also be induced following mitogenic stimulation of human lymphocytes [21]. Thus, it is not surprising that expression of receptors for viruses can be similarly regulated.

Lymphocytes already fixed at different stages of differentiation also show differences in the expression of viral receptors. Clonal lines of T and B cell lymphomas from BALB/c mice have been established and well characterized with respect to their cell surface and enzymatic markers [22,23]. When these cell lines were tested for the presence of EMC virus receptors [15], one of two T-cell lines and one of four B-cell lines were receptor positive. Although the sample size permits no generalization, the results do support the possibility that subsets of lymphocytes at different stages of differentiation may differ in the expression of viral receptors.

A strong correlation was found between receptor expression on cells and their susceptibility to virus infection [15,24]. A large number of established cell lines derived from different species were tested for their ability to bind radiolabeled EMC virus and for susceptibility to EMC virus infection. Cell lines that did not express detectable EMC receptors were not susceptible to infection. In contrast, the majority of the cell lines with demonstrable receptors were capable of being infected. The same correlation was found when freshly prepared murine lymphoid and myeloid cells were tested. Normal and stimulated macrophages and mitogen-stimulated lymphocytes expressed EMC virus receptors and were capable of being infected. In contrast, unstimulated lymphocytes did not bind the virus and were resistant to infection.

Although receptors are necessary, they may not be sufficient to confer susceptibility to infection. Human erythrocytes have a binding site for EMC virus [17]; however, since erythrocytes lack the ability to translate the viral message, they do not support viral replication. The C6 rat glioma line has abundant EMC virus receptors, but is restrictive for viral replication (T. Morishima, personal communication). Studies of differentiating skeletal muscle cells have shown that presumptive myoblasts and postfusion muscle fibers are resistant to group A coxsackieviruses [25]. In contrast, myoblasts that have acquired the ability to fuse are susceptible to these viruses. Although these differences in susceptibility were attributed to changes in the number of receptors, recent studies suggest that a postattachment step may be responsible [26]. From these and other studies, it is clear that the presence and quantity of viral receptors play a major role in susceptibility to infection, but postreceptor events also may be involved in restricting the severity and host range of viral infections.

Studies of virus receptors have used assays of infectivity, radioactive virus binding, or purified virus attachment proteins. These methods are limited by the affinity, stability, and availability of purified virions or virion proteins. Moreover, the multivalent nature of the virus attachment process limits the interpretation of assays used to measure receptor sites and affinities. In order to study the role of receptors in determining virus tropism and genetic variations in receptors, it appears that new methods will be necessary. Antireceptor antibodies offer the possibility of specific, high-affinity probes for these molecules. Immunization with cells or cell membranes and existing hybridoma techniques may be applied successfully to this end. Another approach is to use anti-idiotypic antibodies made against antiviral antibodies whose specificity is for the virus attachment proteins. Some of these anti-idiotypic antibodies may mimic the conformation of the virus protein (since both bind to the antiviral antibody) and thus will have antireceptor activity. This technique has been successfully applied to both hormones and viruses and is discussed elsewhere in Chapter 15 of this volume. A third type of affinity reagent is possible using purified viral attachment proteins or synthetic fragments of these proteins. Whatever methods are used, the availability of specific high-affinity ligands will greatly aid in studying the role of receptors in the determination of tissue tropisms, the genetics of host susceptibility, and the expression of receptor variants in target tissues.

References

1. McCaren LC, Holland JJ, Syverton JT (1959) The mammalian cell-virus relationship. I. Attachment of poliovirus to cultivated cells of primate and non-primate origin. J Exp Med 109:475–485
2. Lonberg-Holm KL, Philipson L (eds) (1981) Virus Receptors: Part 2. Animal Viruses. Chapman and Hall, New York
3. McKusick VA (1980) The anatomy of the human genome. Am J Med 69:267–276
4. Lentz TL, Burrage TG, Smith AL, Crick J, Tignor GH (1982) Is the acetylcholine receptor a rabies virus receptor. Science 215:182–184
5. Jondal M, Klein G, Oldstone MBA, Bokish V, Yefenof E (1976) Surface markers on human B and T lymphocytes. VIII. Association between complement and Epstein–Barr virus receptors on human lymphoid cells. Scand J Immunol 5:401–410
6. Patel P, Menezes J (1981) Epstein–Barr virus (EBV)-lymphoid cell interactions. I. Quantification of EBV particles required for the membrane immunofluorescence assay and the comparative expression of EBV receptors on different human B, T and null cell lines. J Gen Virol 53:1–11
7. Wells A, Koide N, Stein H, Gerdes J, Klein G (1983) The Epstein–Barr virus receptor is distinct from the C3 receptor. J Gen Virol 64:449–453
8. Popovic M, Sarin PS, Robert-Gurroff M, Kalyanaraman VS, Mann D, Minowada J, Gallo RC (1983) Isolation and transmission of human retrovirus (human T-cell leukemia virus). Science 219:856–859

9. Helenius A, Morein B, Fries E, Simons K, Robinson P, Schirrmacher V, Terhorst C, Strominger JL (1978) Human (HLA-A and HLA-B) and murine (H-2K and H-2D) histocompatibility antigens are cell surface receptors for Semliki-forest virus. Proc Natl Acad Sci USA 75:3846-3850
10. Oldstone MBA, Tishon A, Dutko FJ, Kennedy SIT, Holland JJ, Lampert PJ (1980) Does the major histocompatibility complex serve as specific receptor for Semliki forest virus. J Virol 34:256-265
11. Enegren BJ, Burness ATH (1977) Chemical structure of attachment sites for viruses on human erythrocytes. Nature 268:536-537
12. Lonberg-Holm K, Crowell RL, Philipson L (1976) Unrelated animal viruses share receptors. Nature 259:679-681
13. Morishima T, McClintock PR, Aulakh GS, Billups LC, Notkins AL (1982) Genomic and receptor attachment differences between mengovirus and encephalomyocarditis virus. Virology 122:461-465
14. Altrock BW, Arthur LO, Massey RJ, Schochetman G (1981) Common surface receptors on both mouse and rat cells distinguish different classes of mouse mammary tumor viruses. Virology 109:257-266
15. Morishima T, McClintock PR, Billups LC, Notkins AL (1982) Expression and modulation of virus receptors on lymphoid and myeloid cells: Relationship to infectivity. Virology 116:605-618
16. Friend C, Scher W, Holland JG, Sato T (1971) Hemoglobin synthesis in murine virus-induced leukemic cells *in vitro:* Stimulation of erythroid differentiation by dimethylsulfoxide. Proc Natl Acad Sci USA 68:378-382
17. Angel MA, Burness ATH (1977) The attachment of encephalomyocarditis virus to erythrocytes from several animal species. Virology 83:428-432
18. McClintock PR, Billups LC, Notkins AL (1980) Receptors for encephalomyocarditis virus on murine and human cells. Virology 106:261-272
19. Heath J, Bell S, Rees AR (1981) Appearance of functional insulin receptors during the differentiation of embryonal carcinoma cells. J Cell Biol 91:293-297
20. Rees AR, Adamson ED, Graham CF (1979) Epidermal growth factor receptors increase during the differentiation of embryonal carcinoma cells. Nature 281:308-311
21. Krug U, Krug F, Cuatrecasas P (1972) Emergence of insulin receptors on human lymphocytes during *in vitro* transformation. Proc Natl Acad Sci USA 69:2604-2608
22. Kim KJ, Weinbaum FI, Mathieson BJ, McKeever PE, Asofsky R (1978) Characteristics of BALB/c T cell lymphomas grown as continuous *in vitro* lines. J Immunol 121:339-344
23. Kim KJ, Kanellopoulos-Langevin C, Merwin RM, Sachs DH, Asofsky R (1979) Establishment and characterization of BALB/c lymphoma lines with B cell properties. J Immunol 122:549-554
24. McClintock PR, Morishima T, Notkins AL (1983) Expression and modulation of virus receptors: Relationship to infectivity. *In* Scolnick EM, Levine AJ (eds) Tumor Viruses and Differentiation. UCLA Symposium on Molecular and Cellular Biology, Vol. 5, p 299
25. Goldberg RJ, Crowell RL (1971) Susceptibility of differentiating muscle cells of the fetal mouse in culture to coxsackievirus A13. J Virol 7:759-769
26. Schultz M, Crowell RL (1983) Eclipse of coxsackievirus infectivity: The restrictive event for a non-fusing myogenic cell line. J Gen Virol 64:1725-1734

CHAPTER 15
Viral Genes and Tissue Tropism

BERNARD N. FIELDS*

A major feature of viral infections is their localization in specific cells and tissues [1]. This specific localization of different viruses into different tissues, as well as to different cells within those tissues, results in distinct patterns of infectious diseases [2]. Recent studies with the mammalian reoviruses and their infection of newborn mice have identified the specific viral protein that determines tropism (the viral hemagglutinin) and have uncovered approaches to alter regions of the hemagglutinin selectively, thereby altering virulence and changing tropism [3].

The approach our laboratory has taken in these studies is based on two observations: (1) although the three mammalian reovirus serotypes (serotypes 1, 2, and 3) are similar in their capsid structure and genomic organization, they interact with newborn mice in distinct fashions [4,5]; (2) the fact that they contain segmented genomes has allowed us to generate reassortant viruses consisting of genes (dsRNA segments) derived from both parental serotypes [6]. Since each segment can be identified as to its parental origin, it is possible to identify the origin of each gene in hybrid viruses consisting of genes derived from two different parental viruses and thus to identify the role of each of the viral genes in pathogenesis.

To identify the genes responsible for tropism, initially we confirmed earlier studies that suggested that reovirus types 1 (T1) and type 3 (T3) differed in their neurotropism [7,8]. Following intracerebral inoculation into newborn mice, reovirus T1 localized primarily in ependymal cells, producing a nonlethal infection that frequently leads to hydrocephalus. By contrast, reovirus

---

* Harvard Medical School and Brigham and Women's Hospital, Boston, Massachusetts.

T3, produces a highly lethal encephalitis in which the virus has localized in neurons in several regions of the brain.

Studies of the diseases produced in newborn mice following intracerebral inoculation with various reassortant viruses allowed us to identify the S1 gene (the gene encoding the viral hemagglutinin), as the gene solely responsible for those differences in tropism [8,9]. Reassortants containing nine genes from type 3 and the S1 gene from type 1, produce destruction of ependymal cells, accumulate antigen in ependymal cells, and lead to a nonfatal infection, all similar to the parental T1 virus. Reassortants containing nine genes from type 1 and the S1 gene from type 3 produce extensive destruction of neurons, antigen localizes in neurons, and a lethal encephalitis results that is similar to that of the parental T3 virus.

A number of other studies indicate that other features of viral host interactions are properties of the viral hemagglutinin (Table 1) (reviewed in [10]). These include binding to, and agglutination of red blood cells (hence "hemagglutinin"), specificity in humoral immunity (both hemagglutination inhibiting and neutralizing antibodies) and cellular immunity (as detected by cytotoxic T lymphocytes [CTL], T cells that mediate delayed-type hypersensitivity [TDTH], and suppressor T cells [ts]).

Table 1. Role of the S1 dsRNA segment—σ1 protein (the viral hemagglutinin)[a]

I. Cell and tissue tropism

  a) Nervous system
    Type 1 = ependyma
    Type 3 = neurons
  b) Pituitary
    Type 1 = anterior
    Type 3 = intermediate and posterior

II. Specificity of host immune response

  a) Humoral
    Neutralizing antibody
    Hemagglutination inhibition antibody
  b) Cellular
    Cytolytic T lymphocytes (T_{CTL})
    T-cell dependent delayed type hypersensitivity (T_{DTH})
    Suppressor T cells and tolerance (Ts)

III. Interaction with cells

  a) Binding to cellular microtubules
  b) Binding to lymphocytes and to cell membranes
  c) Inhibition of cellular DNA synthesis

[a] From [10] with permission.

To better understand the role of the hemagglutinin, we wondered whether the various properties of the viral hemagglutinin protein were properties of the same, or different regions. In order to study this problem, we isolated a number of monoclonal antibodies directed at the reovirus T3 hemagglutinin [11]. We specifically focused on the T3 hemagglutinin because of our interest in probing the interaction of viruses with neurons.

The results of these studies indicate that the reovirus T3 hemagglutinin has at least three discrete functional and structural domains (Fig. 1) [11,12]. One domain is involved in viral neutralization, a second domain is located at the site that binds to red blood cells, a third is antigenically active but has no known functional correlates. While some of the monoclones appear to overlap these distinct epitopes, in general there is a sharp distinction of the function of the epitopes on the viral HA. In addition to distinct interactions with antibodies, the three epitopes on the viral HA show specificity in the way they interact with various T cells. The same monoclonal antibodies that

ANTIGENIC DOMAINS of the REOVIRUS TYPE 3 HA

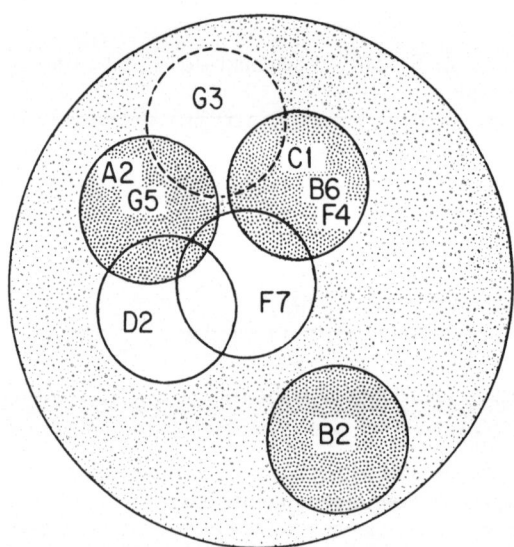

Fig. 1. Model for the antigenic domains on the reovirus type 3 hemagglutinin. Monoclonal antibodies that bind to distinct domains on the hemagglutinin are indicated in circles containing dots. Antibodies that bind to epitopes which overlap these distinct domains are indicated in open circles. The broken circle around the G3 antibody indicates that the RIA data were not conclusive enough to make a definite assignment. The A2 and G5 site is the site responsible for viral neutralization. The C1, B6, F4 site is the site responsible for red blood cell binding. The B2 site has no known function. From Science 220:505–507. Copyright 1983 by the AAAS.

neutralize viral infectivity block the majority of CTL-target interactions, while the others do not [13]. Hence the NT epitope is also the major determinant responsible for recognizing CTL. Studies using HA antigenic variants (see below) led to the same conclusion.

In addition to blocking HA functions, monoclonal antibodies provide important tools for selecting mutants directed at selected regions of biologically important proteins [14]. Such mutants may be used as an alternative to study functions of epitopes on viral polypeptides. Thus our major goal in isolating HA variants was to alter the NT epitope in order to determine whether such alterations might affect the interaction of the reovirus hemagglutinin with neurons in the central nervous system. One of the neutralizing monoclonal antibodies was used to select reovirus antigenic variants that were no longer neutralized by the two neutralizing monoclonal antibodies [15]. The variants were then tested for their neurovirulence by determining their capacity to replicate in mouse brains and cause fatal encephalitis. The results indicate that all of the variants that we have selected in this manner are markedly less virulent than the parental type 3 virus from which they were selected. Detailed analyses of the brain of mice infected by the variants reveal that, while the parental Dearing strain produce a highly lethal encephalitis with lesions widely distributed throughout the brain, the variants induce pathology in certain regions of the brain; i.e., the limbic system, especially hippocampus and hypothalamus (Fig. 2) [16]. Thus the variants display a restricted tropism as compared with the parental virus.

In addition to altering neurotropism, target cells infected with these variants are not lysed as well and are not as capable of cold target inhibition as the parental type 3 infected cells, indicating that the altered hemagglutinin of the variant is not as well recognized by type 3 specific CTL [13]. The pleitropic effect of mutation in the NT epitope, involving critical interactions with neutralizing antibody, CTL, and neurons, suggests that this domain is the portion of the viral HA responsible for interacting with specific receptors on cell surfaces.

Recent studies have extended these findings and used a novel immunologic approach to study viral HA receptor interactions [17]. The similarity of the binding of the viral HA to NT antibody, CTL, and neurons, and its loss of reactivity in the variants suggested to us that the binding of the reovirus type 3 HA to the idiotype on the NT monoclonal antibody might resemble the binding site on CTL and neurons. Investigating the antigen-binding site on the antibody might therefore provide insight into the nature of the binding site of the cell surface receptor for $\sigma 1$. In fact, antisera raised against the idiotype (anti-idiotype) appear to recognize the same binding structure both on the NT monoclonal antibodies and on T cells. The identification of critical cell surface determinants recognized by these specific immunologic probes should shed light on the nature of the viral receptor and thus provide a more detailed explanation for viral tropism [18].

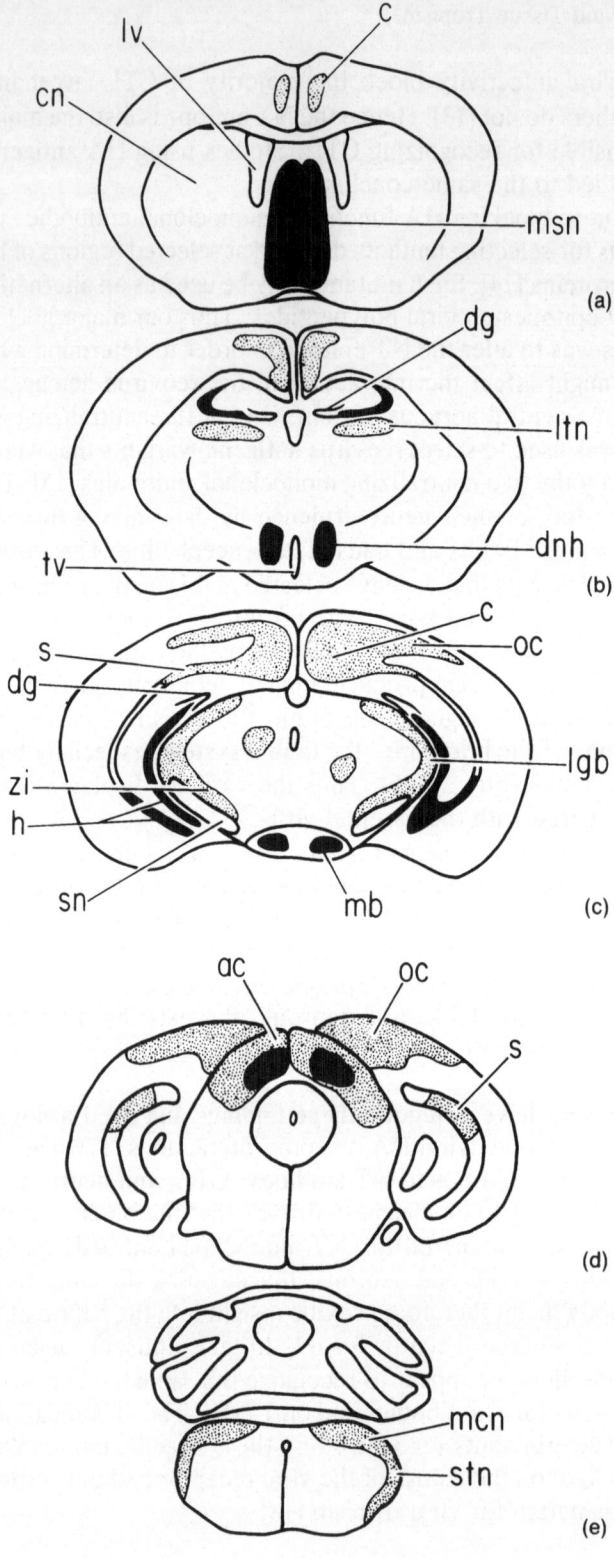

References

1. Smith H, Skehel JJ, Turner ML (1980) The molecular basis of microbial pathogenicity. Verlag Chemie, Basel
2. Mims CA (1976) The pathogenesis of infectious disease. Academic Press, London
3. Fields BN, Greene MI (1982) Genetic and molecular mechanisms of viral pathogenesis: Implications for prevention and treatment. Nature 300:19–23
4. Margolis G, Kilham L, Gonatos N (1971) Reovirus type III encephalitis: Observations of virus–cell interactions in neural tissue. I. Light microscopy studies. Lab Invest 24:91–109
5. Kilham L, Margolis G (1969) Hydrocephalus in hamsters, ferrets, rats and mice following inoculation with reovirus type 1. II. Pathologic studies. Lab Invest 21:189–198
6. Fields BN (1981) Genetics of reovirus. Curr Top Microbiol Immunol 91:1–24
7. Raine CS, Fields BN (1973) Ultrastructural features of reovirus type 3 encephalitis. J Neuropath Exp Neurol 32:19–33
8. Weiner HL, Drayna D, Averill DR, Fields BN (1977) Molecular basis of reovirus virulence: The role of the S1 gene. Proc Natl Acad Sci USA 74:5744–5748
9. Weiner HL, Powers ML, Fields BN (1980) Absolute linkage of virulence with central nervous system cell tropism of reovirus to hemagglutinin. J Infect Dis 141:609–616
10. Fields BN (1982) Molecular basis of reovirus virulence: Brief review. Arch Virol 71:95–107
11. Burstin SJ, Spriggs DR, Fields BN (1982) Evidence for functional domains on the reovirus type 3 hemagglutination. Virology 117:146–155
12. Spriggs DR, Kaye K, Fields BN (1983) Topical and functional analysis of the reovirus type 3 hemagglutinin. Virology 127:220–224
13. Finberg R, Spriggs DR, Fields BN (1982) Host immune response to reovirus: CTL recognize the major neutralization domain of the viral hemagglutinin. J Immunol 129:2235–2238
14. Webster RG, Laver WG (1980) Determination of the number of nonoverlapping antigenic areas on Hong Kong (H3N2) influenza virus hemagglutinin with monoclonal antibodies and the selection of variants with potential epidemiological significance. Virology 104:139–148

Fig. 2. (a to e) Diagrams of coronal brain sections extending from (a) rostral to (e) caudal areas of the brain showing the location of virus-induced lesions in mice infected intracerebrally with either the Dearing strain of reovirus type 3 or a variant Dearing virus. Black areas indicate regions of the brain where Dearing virus as well as the variant viruses induced lesions; dotted areas show regions where only the Dearing virus caused necrosis. From [16] with permission.

Abbreviations: AC, anterior colliculus; C, cingulum; CN, caudate nucleus; DG, dentate girus; DNH, dorsomedial nucleus of hypothalamus; H, hippocampus; LGB, lateral geniculate body; LTN, lateral thalamic nucleus; LV, lateral ventricle; MB, mammilary body; MCN, media cuneate nucleus; MSN, medial septum nucleus; OC, occipital cortex; S, subiculum; SN, substantia nigra; STN, spinal trigeminal nucleus; TV, third ventricle; ZI, zona incerta. From Science 220:505–507. Copyright 1983 by the AAAS.

15. Spriggs DR, Fields BN (1982) Attenuated reovirus type 3 strains generated by selection of hemagglutinin antigenic variants. Nature 297:68–70
16. Spriggs DR, Bronson RT, Fields BN (1983) Hemagglutinin variants of reovirus type 3 have altered central nervous system tropism. Science 220:505–507
17. Nepom JT, Weiner H, Dichter M, Spriggs DR, Gramm C, Powers ML, Fields BN, Greene MI (1982) Identification of a hemagglutinin specific idiotype associated with reovirus recognition shared by lymphoid and neural cells. J Exp Med 155:155–167
18. Nepom JT, Tardieu M, Epstein RL, Weiner HL, Gentsch J, Fields BN, Greene MI (1982) Antiidiotype to a reovirus binding site related idiotype interacts with peripheral lymphocyte subsets. Surv Immunol Res 1:255–261

CHAPTER 16
Utilization of Host Proteins as Virus Receptors

GREGORY H. TIGNOR, ABIGAIL L. SMITH, AND
ROBERT E. SHOPE*

A general construct in microbiology is that parasites and their hosts continuously evolve to ensure the survival of each. A significant portion of virologic research has been devoted to describing strategies used by different viruses to survive despite attempts by the host to prevent their replication. One hypothesis holds that virus attachment to host cell molecules unique to susceptible cells governs virus spread to, and infection of, specific target organs [1]. Our focus is on the implication of recent research developments that support this hypothesis. The observations have been made with viruses from diverse genera and with cells of different functional types. It may be overly simplistic to state that tissue tropisms are readily defined on the basis of attachment *per se*, but attachment is a logical starting point in studying the evolution of virus–host interactions, since it has been shown that both virus and host-cell receptors are genetically controlled [1,2]. An excellent review of general principles related to animal virus attachment has been written recently by Holmes [3]. A key issue in relating host-cell molecules to virus attachment is identification of the "productive receptor" [4]. This is the receptor that leads to infection rather than one that binds virus nonspecifically. Many enveloped viruses bind well to a number of nonproductive substrates including inanimate surfaces, such as polystyrene plates, which are not obviously related to tissue tropisms *in vivo*.

Differential analysis of low- and high-affinity virus–cell interactions may lead to clearer understanding of the pathogenetic importance of virus attachment to host-cell molecules. Multi-affinity binding curves with nonspecific and specific components for viruses have not been routinely reported, although these curves are regularly described in pharmacologic binding as-

---
* Yale University School of Medicine, New Haven, Connecticut.

says. Studies with Rous sarcoma virus (RSV) illustrate the complexities of interpreting adsorption data in the absence of affinity constants for virus binding. Piraino [5] and Crittendon [6] showed that the RSV that bears subgroup A glycoproteins (RSV-A) adsorbed to resistant cells as well as to susceptible cells. Subsequently, Weiss [7] adsorbed virus to resistant and susceptible cells, and after incubation, performed infectious center assays on new monolayers of resistant or susceptible cells. His conclusion was that virus that adsorbed to resistant cells did not penetrate the cell surface, but merely adhered to it. When these cells came into contact with a susceptible monolayer, a reduced, but substantial fraction of the virus was available for interaction with the receptors of the susceptible assay cells. He confirmed that the resistant cells were serving as a physical vector for the infectious particles by demonstrating that the virus that adsorbed to susceptible cells apparently penetrated; the virus became rapidly refractory to antibody inactivation and was not infectious when plated on new monolayers. It is a perfectly tenable hypothesis that, in the case of RSV-A, the specificity of the receptor site is governed by differences in binding affinities which, in turn, affect the rate of penetration. Nevertheless, virus attachment to nonpermissive cells, or the inert molecules of supporting matrices such as collagen, may serve one important pathogenetic function: Virus particles can be concentrated by this mechanism in regions where there are increased densities of specific high-affinity host-cell receptors on target organs.

Experimental studies with rabies virus emphasize the importance of the low–high-affinity binding concept. Rabies virus has evolved as a pathogen of the central nervous system (CNS). Normally the CNS is a sheltered environment with limited access sites for viruses. The morphologic and structural peculiarities of the CNS including the blood–brain and the blood–cerebro-spinal-fluid barriers effectively defend against virus infection since most are non-neurotropic. However, at neuromuscular junctions (NMJ), the peripheral nervous system is open to the microenvironment. Specialized molecules related to neurotransmission are incorporated into nerve terminals at this site. Rabies virus has a special affinity for the NMJ [8] and for the acetylcholine receptor (AChR) or a closely related molecule [9]. Thus, the virus has apparently evolved in such a way that it circumvents normal host barriers to CNS entry by using as a receptor a molecule essential for normal host-cell function. Yet, it was reported during the course of this work [9] that rabies virus bound to other surfaces including both collagen and fibroblasts. Specific ligands, which attached to high density patches of AChR on chick-embryo myotubes, significantly reduced the number of rabies virus-infected myotubes. However, fibroblasts were infected in the presence of these drugs. Thus, rabies virus is capable of binding to at least two different molecules. We postulate that the "productive receptor" is the AChR. Binding of virus to the AChR enhances the probability of virus entering motor nerves because of the high concentration of AChR at the tips of the junctional folds. Both fibroblasts and collagen overlie the NMJ. Therefore, nonspecific or low-affinity binding of virus to collagen or a molecule on fibroblasts, even in the absence of productive infection would also increase the probability of uptake into motor nerves by concentrating the virus at this general region in the vicinity of the NMJ. A diagrammatic view of potential virus binding sites at the NMJ is presented in Fig. 1.

Since the NMJ is open to its microenvironment in the muscle, other viruses that multiply in muscle tissue might similarly enter the nervous system at this site. Several investigators have shown that retrograde transport is a mechanism for the spread of polio, herpes, and pseudorabies viruses, and tetanus and diphtheria toxins in the nervous system (reviewed in [10]). Carbohydrate-containing macromolecules have been localized to synapses by the technique of horseradish peroxidase (HRP) labeling of lectins. Red blood cells coated with concanavalin A attached to axon membranes of chick sympathetic ganglia neurons, and these complexes moved slowly in the retrograde direction at the surface [10]. Similarly, soluble HRP was taken up at active axon terminals [10]. The selectivity of uptake can depend on properties common to all axon terminals; for example, tetanus toxin was taken up presynaptically and transported to all neurons examined. For other neurobiologically active substances, uptake was restricted to certain populations of neurons; for example, nerve growth factor was taken up by sympathetic and large dorsal root ganglion neurons but not by motor neurons [10]. An additional example of selectivity of uptake with respect to the population of neurons is the transport of antibodies raised to dopamine-β-hydroxylase to sympathetic nerve cell bodies, but not in detectable amounts to either sensory or motor neurons [10]. It has been suggested that membrane receptors at the presynaptic axon terminals provide specific binding sites for these substances. Kristensson [10] suggested that the existence of specific receptors at axon terminals may explain why certain viruses attack neurons while other viruses do not.

Among the arboviruses, many of which cause neurotropic disease in man, the study of attachment is complicated by the fact that, in addition to having diverse tissue tropisms, these viruses have a biological cycle that includes both arthropods and vertebrates. One can postulate that the "productive receptor" in the arthropod is different from that in the vertebrate animal. This might require two surface antigens on the virion, one for attachment to the arthropod cell and another for the vertebrate, a tenable hypothesis since most arboviruses have two surface glycoproteins. We have shown that host-cell receptors on mammalian and mosquito cell lines differ in susceptibility to proteolytic enzymes (Smith and Tignor 1980, reviewed in [10]). In general, however, molecules are conserved during the course of evolution. The most economical strategy for a given neurotropic arbovirus would entail the use of a single productive receptor. This receptor would have to be widely distributed in nature but be present at increased concentration in the nervous system. Yet, such a molecule would have to vary from species to species to allow for the host specificities observed in nature. Neurotransmitter or re-

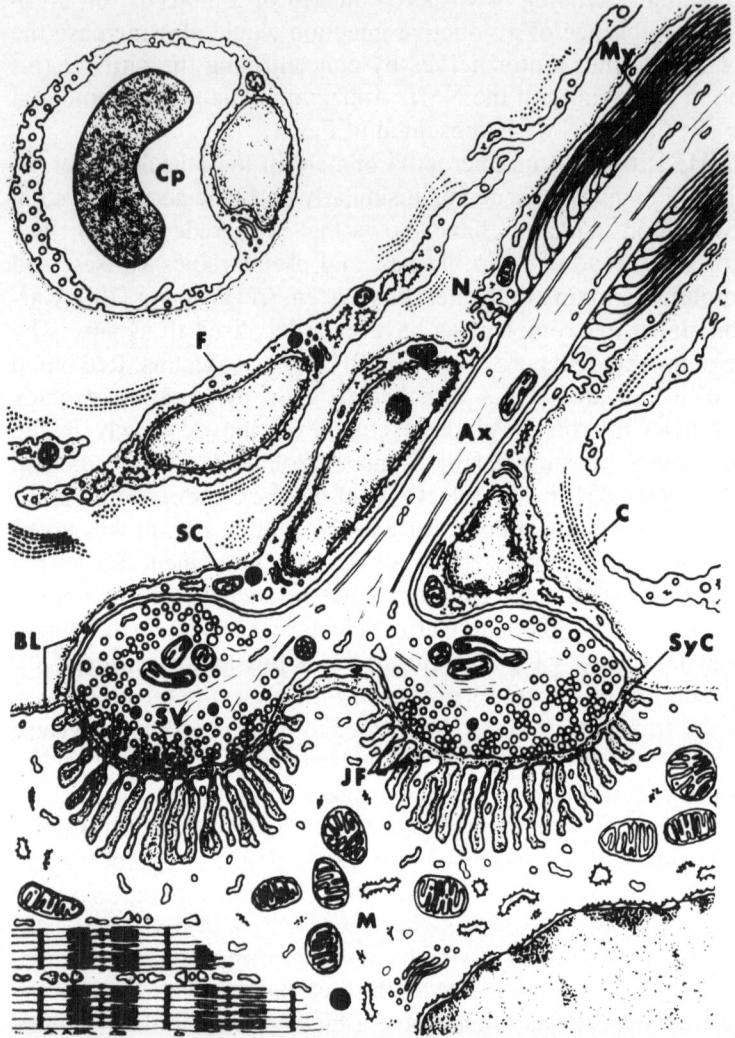

Fig. 1. Diagram illustrating the structures occurring at the neuromuscular junction of vertebrate skeletal muscle. The myelin sheath (My) of the motor nerve axon (Ax) ends preterminally while Schwann cells (SC) continue to envelop the distal axon and nerve ending. The axon terminates in boutons lying within indentations on the myofiber (M) surface. The terminals are filled with synaptic vesicles (SV). The terminal is separated from the muscle surface by a synaptic cleft (SC) while the postsynaptic muscle surface is thrown up into junctional folds (JF). A basal lamina (BL) covers the muscle surface and Schwann cells and occupies the synaptic cleft. The extracellular space contains abundant collagen fibrils (C). Fibroblasts (F) are present external to the axon and nerve terminal. Capillaries (Cp) occur in the intercellular spaces as well. Virus particles introduced into the extracellular spaces are accessible to collagen fibrils, basal lamina, and the surfaces of myofibers, Schwann cells, fibroblasts and capillaries. Sites of accessibility to the neuron are the presynaptic axolemma facing the synaptic cleft and the axon membrane at the distal node of Ranvier (N).

lated molecules fit this category quite neatly. Recent studies with monoclonal antibodies to the AChR, for example, have revealed species-specific differences in the molecule.

We looked for accumulations of viral antigen at neuromuscular junctions using parental and reassortant strains of two California serogroup viruses [11]. These viruses had different pathogenetic potential after peripheral inoculation of adult mice. LaCrosse virus is associated with lethal infection of adult mice after subcutaneous injection, and Tahnya virus does not kill mice similarly inoculated. Virus antigen and the NMJ were colocalized by dual staining of infected muscle tissue with fluorescein-labeled antiviral antibody and rhodamine-conjugated alpha-bungarotoxin, a ligand specific for the AChR. FITC-labeled antibody to parental LaCrosse virus was colocalized with the NMJ early after subcutaneous infection. In later stages of infection, LaCrosse virus apparently multiplied selectively in the *arrector pili,* muscle which is cholinergically innervated; however, the precise molecule to which the virus bound was not identified. Parental Tahyna virus was not found at the NMJ. A LaCrosse–Tahyna reassortant virus bearing the LaCrosse middle-sized gene segment, which codes for two virion surface glycoproteins, was colocalized with the NMJ suggesting that viral genetics play an important role in determining the site of early antigen accumulation. Sindbis virus has also been localized at the NMJ, but preliminary binding data [12] suggest that the virus binds to a molecule on the presynaptic nerve terminal, possibly a catecholaminergic neurotransmitter molecule (see Fig. 1).

We demonstrated ultrastructural changes in the NMJ after rabies virus infection. Degenerative cytopathologic changes usually involved the presynaptic (nerve) element but occasionally the postsynaptic muscle cell as well. Morphologic disarray of the junctional folds themselves was observed in several instances. Immune and/or phagocytic cells were found in the extracellular space at the NMJ. We were unable to resolve with confidence whether rabies virus infection antigenically alters the AChR. Such an alteration could be important as a possible mechanism by which virus infections might initiate autoimmune responses to the AChR, as in myasthenia gravis. Korn and Abramsky [13] reported the onset of myasthenia gravis in five individuals a few weeks after exposure to a variety of viruses or virus immunizations. Additional studies using monoclonal antibodies and electrophoretic protein–protein transfers will undoubtedly lead to more precise definition of the host-cell molecules to which viruses bind at synaptic regions. This definition should clarify the mechanisms governing viral neurotropism in acute and chronic virus infections of both the neuromuscular tissue and the central nervous system.

During the evolution of virus tropisms other than neurotropism, it is a reasonable speculation that viruses would develop high-affinity receptors for a host-cell molecule on immune and/or phagocytic cells. Attachment to and infection of these cell types would enable infecting virions to circumvent a potent host defense against virus invasion. Evidence is accumulating for a relationship between Epstein–Barr virus (EBV) and a complement receptor (C3R), or a closely related molecule, on B lympocytes. Based on serologic surveys, EBV is a highly prevalent agent associated, usually, with subclinical infection. It is causally associated with heterophile-positive infectious mononucleosis cases and is associated also with African Burkitt lymphoma, another lymphoproliferative disease. In both diseases, some of the dividing cells are B lymphoblasts infected with EBV. Disease is a consequence of virus-induced cell proliferation and the host's response to the lymphoproliferation.

In 1976, several lines of evidence were presented for an association between EBV and the C3R. *In vitro* studies using fluorescein- and rhodamine-conjugated antisera colocalized EBV and the C3R on the cell membrane [14]. Kinetics of co-capping of the EBV receptor and C3R were also identical [15].

Several methods have been used to convert nonbinding cells to EBV positive cells. One innovative approach was through receptor implantation, accomplished using Sendai virus envelopes as vehicles [16]. Tritiated thymidine-labeled EBV bound to receptor-implanted cells but not to untreated, receptor-negative cells. In a second method, complement receptors were induced with theophylline [17] with a concomitant appearance of EBV receptors on previously receptor-negative cells. The induced receptors bound C3 with significantly reduced avidity compared with naturally occurring C3R.

Recently, evidence for distinction of the EBV receptor and C3R has been presented [18], despite the fact that they are often coexpressed and closely associated. Not all C3R-positive cells adsorb EBV, suggesting that a subclass of receptors, a C3R-associated molecule, or an as yet unidentified molecule may serve as the EBV receptor [19].

In vivo transformation of epithelial cells by EBV is implicated as the mechanism of induction of undifferentiated nasopharyngeal carcinoma. Recently, receptor-negative cells were shown to bind EBV after treatment with either inactivated Sendai virus or reconstituted Sendai virus envelopes [20]. Since epithelial cells are normally EBV receptor-negative, the authors postulate a role for "helper" viruses that could mediate EBV binding as the first step in transformation and carcinoma induction.

With EBV, as with rabies virus, the final identity of the host-cell receptor awaits resolution by biochemical analyses. These attempts will be accompanied by additional difficulties and controversies since high-affinity binding properties may depend on tertiary structure, which cannot be preserved during the purification procedures currently in use. Nevertheless, attempts to develop a functional characterization provide significant guideposts for directing future pathogenetic studies.

Acknowledgments. These studies were supported in part by NIH Grants AI-11132 and AI-12541. The line drawing was provided by Thomas Lentz of the Section of Cell Biology at the Yale University School of Medicine.

References

1. Crowell RL (1976) Comparative generic characteristics of picornavirus-receptor interactions. *In* Beers RF Jr, Bassett EG (eds) Cell Membrane Receptors for Viruses, Antigens and Antibodies, Polypeptide Hormones, and Small Molecules. Raven Press, New York, p 202
2. Green H (1974) The gene for the poliovirus receptor. N Engl J Med 290:1018–1019
3. Holmes KV (1981) The biology and biochemistry of cellular receptors for enveloped viruses. *In* Lonberg-Holm K, Philipson L (eds) Virus Receptors. Part 2. Chapman and Hall, London, p 85
4. Dimmock NJ (1982) Initial stages in infection with animal viruses. J Gen Virol 59:1–22
5. Piraino F (1967) The mechanism of genetic resistance of chick embryo cells to infection by Rous sarcoma virus (BS-RSV). Virology 32:700–707
6. Crittenden LB (1968) Observations on the nature of a genetic resistance to avian tumor viruses. J Natl Cancer Inst 41:145–153
7. Weiss RA (1976) Receptors for RNA tumor viruses. *In* Beers RF Jr, Bassett EG (eds) Cell Membrane Receptors for Viruses, Antigens and Antibodies, Polypeptide Hormones, and Small Molecules. Raven Press, New York, p 237
8. Watson HD, Tignor GH, Smith AL (1981) Entry of rabies virus into the peripheral nerves of mice. J Gen Virol 56:371–382
9. Lentz TL, Burrage TG, Smith AL, Crick J, Tignor GH (1982) Is the acetylcholine receptor a rabies virus receptor? Science 215:182–184
10. Kristensson K (1978) Retrograde transport of macromolecules in axons. Ann Rev Pharmacol Toxicol 18:97–110
11. Tignor GH, Shope RE, Bishop DHL, Smith AL, Burrage TG (1983) California serogroup gene structure-function relationships: Virulence and tissue tropisms. *In* Calisher CH (ed) California Serogroup Viruses. Alan R. Liss, New York, in press
12. Smith AL (1978) *In vitro* characterization of host cell receptors in a Sindbis virus model of neurovirulence. Thesis, Yale University, New Haven, Connecticut
13. Korn IL, Abramsky O (1981) Myasthenia gravis following viral infection. Eur Neurol 20:435–439
14. Yefenof E, Klein G, Jondal M, Oldstone MB (1976) Surface markers on human B and T lymphocytes. IX. Two-color immunofluorescence studies on the association between EBV receptors and complement receptors on the surface of lymphoid cell lines. Int J Cancer 17:693–700
15. Jondal M, Klein G, Oldstone MB, Bokish V, Yefenof E (1976) Surface markers on human B and T lymphocytes. VIII. Association between complement and Epstein–Barr virus receptors on human lymphoid cells. Scand J Immunol 5:401–410
16. Volsky DJ, Shapiro IM, Klein G (1980) Transfer of Epstein–Barr virus receptors to receptor-negative cells permits virus penetration and antigen expression. Proc Natl Acad Sci USA 77:5453–5457
17. Magrath I, Freeman C, Santaella M, Gadek J, Frank M, Spiegel R, Novikovs L (1981) Induction of complement receptor expression in cell lines derived from human undifferentiated lymphomas. II. Characterization of the induced complement receptors and demonstration of the simultaneous induction of EBV receptor. J Immunol 127:1039–1043
18. Hutt-Fletcher LM, Fowler E, Lambris JD, Feighny RJ, Simons JG, Ross GD

(1983) Studies of the Epstein–Barr virus receptor found on Raji cells. II. A comparison of lymphocyte binding sites for Epstein–Barr virus and C3d. J Immunol 130:1309–1312
19. Sugden B (1982) Epstein–Barr virus: A human pathogen inducing lymphoproliferation in vivo and in vitro. Rev Infect Dis 4:1048–1061
20. Khelifa R and Menezes J (1983) Sendai virus envelopes can mediate Epstein–Barr virus binding to and penetration into Epstein–Barr virus-receptor negative cells. J Virol 46:325–332

CHAPTER 17
Host Range and Tissue Tropisms: Antibody-Dependent Mechanisms

JAMES S. PORTERFIELD AND M. JANE CARDOSA*

The classic concept of viral infection involves an initial interaction between determinants on the surface of a virus and cellular receptor binding sites of appropriate viral specificity, followed by stages of penetration, uncoating, and replication of the viral genome. With most infections, antiviral antibody will inhibit the initial binding step, and in so doing, will reduce viral infectivity. However, in certain cell–virus interactions, the presence of antiviral antibody increases infectivity, a phenomenon which is known as *antibody-dependent enhancement*, or ADE, of viral infectivity [1–4]. The explanation for this apparent paradox is that this type of infection is mediated, not by specific viral receptors, but by cellular receptors for components of the immune system that function as accessory viral receptors by binding infectious complexes of antiviral antibody with virus, this unusual binding step being followed by later replicative stages. ADE is thus conditional upon the presence on host cells of the necessary accessory receptors, but it is also determined by the nature of the infecting virus, by the specificity of the antiviral antibodies, and by the conditions under which the cells, virus, and antibody are brought together.

Cellular Receptors Mediating ADE

There are two principal classes of receptors for components of the immune system, namely, cellular receptors for the Fc portion of immunoglobulin molecules (Fc receptors, or FcR), and receptors for components of the complement system (complement receptors, or CR). FcR are present on a

---

* Sir William Dunn School of Pathology, Oxford, England.

variety of immunocompetent cells, and CR are found on an even wider range of cell types [5]. Both classes are heterogeneous, both are variably expressed at the cell surface and may be inducible following viral infection or injury [6].

The concept that FcR might be capable of mediating ADE was first put forward by Halstead and his associates [2,7] in the context of hemorrhagic dengue and the dengue shock syndrome. Using preparations of primary peripheral blood white cells of human or primate origin, they were able to demonstrate substantially enhanced yields of dengue virus in the presence of subneutralizing concentrations of antiviral IgG, but not IgM or F(ab')$_2$ fragments of antibody. Enhancement was associated with cells of the monocyte–macrophage lineage, which are rich in FcR. However, the experimental system was difficult, probably because of heterogeneity in the primary cells being used. The substitution of continuous lines of macrophage-like cells for the primary cell cultures greatly simplified experimentation, and made possible a number of studies on the mechanism underlying ADE [3,8]. Mouse macrophages are now well characterized and are known to have at least two FcR, readily distinguishable by a monoclonal antibody, 2.4G2, which binds strongly to the trypsin-resistant FcRII. In a model system involving the mouse macrophage line P388D1, the flavivirus West Nile virus (WNV) and anti-WNV antibodies, pretreatment of cells with 2.4G2 antibody blocked ADE, clearly implicating FcRII in this enhancement [8]. Further evidence for the role of FcR has come from the use of hybrid cells, ADE being demonstrable only in those hybrids that contain FcR [9] and from observations that aggregated IgG will block enhancement [10].

Complement is well known to potentiate the neutralization of viruses by antiviral antibody, and complement also has a direct virolytic effect upon some enveloped viruses. It is, therefore, somewhat surprising to discover that a combination of antiviral IgM with complement can mediate enhanced viral replication. The evidence for this comes from the same WNV model system in P388D1 cells, which are known to carry CR as well as FcR, and in which two quite distinct pathways for ADE are possible [11,12]. The first pathway, which can result in 20- to 100-fold enhancement, involves FcR and antiviral IgG, and is independent of the presence of complement. The second pathway, which can result in 10- to 20-fold enhancement, involves CR and antiviral IgM with complement and is abolished by heat treatment or cobra venom treatment of the complement source. As previously noted, the monoclonal antibody 2.4G2, which binds to FcRII, blocks the IgG-dependent ADE, but has no effect upon the CR-mediated ADE, which is IgM dependent. Conversely, a different monoclonal antibody, M1/70, with specificity for CR3, blocks IgM enhancement, but has no effect upon FcR-mediated IgG enhancement. Although the two enhancement pathways have been separated under these experimental conditions, it is quite possible that the two mechanisms operate synergistically when both immunoglobulin classes are present, as may be the case early in the immune response to viral infection.

Japanese workers have reported substantial enhancement produced in BHK-21 cells infected with the alphavirus Getah virus, or the flavivirus, Japanese encephalitis virus, in the presence of appropriate antiviral antisera [13]. However, we have been unable to confirm these observations in a number of experiments in BHK-21 cells infected with Getah, West Nile, or Bunyamwera viruses, although all three viruses are capable of producing significant ADE when the same virus–antibody mixtures were tested in P388D1 cells.

Viruses Producing ADE

Dengue and West Nile viruses are both members of the genus flavivirus, in the family Togaviridae, and ADE has also been reported with Murray Valley encephalitis, Japanese encephalitis, Kunjin, Uganda S, yellow fever, and Zika viruses in the same genus, and with the alphaviruses Getah, Semliki forest, Sindbis, and Western equine encephalitis viruses in the same family [1,2,4,11,13]. The family Bunyaviridae contains four genera, in each of which at least one virus has been shown to produce ADE. Thus, in the Bunyavirus genus, Bunyamwera, Batai, Maguari, Lokern, California encephalitis, Tahyna, Trivittatus, and a number of recombinant viruses have all shown ADE [14]. It therefore seems probable that ADE is a general phenomenon as far as all viruses in the families Togaviridae and Bunyaviridae are concerned.

In other families, studies have been limited to relatively few viruses, but ADE has been reported with the following: rabbit pox virus [1], family Poxviridae; rabies virus and fish rhabdoviruses, family Rhabdoviridae [15]; Reovirus type 3, family Reoviridae; murine cytomegalovirus, family Herpesviridae [16]; and feline infectious pancreatitis virus, family Coronaviridae [17]. Failure to detect ADE has been reported with only two viruses, Mengo virus, family Picornaviridae, and herpes simplex virus, family Herpesviridae [4].

Without doubt the list of viruses showing ADE will be extended as further viruses are tested.

Antibody Specificity of ADE

Are enhancing antibodies distinct from the antibodies measured by neutralization (N), hemagglutination inhibition (HI), complement fixation (CF), or any other serological test? Quantitatively, enhancing antibodies are detectable at substantially lower serum concentrations than are N, HI, or CF antibodies. In a study involving a range of different flavivirus antisera, there was some correlation between enhancing antibody titers and cross-reactivity detected by HI, but little correlation between enhancing and N-antibody titers [18]. Studies with monoclonal antibodies provide more definitive evidence. Three monoclonal antibodies with specificity directed against the envelope glycoprotein of WNV were all potent at producing ADE, but only one had significant N activity, and a different one had substantial activity against the viral hemagglutinin [19]. These, and other studies with monoclonal antibodies [10,20] support the view that any antibody with specificity against a virion surface epitope, but not antibodies against internal nucleocapsid antigens, would be expected to produce ADE, provided the antibody retained an intact Fc-terminus, and provided the host cells carried FcR or CR of appropriate receptivity. It is worth noting that the same monoclonal antibody could enhance viral replication when tested in one macrophage cell line, but fail to enhance in a second macrophage line that carried fewer FcR, or FcR of inappropriate IgG subclass specificity.

To date, there are no reports of ADE produced by IgA or IgE antibodies.

In addition to the heterogeneity of antibodies with respect to viral antigens, there are also some interspecies constraints that can affect ADE. Thus, while antiviral antibodies prepared in avian species will produce ADE when tested in avian cells, such avian antibodies will not produce ADE when tested in mammalian cells [1]. There are major differences in the strength of binding of IgGs of different mammalian species to mouse FcR [21], and the species homology, or lack of it, between the source of antiviral IgG and the host cell providing the FcR, can materially affect ADE. While these considerations may have profound effects in experimental procedures, they are obviously not relevant to the pathogenesis of viral infections *in vivo*.

Temporal Relationships

Prolonged pre-incubation of a virus–antibody mixture favors neutralization of infectivity, whereas enhancement is more likely if mixtures are applied to appropriate FcR-bearing cells without pre-incubation [1,4]. Similarly, pre-incubation at 37°C favors neutralization, whereas serum–virus mixtures that have been held in the cold are more likely to enhance infectivity.

ADE *in vivo*

Is ADE simply an interesting laboratory artifact or do the mechanisms underlying ADE contribute to viral pathogenesis in nature? Much of the laboratory work on ADE has stemmed from studies on dengue and dengue hemorrhagic fever [2,7], but there are other viral infections in which ADE may contribute to disease. The rabies "early death" phenomenon [22] may be a manifestation of ADE, as may a similar antibody-dependent early death in cats infected with the virus of feline infectious peritonitis, a member of the family Coronaviridae [17]. The severe reactions seen in some children who have been vaccinated against measles and respiratory syncytial viruses and have subsequently suffered natural infections with these agents may be attributable to the absence of antibodies against the fusion factor in these viruses, but it is possible that non-neutralizing antibodies may potentiate infection of FcR-bearing cells in the respiratory tract, or of endothelial cells. Since it is known that several viruses in the family Herpesviridae are capable of inducing FcR [23], sequential infections, first with a herpesvirus, and shortly after with a different virus, may result in enhanced replication of the second with consequent disease. Other stimuli that activate macrophages can enhance expression of FcR [24], and may thus also contribute to ADE. Whether or not CR are also inducible by viral infection is not known. Much remains to be done to explore the contribution of CR, and other cell surface receptors, such as the mannose receptor, to viral pathogenesis.

References
1. Hawkes RA, Lafferty KJ (1967) The enhancement of viral infectivity by antibody. Virology 33:250–261
2. Halstead SB, O'Rourke EJ (1977) Dengue viruses and mononuclear phagocytes. 1. Infection enhancement by non-neutralizing antibody. J Exp Med 146:201–217
3. Peiris JSM, Porterfield JS (1979) Antibody-mediated enhancement of flavivirus replication in macrophage-like cell lines. Nature 282:509–511
4. Peiris JSM, Porterfield JS (1981) Antibody-dependent enhancement of plaque formation on cell lines of macrophage origin—a sensitive assay for antiviral antibody. J Gen Virol 57:119–125
5. Unkeless JC, Fleit H, Mellman IS (1981) Structural aspects and heterogeneity of immunoglobulin Fc receptors. Adv Immunol 31:247–270
6. Ryan US, Schultz RD, Ryan DR (1981) Fc and C3b receptors on pulmonary endothelial cells: Induction by injury. Science 24:557–558
7. Halstead SB (1980) Immunological parameters of Togavirus disease syndrome. *In* Schlesinger RW (ed) The Togaviruses: Biology, Structure, Replication. Academic Press, New York, p 107
8. Peiris JSM, Gordon S, Unkeless JC, Porterfield JS (1981) Monoclonal anti-Fc receptor IgG blocks antibody enhancement of viral replication in macrophages. Nature 289:189–191
9. Peiris JSM, Gordon S, Porterfield JS, Unkeless JC (1981) Antibody mediated enhancement of virus replication in macrophage cell lines—its dependence on the Fc receptor. *In* Förster O, Landy M (eds) Heterogeneity of Mononuclear Phagocytes. Academic Press, London, p 469
10. Brandt WE, McCown JM, Gentry MK, Russell PK (1982) Infection enhancement of dengue-2 virus in the U-937 human monocyte cell line by antibodies to flavivirus cross-reactive determinants. Infect Immun 36:1036–1041
11. Porterfield JS (1982) Immunological enhancement and the pathogenesis of dengue haemorrhagic fever. J Hyg 89:355–364
12. Cardosa MJ, Porterfield JS, Gordon S (1983) Complement receptor mediates enhanced flavivirus replication in macrophages. J Exp Med 158:258–263
13. Kimura T, Ueba N, Minekawa Y (1981) Studies on the mechanism of antibody-mediated enhancement of Getah virus infectivity. Biken J 24:39–45
14. Millican D, Porterfield JS (1982) Relationship between glycoproteins of the viral

envelope of Bunyaviruses and antibody dependent plaque enhancement. J Gen Virol 63:233–236
15. Clerx JPM, Horzinek MC, Osterhaus ADME (1978) Neutralisation and enhancement of infectivity of non-salmonid fish rhabdoviruses by rabbit and pike immune sera. J Gen Virol 40:297–308
16. Inada T, Chong KT, Mims CA (1983) Enhancing antibodies in murine cytomegalovirus infection. In press.
17. Weiss RC, Scott FW (1981) Antibody-mediated enhancement of disease in feline infectious peritonitis: Comparisons with dengue haemorrhagic fever. Comp Immun Microbiol Inf Dis 4:175–189
18. Halstead SB, Porterfield JS, O'Rourke EJ (1980) Enhancement of dengue virus infection in monocytes by flavivirus antisera. Am J Trop Med Hyg 29:638–642
19. Peiris JSM, Porterfield JS, Roehrig JT (1982) Monoclonal antibodies against the flavivirus West Nile. J Gen Virol 58:283–289
20. Chanas AC, Gould EA, Clegg JCS, Varma MGR (1982) Monoclonal antibody to Sindbis virus glycoprotein E1 can neutralize, enhance infectivity and independently inhibit haemagglutination or haemolysis. J Gen Virol 58:37–46
21. Haeffner-Cavaillon N, Klein M, Dorrington KJ (1979) Studies on the Fc receptor of the murine macrophage-like cell line P388D1. J Immunol 123:1905–1914
22. Prabhakar BS, Nathanson N (1981) Acute rabies death mediated by antibody. Nature 290:590–591
23. Yee C, Costa J, Hamilton V, Klein G, Rabson AS (1982) Changes in the expression of Fc receptor produced by induction of Epstein–Barr virus in lymphoma cell lines. Virology 120:376–382
24. Ezekowitz RAB, Bampton M, Gordon S (1983) Macrophage activation selectively enhances expression of Fc receptors for IgG2a. J Exp Med 157:807–812

CHAPTER 18
Enveloped Virus Maturation at Restricted Membrane Domains

RICHARD W. COMPANS*

Synthesis and assembly of the envelope proteins of lipid-containing viruses require the biosynthetic and transport processes involved in cellular membrane biogenesis, and such viruses have therefore been used extensively for investigation of these processes. With the exception of poxviruses, which assemble their membranes *de novo* in the cytoplasm, the assembly of enveloped viruses takes place on preformed cellular membranes. The precise location at which assembly occurs is a distinctive characteristic that is highly conserved among structurally similar viruses. Herpesvirus maturation occurs by budding at the inner nuclear membrane. Coronaviruses are assembled at the rough endoplasmic reticulum, and bunyaviruses bud at membranes of the Golgi complex. Virus particles of several other families are formed by budding at the plasma membrane. Polarized epithelial cells exhibit distinct apical and basolateral surface domains separated by tight junctions, and it has been observed that assembly of enveloped viruses occurs at one or the other of these membrane domains. Thus, with the exception of mitochondrial membranes, any membrane of the cell is known to be capable of serving as a virus maturation site.

What determines the cellular membrane selected for maturation by a particular enveloped virus? In those cases that have been investigated, viral envelope proteins accumulate at the membrane where virus budding occurs. The mechanisms by which viral envelope proteins become localized at specific membrane sites are likely to involve normal cellular processes for sorting proteins into different compartments. Current research in this area is centered around several key questions: (1) What is the precise nature of the information that determines sorting of viral proteins into distinct membrane

---

* University of Alabama in Birmingham, Alabama.

compartments? (2) Where do the sorting events occur, and what are the cellular mechanisms for sorting of membrane proteins? (3) What are the biological consequences of virus maturation at restricted membrane domains?

Polarized Expression of Viral Glycoproteins

In the polarized line of Madin-Darby canine kidney cells (MDCK), which resemble differentiated cells of kidney distal tubule epithelia, influenza and parainfluenza viruses have been observed to assemble at the free apical surface, whereas vesicular stomatitis virus (VSV) and several retroviruses are assembled at basolateral membranes [1,2]. In each case, the viral surface glycoproteins are localized at the membrane domain from which the virus buds [2,3], suggesting that the site of glycoprotein insertion determines the viral maturation site. An alternate possibility would involve directional transport of viral matrix proteins (internal, nonglycosylated envelope proteins), and subsequent localization of glycoproteins at the same site by specific transmembrane interactions with the matrix proteins. This possibility is unlikely, however, since glycoprotein expression was shown to be polarized in the absence of any other viral gene products. Cloned DNA copies of influenza virus HA glycoprotein genes inserted into SV40 vectors yield fully glycosylated HA molecules, which are expressed at cell surfaces. When such recombinant viruses were used to infect primary African Green monkey kidney epithelial cells, in which influenza virus maturation is polarized at apical surfaces, the expression of the HA glycoprotein was found to be restricted to apical cell surfaces [4]. These observations indicate that the information for directional transport (sorting) of the HA glycoprotein is contained in the HA molecule itself, and does not depend on any other viral component. Further, polarized virus maturation and glycoprotein insertion continue under conditions in which glycosylation of HA is completely inhibited by the drug tunicamycin [5,6], demonstrating that the information for glycoprotein sorting resides in the polypeptide portion of the molecule, and not the carbohydrate portion.

Although similar information is not yet available for other viral glycoproteins, it is likely that the site of accumulation of viral glycoproteins will prove to be of general importance as a determinant of budding sites of enveloped viruses. In the case of a virus that forms at an intracellular membrane, the E1 glycoprotein of mouse hepatitis virus, a coronavirus, has been observed to be restricted to the perinuclear region of infected cells [7]. At least two exceptional cases must be considered, however, in which enveloped viruses appear to assemble in the absence of the viral glycoproteins. These include avian retroviruses as well as temperature-sensitive glycoprotein mutants of vesicular stomatitis virus, both of which yield virus particles that appear to be devoid of surface glycoproteins [8,9]. The assembly processes of such virus particles should be investigated further.

The finding that viral glycoproteins are localized in distinct plasma membrane domains of epithelial cells indicates that cellular transport processes result in their vectorial transport and insertion into a particular membrane site. Once inserted, barriers must exist to prevent lateral diffusion in the plane of the membrane, which would otherwise be expected to result in a randomized distribution around the cell surface. The junctional complexes that are present in regions of contact between adjacent epithelial cells probably form such a barrier. However, virus maturation was also found to be polarized at apical or basal surfaces even in individual MDCK cells in contact with a substrate, in which no tight junctions could exist [10]. How the mobility of surface components is restricted in such cells remains to be determined.

Because of the extensive structural information available about many viral glycoproteins, it is of particular interest to consider where, within the amino acid sequence, the signals for intracellular sorting and directional movement reside. Since the movement of glycoprotein molecules occurs relative to cellular membranes in which they are embedded, such "sorting signals" might be found in proximity to the hydrophobic membrane-anchoring sequences, either within the hydrophobic domain itself or in the adjacent cytoplasmic or external regions. The sites at which such information resides can probably be determined by introducing specific alterations into the amino acid sequences of glycoproteins by modifying genes at the level of cloned DNA. This approach has shown that modification of the cytoplasmic domain of the VSV G protein affects the rate of its transport to the cell surface [11].

Intracellular Pathways for Membrane Glycoprotein Transport

As discussed above, it is likely that maturation sites of enveloped viruses are determined, at least in the majority of cases, by the sites of insertion and accumulation of their envelope proteins. The accumulated evidence indicates that the intracellular pathway followed by viral and other membrane glycoproteins resembles the pathway of exocytosis of secretory proteins: synthesis on membrane-bound polyribosomes in the endoplasmic reticulum, movement through the Golgi complex, and transport to the cell surface in vesicles that fuse with the plasma membrane [12,13]. Glycosylation is initiated by transfer of lipid-linked oligosaccharides to nascent polypeptide chains in the rough endoplasmic reticulum, and many of the subsequent processing events in the formation of complex carbohydrate chains occur in the various compartments of the Golgi complex. In the case of viral glycoproteins synthesized in doubly infected MDCK cells, proteins destined for opposite membrane domains share a similar intracellular pathway of transport until they reach the Golgi complex [14]. Several observations suggest that such proteins are sorted into distinct sets of transport vesicles within the Golgi complex, which are then targeted to apical or basolateral cell surfaces.

The sodium ionophore monensin blocks the exit from the Golgi complex of secretory proteins as well as most membrane glycoproteins [15]; however, the ionophore differentially affects the transport of viral glycoproteins destined for apical versus basolateral membranes of MDCK cells [16,17]. Influenza virus glycoprotein transport, and virus maturation at apical cell surfaces, are unaffected by the ionophore, whereas transport of the G protein of VSV to basolateral membranes is completely inhibited. The terminal stages of glycosylation of both viruses are inhibited [17], indicating that the glycoproteins may not reach the distal cisternae of the Golgi complex. Nevertheless, the influenza viral glycoproteins are able to reach the cell surface efficiently, suggesting that exit from the Golgi complex may occur by distinct pathways: a monensin-resistant pathway utilized by influenza virus, and a monensin-sensitive pathway utilized by VSV.

Simultaneous infection of MDCK cells with influenza virus and VSV results in continued polarity of virus maturation, at least during early stages of infection [18]. Influenza virus budding is found at the free apical surfaces, and VSV is assembled at basolateral membranes. VSV readily incorporates heterologous envelope proteins to form phenotypically mixed virus particles and pseudotypes, but such particles are not detected in doubly infected MDCK cells at early times after infection, indicating that viral glycoproteins are segregated into different cell surface domains [18]. These observations also support the conclusion that the viral glycoproteins are sorted into distinct sets of transport vesicles, since vesicles containing mixtures of glycoproteins would be expected to result in simultaneous insertion of both types of viral glycoproteins into apical or basolateral membranes.

Very little information is available on the exact nature of the vesicles involved in transport from the Golgi complex to the cell surface, or how the sorting of proteins into different vesicle populations might occur. Coated vesicles have been implicated in various stages of membrane traffic within the cell, including the movement of viral envelope proteins [19,20]. One attractive possibility is that different subpopulations of coated vesicles could form at the Golgi complex, possessing receptors that recognize the sorting signals in viral glycoproteins. Such vesicles could effect the subsequent movement of membrane proteins to various compartments. Further information on transport pathways of viral glycoproteins should provide considerable insight into the process of sorting of cellular membrane proteins in general.

Consequences of Restricted Viral Maturation Sites

Virus maturation at restricted membrane domains, or at intracellular sites, limits the availability of virions or viral antigens for interaction with components of the immune system. This has been demonstrated for VSV infection of MDCK cells, in which antibody was unable to prevent the spread of virus from cell to cell and the resultant formation of plaques [21]. In contrast, influenza virus plaque formation in MDCK cells, or VSV plaque formation in nonpolarized cell lines such as BHK21, was completely prevented by the respective antisera. Since VSV maturation in MDCK cells occurs beneath tight junctions, which prevent the passage of large molecules such as antibodies into the intercellular spaces, the progeny virions are able to infect adjacent cells without exposure to antibody. It will be of interest to investigate the extent to which viral maturation sites may play a role in limiting the accessibility of viral components to antibody or cytotoxic cells in natural disease processes.

The site of insertion of viral glycoproteins at cell surfaces may also affect the type of cytopathology resulting from virus–cell interaction. Striking differences have been observed in the responses of various cell types to infection by paramyxoviruses [22]. Many cell types are highly susceptible to virus-induced cell fusion, whereas other cell types are resistant to fusion and the cells continue to appear morphologically normal. The latter are cells of epithelial morphology, in which parainfluenza virus assembly occurs at free apical surfaces. Paramyxovirus-induced cell fusion probably involves viral fusion glycoproteins acting to form a bridging structure between two adjacent cells, and the ability of virions or surface glycoproteins to form contacts between adjacent cells in monolayers of epithelial cell types is prevented by restriction of viral glycoproteins to apical surfaces above tight junctions. The role of intact junctions in resistance to cell fusion was demonstrated by treatment of MDCK cells with EGTA, which causes disruption of the junctional complexes [23]. Intact MDCK monolayers exposed to concentrated Sendai virus showed no fusion, whereas cells pretreated with EGTA showed formation of large syncytia. In the EGTA-treated cells, the virus was presumably able to penetrate into spaces between cells and induce fusion between adjacent cell membranes.

There is little information available on the possible influence that the sites of virus assembly in infected tissues may exert on the pathogenesis of viral diseases. Numerous examples exist where structurally similar viruses cause distinctly different natural infections. Parainfluenza viruses cause infections restricted to the respiratory tract, whereas measles virus, which is structurally similar, causes a generalized infection. Human influenza viruses are also restricted to the respiratory tract, whereas in avian species, influenza virus multiplies to high titers in the intestinal tract as well. In MDCK cells, all of these agents were found to be assembled at apical cell surfaces [24,23], indicating that the differences in disease processes are not reflected by a difference in viral maturation sites in this cell system. The biological properties of avian influenza viruses may be explained by their stability to low pH, enabling the virions to pass through the digestive tract and retain infectivity [25]. Since virus maturation sites in the MDCK cell system may not reflect the process of infection *in vivo,* it will be of interest to further investigate the possible consequences of restricted virus maturation sites on disease processes.

References

1. Rodriguez Boulan E, Sabatini DD (1978) Asymmetric budding of viruses in epithelial monolayers: A model system for study of epithelial polarity. Proc Natl Acad Sci USA 75:5071–5075
2. Roth MG, Srinivas RV, Compans RW (1983) Basolateral maturation of retroviruses in polarized epithelial cells. J Virol 45:1065–1073
3. Rodriguez Boulan E, Pendergast M (1980) Polarized distribution of viral envelope proteins in the plasma membrane of infected epithelial cells. Cell 20:45–54
4. Roth MG, Compans RW, Giusti L, Davis AR, Nayak DP, Gething M-J, Sambrook J (1983) Influenza virus hemagglutinin expression is polarized in cells infected with recombinant SV40 viruses carrying cloned hemagglutinin DNA. Cell 33:435–442
5. Green RF, Meiss HK, Rodriguez Boulan E (1981) Glycosylation does not determine segregation of viral envelope proteins in the plasma membrane of epithelial cells. J Cell Biol 89:230–239
6. Roth MG, Fitzpatrick JP, Compans RW (1979) Polarity of influenza and vesicular stomatitis virus maturation in MDCK cells: Lack of a requirement for glycosylation of viral glycoproteins. Proc Natl Acad Sci USA 76:6430–6434
7. Doller EW, Holmes KV (1980) Different intracellular transportation of the envelope glycoproteins E1 and E2 of the coronavirus MHV. Abstr Am Soc Micro, p 267
8. De Guili C, Kawai S, Dales S, Hanafusa H (1975) Absence of surface projection on some noninfectious forms of RSV. Virology 66:253–260
9. Schnitzer TJ, Dickson C, Weiss RA (1979) Morphological and biochemical characterization of viral particles produced by the ts045 mutant of vesicular stomatitis virus at restrictive temperature. J Virol 29:185–195
10. Rodriguez Boulan E, Paskiet KT, Sabatini DD (1983) Assembly of enveloped viruses in Madin-Darby canine kidney cells: Polarized budding from single attached cells and from clusters of cells in suspension. J Cell Biol 96:866–874
11. Rose JK, Bergmann JE, Gallione CJ, Florkiewicz RZ (1983) Changes in the cytoplasmic sequence of VSV G protein alter its rate of transport to the plasma membrane. J Cell Biochem Supp 7B:356
12. Bergeron JJM, Kotwall GJ, Levine G, Bilan P, Rachubinski R, Hamilton M, Shore GC, Ghosh HP (1982) Intracellular transport of the transmembrane glycoprotein G of vesicular stomatitis virus through the Golgi apparatus as visualized by electron microscope radioautography. J Cell Biol 94:36–41
13. Bergmann JE, Tokuyasu KT, Singer SJ (1981) Passage of an integral membrane protein, the vesicular stomatitis virus glycoprotein, through the Golgi apparatus en route to the plasma membrane. Proc Natl Acad Sci USA 78:1746–1750
14. Rindler MJ, Ivanov IE, Rodriguez Boulan E, Sabatini DD (1981) Simultaneous budding of viruses with opposite polarity from doubly-infected MDCK cells. J Cell Biol 91:118a
15. Tartakoff AM (1983) Perturbation of vesicular traffic with the carboxylic ionophore monensin. Cell 32:1026–1028
16. Alonso FV, Compans RW (1981) Differential effect of monensin on enveloped viruses that form at distinct plasma membrane domains. J Cell Biol 89:700–705
17. Alonso-Caplen FV, Compans RW (1983) Modulation of glycosylation and transport of viral membrane glycoproteins by a sodium ionophore. J Cell Biol 97:659–668

18. Roth MG, Compans RW (1981) Delayed appearance of pseudotypes between vesicular stomatitis virus and influenza virus during mixed infection of MDCK cells. J Virol 40:848–860
19. Farquhar MG (1983) Multiple pathways of exocytosis, endocytosis, and membrane recycling: Validation of a Golgi route. Fed Proc 42:2407–2413
20. Rothman JE, Pettegrew HG, Fine RE (1980) Transport of the membrane glycoprotein of the vesicular stomatitis virus to the cell surface in two stages by clathrin-coated vesicles. J Cell Biol 86:162–171
21. Roth MG, Compans RW (1980) Antibody-resistant spread of vesicular stomatitis virus infection in cell lines of epithelial origin. J Virol 35:547–550
22. Holmes KV, Choppin PW (1966) On the role of the response of the cell membrane in determining virus virulence. Contrasting effects of the parainfluenza SV5 in two cell types. J Exp Med 124:501–520
23. Compans RW, Roth MG, Alonso FV, Srinivas RV, Herrler G, Melsen LR (1982) Do viral maturation sites influence disease processes? In Mackenzie JS (ed) Viral Disease in Southeast Asia and the Western Pacific, Academic Press, Australia, p 328
24. Basak S, Melsen LR, Alonso-Caplen FV, Compans RW (1983) Maturation sites of human and avian influenza viruses in polarized epithelial cells. In Laver WG (ed) The Origin of Pandemic Influenza Viruses. Elsevier, New York, pp 139–145
25. Webster RG, Yakhno M, Hinshaw VS, Bean WJ, Murti KG (1978) Intestinal influenza: Replication and characterization of influenza viruses in ducks. Virology 84:268–278

CHAPTER 19

Viruses and Differentiation: The Molecular Basis of Viral Tissue Tropisms

ARNOLD J. LEVINE*

Viruses often have complicated life cycles in their natural hosts, replicating in a series of specific tissue types in a defined order that may even vary depending upon the route of inoculation. These viruses show distinct preferences for efficient replication in selected differentiated cell types and this has been termed tissue tropism. In order to understand viral pathogenesis at the molecular, cellular, or organismic levels, one must consider the differential regulation of viral genetic information in the great diversity of host-cell backgrounds that compose the organism. Not only can the stage of differentiation of a host cell affect viral gene expression [1], some viruses are capable of reprogramming or altering cellular gene expression, or the regulation of cell division, and in doing so produce a disease state [2]. The subject of this brief communication is to review the mechanisms that underlie virus-specific tissue tropisms and latency and to consider how these can result in defined pathology.

The selective expression of cellular receptors for virus adsorption on different tissue types can result in a limited spectrum of virus replication and tissue destruction [3]. Variant or mutant viruses may acquire the ability to adsorb to and replicate in new cell types [4], resulting in pathologies not normally associated with the parent or wild type virus. Latency can be defined as the presence of a viral agent in a noninfectious form that may be activated by external agents. Latency is a hallmark of the Herpesviruses [5] and the five human Herpesviruses show distinct tissue tropisms. Herpes simplex type 1, 2 (HSV), and Varicella-Zoster viruses (VZV) remain latent in the trigeminal, sacral, and vertebral ganglia, respectively, while Epstein–Barr virus (EBV) and cytomegalovirus (CMV) are lymphotropic agents.

* State University of New York at Stony Brook, New York.

Herpesviruses, like HSV-1, regulate their genetic information in a temporal fashion during productive infection so that three classes of viral gene products are now recognized: immediate early, delayed early, and late functions [6]. An immediate early function is required to regulate, in a positive fashion, delayed early gene expression, which is in turn utilized to regulate late viral genes [7]. It is not yet clear whether the latent viral genome expresses only the immediate early genes, but alterations in a single immediate early function (ICP-4) that is required for subsequent viral gene expression, could then maintain the latent and noninfectious state. If this hypothesis is correct, the way in which external stimuli, such as ultraviolet light, trauma, steroid hormones, etc., restore normal function to ICP-4 would then be a most interesting subject for future investigation.

Several regulatory signals for transcriptional initiation events of viral genes are now recognized [8]. The 5' ends of viral transcripts, where transcription begins (termed nucleotide number +1), contain a cap structure with the general formula $m^7GpppXm$ [9]. Some 20–30 nucleotides 5' or prior to the cap site (-20 to -30 nucleotides), a TATA-like sequence is commonly found [10]. This signal appears to measure the distance between the TATA box and the initiation or start site (cap site) for ribonucleotide synthesis [11]. A third important element in transcriptional initiation is the gene enhancer. Gene enhancers are nucleotide sequences usually located several hundred base pairs prior to the cap site (-100 to -300 base pairs). The enhancer is frequently, but not always, composed of a repeated sequence of nucleotides [12]. This type of signal can be cloned into a plasmid and promote the transcription of a foreign gene contained in the plasmid, such as the globin gene [13]. In polyoma and SV40, the gene enhancer nucleotide sequences reside in a region of the chromosome that is not packaged by nucleosomes [14,15]. Gene enhancer signals appear to act in a fashion independent of the distance from the cap site. As long as a gene enhancer is located on the same plasmid or DNA molecule (*cis* acting), [13] it is functional. For this reason, researchers have speculated that gene enhancers may represent entry sites for RNA polymerase II onto the DNA molecule to be transcribed. The region of the polyoma virus gene enhancer and transcriptional signals are detailed in Fig. 1.

Polyoma virus fails to express its early proteins or tumor antigens in developmentally immature mouse cells, embryonal carcinoma cells. These tumor antigens and infectious polyoma virus are produced in most differentiated murine cell types [16,17]. Polyoma mutants that synthesize tumor antigens and infectious virus in embryonal carcinoma cells have been isolated [17–19]. These mutants contain alterations that have been localized in the polyoma gene enhancer region of the genome (see Fig. 1) [1]. Furthermore, polyoma mutants selected for replication in two different embryonal carcinoma cell lines (F9, PCC-4) contain different mutations (Fig. 1) permitting transcription; these mutations are specific for the cell line in which they were selected. It would appear then that cells in different stages of development

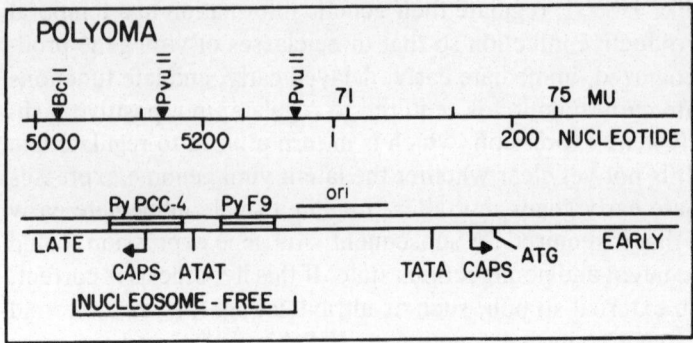

Fig. 1. Detail of the region of the polyoma virus genome involved in transcriptional regulation and host-range mutations. The figure represents a portion of the polyoma genome between map units (mu) 65–75 or nucleotide numbers 5000–200 (origin is nucleotide 1) that contains the origin of DNA replication (*ori*). To the right of this origin region are the early genes (tumor antigens) and their TATA box, cap sites, and ATG (methionine) for the N-terminal end of the tumor antigens. To the left of the origin sequences are the late or structural genes of polyoma, TATA box, and cap signals. In the nucleosome free region is the gene enhancer region for regulating early gene transcription. Mutants of polyoma that replicate in embryonal carcinoma cell lines F9 or PCC-4 map in this region. Wild type virus will not replicate in these stem cells and so the mutants confer a new host range upon this virus permitting it to replicate in primitive embryonic cells.

contain functions that must recognize *cis*-acting signals required for transcription of the viral genome. Thus, tissue specificities of viruses can be regulated at the transcriptional level by alterations in gene enhancer signals. This is obviously an important issue for agents that cause congenital birth defects and show specific tissue tropisms.

The genetic background of the host is equally important in viral-induced tissue pathology. Genetic resistance at the *Fv-1* locus of mice [20], which restricts replication of certain murine leukemia viruses, can be overcome by virus mutations [21]. A single genetic locus like *Fv-1* can have a dramatic effect upon leukemia incidence in AKR mice [22]. Similarly, histocompatibility-linked loci can specify alleles that affect the immune system and result in an altered incidence of murine leukemia [23]. Thus, genetic alterations regulating both the virus and the host cell can impact upon the extent and type of tissue pathology resulting from virus infections.

The DNA and RNA tumor viruses appear to be able to reprogram the gene expression of their host cells, giving rise to the transformed phenotype. Transformed cells often have altered sizes, shape, and growth-regulating properties. These viruses synthesize transforming proteins (*onc* gene products or tumor antigens) whose action results in an alteration in the levels or extent of cellular gene expression [24]. Many different transformed cells express higher levels of a cellular protein of 53,000 daltons, termed p53, than their normal counterparts (for a review of this subject, see [25]. Levels of p53 in SV40-transformed cells are regulated at the protein stability or posttranslational level with p53 protein having a shorter half-life (20–30 minutes) in normal cells than in viral transformed cells (greater than 24 hours) [26]. In embryonal carcinoma cells that are stimulated to differentiate into nontransformed endoderm cells, the p53 levels fall to about one-tenth. The mRNA levels for p53 also fall to one-tenth, indicating that a transcriptional or mRNA stability level of regulation is operating in this case [27]. Thus, the tumor viruses produce gene products that modulate cellular functions using several different mechanisms, resulting in cellular pathology and alterations in growth control.

This review of the mechanisms that underlie virus-induced tissue-specific pathology points out the need to understand both viral and cellular regulatory mechanisms operating in each situation. Both the virus and host cell contribute components to the resultant pathology. Furthermore, genetic alterations in either the virus or host cell can change the dynamics of this interaction, resulting in the creation of a new syndrome or a loss of the expected pathology associated with the virus. The regulatory elements that impact upon tissue-specific pathology act upon the levels of transcription, RNA processing, stability and transport, translation, protein stability, and processing or transport into cellular compartments. From what we know, we should not be surprised to observe viruses that were once thought to be thoroughly understood creating new disorders and tissue-specific pathologies. The genetic alterations that give rise to these situations will allow researchers to localize new regulatory signals and identify their functions. These observations serve to emphasize a well-understood biological principle, i.e., life processes or genetic processes are dynamic, continuing over time to provide new opportunities for growth and replication of the organism. The challenge to the virologist and pathologist will be to understand the mechanisms involved and respond accordingly.

References

1. Levine AJ (1982) The nature of the host range restriction of SV40 and polyoma viruses in embryonal carcinoma cells. Curr Top Microbiol Immunol 101:1–30
2. Maltzman W, Levine AJ (1981) Viruses as probes for development and differentiation. Adv Cancer Res 26:65–116
3. Holland JJ (1961) Receptor affinities as major determinants of enterovirus tissue tropisms in humans. Virology 15:312–326
4. Yoon JW, Onodera T, Notkins AL (1978) Virus-induced diabetes mellitus. XV. Beta cell damage and insulin-dependent hyperglycemia in mice infected with coxsackie virus B4. J Exp Med 148:1068–1080
5. Stevens JG (1975) Latent herpes simplex virus and the nervous system. Curr Top Microbiol Immunol 70:31–50
6. Honess RW, Roizman B (1974) Regulation of herpesvirus macro-molecular synthesis. I. Cascade regulation of the synthesis of three groups of viral proteins. J Virol 14:8–19
7. Honess RW, Roizman B (1975) Regulation of herpesvirus macromolecular syn-

thesis: Sequential transition of polypeptide synthesis requires functional viral polypeptides. Proc Natl Acad Sci USA 72:1276–1280
8. Shenk T (1981) Transcriptional control regions: Nucleotide sequence requirements for initiation by RNA polymerase II and III. Curr Top Microbiol Immunol 93:25–46
9. Shatkin AJ (1978) Capping of eukaryotic RNA's. Cell 9:646–653
10. Corden J, Wasylyk B, Buchwalder A, Sasson-Gorsi P, Kedenger C, Chambon P (1980) Promoter sequences of eukaryotic protein coding genes. Science 209:1406–1414
11. Ghosh PK, Lebowitz P, Frisque AJ, Frisque RJ, Glutzman Y (1981) Identification of a promoter component involved in positioning the 5' termini of the simian virus 40 early mRNA's. Proc Natl Acad Sci USA 78:100–104
12. Gruss P, Dhar R, Khoury G (1981) The SV40 tandem repeats as an element of the early promoter. Proc Natl Acad Sci USA 78:9430–9437
13. Banerji J, Rusconi S, Schaffner W (1981) Expression of a B-globin gene is enhanced by remote SV40 DNA sequences. Cell 27:299–308
14. Saragosti S, Mayne G, Yaniv M (1980) Absence of nucleosomes in a fraction of SV40 chromatin between the origin of replication and the region codings for the late leader RNA. Cell 20:65–73
15. Scott WA, Wigmore DJ (1978) Sites in SV40 chromatin which are preferentially cleaved by endonucleases. Cell 15:1511–1519
16. Fujimura FK, Silbert P, Eckhart W, Lenney E (1981) Polyoma virus infection of retinoic acid induced differentiated teratocarcinoma cells. J Virol 39:306–312
17. Vasseur M, Kress C, Montreau N, Blangy D (1980) Isolation and characterization of polyoma virus mutants able to develop in embryonal carcinoma cells. Proc Natl Acad Sci USA 77:1069–1072
18. Fujimura FK, Deininger PL, Friedman T, Lenney T (1981) Mutation near the polyoma DNA replication origin permits productive infection of F9 embryonal carcinoma cells. Cell 23:809–814
19. Sekikawa, K, Levine AJ (1981) Isolation and characterization of polyoma host range mutants that replicate in nullipotent embryonal carcinoma cells. Proc Natl Acad Sci USA 78:1100–1104
20. Pincus T, Harley JW, Rowe WP (1971) A major genetic locus affecting resistance to infection with murine leukemia viruses. I. Tissue culture studies of naturally occurring viruses. J Exp Med 133:1219–1233
21. Gautsch J, Elder J, Schindler S, Jensen F, Lerner RA (1978) Structural markers on core protein p30 of murine leukemia virus: Functional correlation with Fv-1 tropism. Proc Natl Acad Sci USA 75:4170–4175
22. Rowe WP (1977) Leukemia virus genomes in the chromosomal DNA of the mouse. Harvey Lect Ser 71:173–192
23. Lilly F, Pincus A (1973) The Fv-1 locus. Adv Cancer Res 17:231–277
24. Levine AJ (1982) Transformation-associated tumor antigens. Adv Cancer Res 37:75–109
25. Klein G (1982) The transformation-associated cellular p53 proteins. *In* Advances in Viral Oncology II. Raven Press, New York, p 81
26. Oren M, Maltzman W, Levine AJ (1981) Post-translational regulation of the 54K cellular tumor antigen in normal and transformed cells. Mol Cell Biol 1:101–110
27. Oren M, Reich N, Levine AJ (1982) The regulation of the cellular p53 tumor antigen in teratocarcinoma cells and their differentiated progeny. Mol Cell Biol 2:443–449.

Virus Maturation and Diversity

20. Continuum of Change in RNA Virus Genomes
    JOHN J. HOLLAND.. 137

21. Reassortment Continuum
    PETER PALESE 144

22. Immune Selection of Virus Variants
    JANICE E. CLEMENTS AND OPENDRA NARAYAN 152

23. Antigenic Variants of Viruses and their Relevance to Clinical Disease
    BELLUR S. PRABHAKAR AND ABNER LOUIS NOTKINS 158

CHAPTER 20
Continuum of Change in RNA Virus Genomes

JOHN J. HOLLAND*

The mechanisms of evolution of RNA virus genomes have recently received less attention than the evolution of DNA genomes (or of RNA retroviruses able to reverse-transcribe their RNA genomes into DNA proviruses). The latter two virus groups can employ all of the myriad mutational and recombinational mechanisms of DNA evolution, which have been intensively investigated during the past decade. On the other hand the genomes of "ordinary," nonretrovirus RNA viruses rarely, if ever, are reverse-copied into DNA, so they can neither integrate onto host chromosomes, nor recombine or transpose with other DNA elements. Yet the earlier studies of RNA virus genetics [1] showed that they are extremely mutable. The author and his colleagues recently published an extensive review of the literature documenting that RNA viruses not only exhibit very high mutation rates, but that their genomes undergo extremely rapid rates of evolution [2]. Space limitations prevent referencing herein most of the numerous studies upon which these conclusions are based, but the reader can obtain these from the above review.

Virus RNA genome mutation frequencies exceed by many orders of magnitude the mutation rates of the DNA of their eukaryotic hosts. Because of proofreading mechanisms, error rates in DNA synthesis are usually very low despite considerable variability. The rate of DNA base substitutions can be as low as 10^{-8} to 10^{-11} per base pair per replication [3]. Recent measurements of mutation *in vivo* of human peripheral blood lymphocytes [4] to the HGPRT$^-$ phenotype showed a mutation frequency of the order of 10^{-6}, well within the expected frequency range for most DNA genes.

---

* University of California at San Diego, La Jolla, California.

By contrast, the studies of Domingo et al. [5] showed that the RNA base misincorporation rate per genome doubling of Qβ RNA phage is between 10^{-3} and 10^{-4} at given base positions. This is more than a millionfold greater error rate than is observed in the higher-fidelity DNA replication systems. This disparity occurs largely because the fidelity of DNA replication is maintained by proofreading mechanisms, whereas RNA replication apparently lacks proofreading systems. No investigator has yet described a proofreading system for RNA replication and it seems likely that none exist. There appears to be no *a priori* reason why larger RNA viruses could not have evolved some proofreading ability; so perhaps it is disadvantageous for them to do so. In fact it is argued below that high error frequency with its accompanying genetic plasticity constitutes a crucial asset of RNA viruses.

Following Granoff's [1] demonstration of a 3%–7% frequency of small plaque mutants in clonal pools of Newcastle disease virus, numerous other investigators have reported mutation frequencies generally ranging between 10^{-1} and 10^{-5} among a variety of different RNA viruses (reviewed in [2]). These frequencies vary depending upon phenotype scored, stringency of assay criterion, etc., but they are uniformly high, as would be expected following high rates of nucleotide substitution during replication. No investigators have yet measured directly the rates of base misincorporation for an RNA virus of eukaryotic cells, but we are presently attempting to quantitate this for VSV. It will be surprising if it is not quite high, based upon accumulated genetic data [6].

The rates of evolution (i.e., rates of accumulation of mutations) of RNA genomes can enormously exceed the rates of evolution of the DNA of their eukaryotic hosts. High mutation frequencies favor but do not necessarily assure high rates of genomic evolution. "Wild type" RNA viruses often exhibit great stability when passaged numerous times in different laboratories (see [7], for example). This genomic stability of wild type (or of mutant) viruses is probably due to greater competitive ability of a certain variant or variants in the presence of a vast array of less competitive mutant genomes. Domingo et al. [5] deduced that each viable genome in a Qβ RNA phage population is mutated at one or more sites from the "average" "wild type" sequence in that population. However, they showed that mutants with a relative replication rate of 0.8 to 0.9 are outcompeted by the faster growing "wild type" virus. Therefore, equilibrium pools of these "wild type" RNA viruses will contain a mixture of mutants whose weighted average sequence comprises the "defined" structure of the Qβ genome as determined by RNA sequencing methods. This "probabilistic" nature of an RNA virus genome is inevitable when the number of bases in the genome exceeds the reciprocal of the base-substitution frequency. For example, if the base-substitution rate per nucleotide incorporated into the 11 kilobase VSV genome is 10^{-3} to 10^{-4}, then nearly every progeny genome will be different from its parental genomes and from nearly all of its sister genomes in at least one base. This will be true even when cloned "wild type" parental virus is replicated and even when phenotypic characteristics, genome oligonucleotide maps, and gene sequences "confirm" the presence of only "wild type" phenotype and gene structure. The situation will of course be much more uncertain whenever equilibrium is upset such that a "wild type" or mutant "consensus sequence" no longer dominates the replicating genome population pool. Conditions that upset equilibrium dominance by a single variant should (and in fact, do) lead to extremely rapid evolution of RNA virus genomes.

Many conditions do lead to rapid emergence of multiple mutants of RNA viruses (see [2] for references). The drifts and shifts of influenza virus are well known but not exceptional for RNA viruses. For example, Kew et al. [8] documented dozens of mutations as poliovirus spread among humans in epidemics, and even as virus shedding persisted in an individual human and his contacts. In a controlled laboratory setting we have documented rapid, continuous, and extensive evolution of the genome of originally cloned vesicular stomatitis virus during nearly ten years of persistent infection of BHK_{21} cells *in vitro,* and during serial undiluted passages [9]. These and many other studies suggest that rates of base substitutions per average site per year in viral RNA genomes can exceed 10^{-3}. This exceeds by more than a millionfold the average 10^{-9} nucleotide substitutions per site per year of the chromosomal DNA of their eukaryotic hosts. This enormous disparity in DNA and RNA evolution rates poses numerous interesting questions.

The relative contribution of recombination to RNA virus genomes is still unclear. During the past few years, rather definitive molecular evidence for intratypic recombination between picornavirus genomic RNAs has been presented [10,11]. Also unequivocal sequence evidence proves that both intra- and intermolecular recombination events were involved in the generation of a unique "mosaic" genome DI particle of influenza virus [12]. Most virologists favor "copy choice" (replicative error) models for RNA virus recombination events—including those involved in DI particle generation, but no molecular mechanisms have been proved. If replicase errors are involved, then RNA genome recombination can be viewed as just another manifestation of the high replicative mutation rates of RNA viruses. In any case, bizarre mosaic-type DI genomes such as those described by Fields and Winter [12] indicate that recombination events can sometimes occur at high frequency and it would not even be surprising to find some recombinations between different types of RNA virus or between viral RNA genomes and cellular RNA. However it will always be difficult to assess the relative contribution of recombination to RNA genome evolution because the extremely high rates of mutation will tend to mask recombination events whenever they occur at precise points of homology between two genomes of similar sequence in a replicating virus pool. One type of recombination event that can help drive rapid RNA genome evolution is the generation of DI particles as discussed below.

DI particles can drive rapid RNA genome evolution by "coevolving" with homologous virus populations and upsetting their population equilibria. DI particle-resistant (Sdi⁻) mutants of a number of RNA viruses have now been reported (see [2] for review). We have recently observed [13] that numerous diverse Sdi⁻ mutants of VSV are selected continuously in a stepwise manner during persistent infections and during serial undiluted passages. Concurrently with the appearance and disappearance of different Sdi⁻ mutants of infectious VSV, new DI particles with altered interference interactions with these Sdi⁻ mutants also appear and disappear. Sdi⁻ mutations appear to result from alterations in several viral proteins involved in replication-encapsidation of virus and DI genome. In recent sequencing studies [14], we observed a remarkable stepwise accumulation of stable base substitutions within the 5'-terminal 47 nucleotides of Sdi⁻ mutants isolated at intervals during VSV evolution during undiluted passages or during persistent infections. A lower rate of evolution was observed at the 3' terminus and within coding cistrons. The terminal changes appear to be stepwise compensatory changes at replicase initiation-encapsidation sites, and they are not strictly required to obtain the Sdi⁻ phenotype. This may be the major reason why undiluted passages or persistent infections by VSV generate rapid genome evolution whereas dilute passages lead to very low rates of genome evolution [9].

In theory, any factor(s) able to upset the equilibrium dominance of particular RNA virus genome variants in a virus population should promote rapid evolution of the genome (e.g., immune selection, spread to different tissue types, to different host species, etc.). However replicative competition among changing DI particle populations and changing Sdi⁻ mutant populations should particularly favor rapid RNA genome evolution because the equilibrium can shift constantly and rapidly, and because the selection process acts intracellularly upon those virus gene products (replication-encapsidation proteins), which are directly involved in replicative errors. It is not yet known how much viral replicase error is tolerable, nor how extensively base-misincorporation levels may fluctuate between mutants of the same virus, but Pringle *et al.* [15] have characterized a "mutator" gene mutant of VSV, and these may be rather common and important in promoting high rates of genome evolution.

High levels of replicative error are an asset for RNA genomes. Extremely high mutation rates during replication impose some obvious limitations upon RNA virus genomes. The most obvious is the limitation of genome size. Extremely large viral RNA genomes do not exist (and cannot exist) because the probability of lethal error per replication is a function of both genome size and replicase error rate. As mentioned above, RNA viruses undoubtedly could improve replicase fidelity (and thereby increase genome size). However, they have not done so, and the average size of viral RNA genomes (about 10 kilobases) probably represents an optimum size that allows sufficient genetic information to be coupled to high error rates. For segmented genomes the allowable additive size, or allowable error rate (or both), can be increased somewhat. Interestingly, Drake [3] showed that for procaryotes the mutation rate decreases as genome size increases, so that the mutation rate per genome is rather constant. This also suggests that high error rates are favored to the extent that genome size permits.

A few examples of the many possible advantages of RNA virus replicative infidelity and high evolution rates are: (1) *Escape from immune systems.* Influenza virus, the only pandemic disease agent of man, is an obvious case, but all RNA viruses probably undergo constant minor antigenic variation in the face of immune responses. For example, Prabhakar *et al.* [16] showed very high frequency of antigenic variants among clinical isolates of Coxsackie B4. This event is discussed in detail in chapter 23. (2) *Broad and/or variable host-species ranges and host-tissue tropisms.* The overwhelming majority of viruses that replicate in both insects and mammals, or insects and plants, are RNA viruses rather than DNA viruses. Presumably, only RNA viruses that diverge extremely rapidly are able to evolve genomes capable of efficient replication in the appropriate cells of host and vector to allow regular transmission back and forth. In fact, each transmission from arthropod to plant or mammal (or back to arthropod) might regularly involve the creation of RNA population disequilibrium and the selection of virus variants more suited to the new host. This possibility is difficult to test directly, but the rapid selection of multiple mutants of VSV after infection of whole *Drosophila* flies [17], or of cultured *Drosophila* cells [18], strongly suggests that this occurs. Similar events might occur as rabies virus adapts to a new species following bite transmission, or as it moves from muscle to neurons, or from neurons to salivary glands, etc. Many such viral adaptations need to be explored before we can assess the role of RNA genome hypermutability in virus transmission and pathogenesis. However, it is clear from their much greater numbers and diversity of hosts that RNA viruses are more "successful" than DNA viruses in plant and animal hosts. This is not true for procaryotic hosts, where DNA viruses are apparently more abundant and successful, possibly because they face less complexity in the procaryotic world.

How many RNA virus variants can there be? We do not know the limits of RNA virus sequence change but comparative sequence studies of RNA viruses show that in some of them well over half of the genome positions can be substituted without sacrificing viability. If we assume very conservatively that only 20% of the positions in a 10 kilobase RNA virus genome can be substituted, there are 4^{2000} possible permutations, of which only certain combinations will yield a viable, competitive genome. Of course the viable combinations will vary with host, temperature, cell type, etc. This staggering number assures that there can never be enough time or space (or host cells) in our universe to produce and test even a tiny fraction of all possible viable, competitive variants. Instead, only a relatively minuscule (but still astronomical) number of RNA virus genome variants is drawn from the cellular grabbags each year by all existing viral RNA genomes on earth, as they undergo rapid and unpredictable evolution. In the future years, some of these are likely to provide unpleasant surprises when they affect our plant crops, domestic animals, and burgeoning human population. The relentless divergence of nucleotide sequences is an unavoidable process for all living things as the molecular clock maintains its rather steady beat [19]. At a given moment, the possibilities for a significant evolutionary change in any life form are limited by its existing genomic sequence arrangements, and the constraints imposed by the vital functions they encode. But for RNA genomes the molecular clock hands spin at dizzying speed.

References

1. Granoff A (1961) Induction of Newcastle disease virus mutants. Virology 13:402–408
2. Holland J, Spindler K, Horodyski F, Grabau E, Nichol S, VandePol S (1982) Rapid evolution of RNA genomes. Science 215:1577–1585
3. Drake JW (1969) Comparative rates of spontaneous mutation. Nature 221:1132
4. Morley AA, Trainor KJ, Seshadri R, Ryall R (1983) Measurement of *in vivo* mutations in human lymphocytes. Nature 302:155–156
5. Domingo E, Sabo D, Taniguchi T, Weissmann C (1978) Nucleotide sequence heterogeneity of an RNA phage population. Cell 13:735–744
6. Pringle CR (1982) The genetics of vesiculoviruses. Arch Virol 72:1–34
7. Clewley JP, Bishop DHL, Kang CY, Coffin J, Schitzlein WM, Reichmann ME (1977) Oligonucleotide fingerprints of RNA species obtained from rhabdoviruses belonging to the vesicular stomatitis virus subgroup. J Virol 23:152–166
8. Kew OM, Nottay BK, Hatch MH, Nakano JH, Obijeski JF (1981) Multiple genetic changes can occur in the oral poliovaccines upon replication in humans. J Gen Virol 56:337–347
9. Spindler KR, Horodyski FM, Holland JJ (1982) High multiplicities of infection favor rapid and random evolution of vesicular stomatitis virus. Virology 119:98–108
10. King AMQ, McCahon D, Slade WR, Newman JWI (1982) Recombination in RNA. Cell 29:921–928
11. Tolskaya EA, Ramonova LA, Kolensnikova MS, Agol VI (1983) Intertypic recombination in poliovirus: Genetic and biochemical studies. Virology 124:121–132
12. Fields S, Winter G (1982) Nucleotide sequences of influenza virus segments 1 and 3 reveal mosaic structure of a small virus RNA segment. Cell 28:303–313
13. Horodyski FM, Nichol ST, Sprindler KR, Holland JJ (1983) Properties of DI particle-resistant mutants of vesicular stomatitis virus isolated from persistent infections and from undiluted passages. Cell 33:801–810
14. O'Hara PJ, Horodyski FM, Nichol ST, Holland JJ (1984) Vesicular stomatitis virus mutants resistant to defective interfering particles accumulate stable 5'-terminal and fewer 3'-terminal mutations in a stepwise manner. J Virol 49:793–798
15. Pringle CR, Devine V, Wilkie M, Preston CM, Dolan A, McGeoch DJ (1981) Enhanced mutability associated with a temperature sensitive mutant of vesicular stomatitis virus. J Virol 39:377–389

16. Prabhakar BS, Hospel MW, McClintock PR, Notkins AL (1982) High frequency of antigenic variants among naturally occurring human Coxsackie B4 virus isolates identified by monoclonal antibodies. Nature 300:374–376
17. Printz P (1970) Adaptation du virus de la stomatite vésiculaire à *Drosophila melanogaster*. Ann Inst Pasteur, Paris 119:520–537
18. Mudd JA, Leavitt RW, Kingsbury DT, Holland JJ (1973) Natural selection of mutants of vesicular stomatitis virus by cultured cells of *Drosophila melanogaster*. J Gen Virol 20:341–351
19. Jukes TH (1980) Silent nucleotide substitutions and the molecular evolutionary clock. Science 210:973–978

CHAPTER 21
Reassortment Continuum

PETER PALESE*

Among the animal viruses only RNA-containing viruses are known to possess segmented genomes. Specifically, five out of approximately a dozen different RNA virus families contain viruses with multiple RNA segments. These viral families comprise the arenaviruses [1], the birnaviruses [2], the bunyaviruses [3], the influenza viruses [4], and the reoviruses [5] (Table 1). Although these viruses share few biological properties, have different host-cell specificities, and replicate in quite different manners, they all have the capacity to reassort their genes during replication. For example, when two different influenza A viruses infect the same cell, the eight RNA segments may be rearranged in such a way as to give rise to 254 possible different reassortants (plus the two parent strains). This exchange of RNA segments (minichromosomes) during viral replication is called reassortment, although it is often (incorrectly) referred to in the literature as recombination, which implies the breakage and linkage of nucleotide bonds. True recombination has been observed with RNA viruses such as picornaviruses [6] and it may also occur among members of the above-mentioned virus families, but the present paper does not deal with this aspect of genetic interaction of segmented RNA viruses. Rather, reassortment will be discussed here as it relates to the replication of viruses in nature, and to possible applications in the laboratory.

Special emphasis is given to the influenza viruses as a model of reassortment, since this system has been known for the longest time and much has been learned about it. It should also be noted that RNA tumor viruses are

---

* Mount Sinai School of Medicine, New York, New York.

Table 1. RNA-containing animal viruses with segmented genomes

Virus family	Number and type of RNA segments	Reference
Arenaviruses	2, single-stranded (minus sense)	[1]
Birnaviruses	2, double-stranded	[2]
Bunyaviruses	3, single-stranded (minus sense)	[3]
Influenza viruses	7 or 8, single-stranded (minus sense)	[4]
Reoviruses	10–12, double-stranded	[5]

not considered, because these viruses contain two RNA segments of identical information; genetic exchange, in this case, involves transfer of portions of the RNA rather than of an entire segment. Furthermore, we do not discuss nonanimal virus systems, but it is likely that the conclusions drawn for animal viruses with segmented genomes also apply to the $\phi 6$ phages [2], which contain three double-stranded (ds) RNAs, to the segmented ds RNA viruses of fungi [2], and to the bi- and tripartite single-stranded (ss) RNA viruses of plants as well as to the recently discovered geminiviruses, which code for two different circular ss DNAs [7].

Mechanism of Reassortment

The discovery of reassortment followed the identification of precise genetic markers for animal viruses. Early work by Burnet on influenza viruses suggested a rapid exchange of genetic properties following coinfection of cells with two different strains [8]. Genetic and biochemical work of many groups then established the fact that influenza A viruses direct the synthesis of ribonucleoprotein complexes within the cell. The individual ribonucleoprotein complexes (RNPs), which consist of ss RNA and four viral proteins (the three P proteins and the nucleoprotein), are then packaged into a budding virus particle. It appears that the assembly involves "active" recognition of RNPs resulting in the packaging of at least one equivalent of all eight RNAs. Thus the assembly of influenza virus genes is probably different from a process such as randomly drawing tickets out of a hat. However, whether more than eight RNA segments are actually packaged into a single influenza virus particle is still not known. Since influenza virus preparations contain many noninfectious and morphologically heterogeneous particles, such data are difficult to obtain and many questions remain about the precise mechanism of RNP packaging in influenza viruses. Out of crosses between influenza A/PR/8/34 and A/HK/8/68 viruses, more than 60 different reassortants have been obtained [4; Schulman and Palese, unpublished]. Despite this high frequency of reassortants with different genotypes, further analysis suggested that reassortment may not be completely random with respect to all gene combinations [9]. For example, many more reassortants were isolated that had both PB1 and PB2 genes derived from the same parent rather than from different parents. This may be explained by a differential interaction of RNA segments and/or by a requirement for effective protein–protein interaction in a particular reassortant, which leads to better growth or another selective advantage. Also, it has been found that there is no intertypic reassortment among influenza A, B and C type viruses; genetic reassortment occurs only among different strains of the same type. The reason for this finding remains obscure, since at least influenza A and B virus genes show extensive sequence homologies [4].

The segmented nature of the reovirus genome is also responsible for the emergence of reassortants between different reovirus strains. In this case, however, intertypic reassortment has been observed. Coinfection of cells with two different reovirus serotypes results in a high percentage of intertypic reassortants.

What are the factors that may influence the frequency of reassortment between genes of different viral strains? First, both viruses must be able to infect the same cell. Second, interference between strains (or mutants) has been observed, and clearly may affect the frequency of genetic reassortment between strains. (A careful study concerning this factor has recently been done in the rotavirus system [10].) Thirdly, in some systems the formation of viral aggregates may increase the likelihood of viable reassortants; this is the case with influenza viruses. Finally, as already suggested, certain gene combinations may result in nonviable viruses or in reassortants that are difficult to isolate and/or to maintain.

Reassortment in Nature

Since reassortment between the genes of different segmented RNA viruses had been demonstrated in the laboratory, the question arose as to whether or not this phenomenon also occurs during "normal" propagation of strains in nature. Again, influenza viruses serve as a useful model. There is indirect evidence that new pandemic influenza viruses may originate from reassortment between animal strains and human viruses [11]. For example, the 1968 influenza viruses, carrying an H3 hemagglutinin and an N2 neuraminidase, most likely resulted from a reassortment event involving earlier human H2N2 viruses and animal influenza virus strains carrying an H3 hemagglutinin. Similar arguments involving reassortment between an animal and a human influenza virus can also be made to explain the origin of the H2N2 strains. Thus, reassortment provides at least one mechanism by which influenza viruses may acquire novel surface antigens.

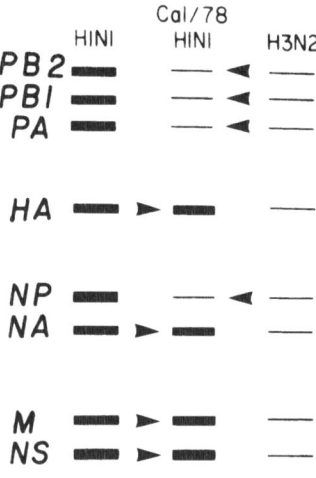

Fig. 1. Reassortment of genes between two influenza A viruses. In 1978 reassortant viruses of the A/Cal/10/78 prototype were found circulating in the human population. Those strains derived four RNAs (PB2, PB1, PA, and NP genes) from an H3N2 virus (light bars) and four RNAs (HA, NA, M, and NS genes) from an H1N1 strain (heavy bars). Arrows denote derivation of genes. Modified from [12].

In addition to these events leading to an antigenic shift of the virus, reassortment of genes can also occur when two different human influenza viruses cocirculate in the human population. Such a situation must have occurred when reassortant viruses of the A/Cal/78 prototype appeared in 1978 [12]. These reassortant viruses were shown to derive the three P genes and the NP gene from H3N2 viruses, and hemagglutinin (HA), neuraminidase (NA), M and NS genes from cocirculating H1N1 viruses (Fig. 1). In this case, direct biochemical evidence was obtained to show that reassortment of genes occurs in nature while viruses are passaged along in the population. A similar biochemical analysis demonstrated that avian influenza viruses also undergo reassortment during passage in feral birds [13]. Thus there is ample evidence to demonstrate that reassortment contributes to genetic changes in influenza viruses replicating under natural conditions and that reassortment may lead to epidemiologically successful variants.

Other segmented viruses, such as bunyaviruses, have also been shown to undergo reassortment in a natural epidemiological setting [14] and preliminary evidence suggests that the same holds true for reoviruses [15]. It thus appears that reassortment is a "natural" phenomenon among segmented RNA viruses, but it should be noted that reassortment is not the only factor which determines the epidemiological significance of a viral strain. In terms of biological properties, it is difficult to predict what the outcome of a reassortment event may be. This can be gleaned from laboratory experiments; reassortment of genes derived from a highly pathogenic and a nonpathogenic strain may lead to strains of intermediate pathogenicity but can also result in a strain even more pathogenic than the pathogenic parent. This has clearly been demonstrated by crossing the weakly mouse-neurotropic fowl plague virus and the Hong Kong influenza virus strain, which does not multiply in mouse brains at all. Several reassortments derived from these parents are, in fact, highly mouse neurotropic [16].

Reassortment: A Tool for Genetic Analysis of Viruses and Virulence Factors

Mapping of the genome of influenza viruses and of those of other segmented RNA viruses has been accomplished by using reassortants derived from parents with different genetic markers [1,17,18]. If separation of RNAs and proteins is sufficient to permit identification of the derivation of genes and gene products in the reassortants, genetic maps can be established for these viruses. The task is made easier by "loading" the reassortment experiment with genes from one parent through high multiplicities of infection of one strain and/or through ultraviolet or thermal inactivation of the other parent.

Such mapping experiments were helpful in identifying the function of particular influenza virus genes and their gene products. For example, the Hong Kong (H3N2) influenza virus does not form plaques in the MDBK cell line. On the other hand, the WSN virus strain plaques easily in this system. Analysis of many reassortants between the two strains revealed that transfer of a single WSN virus gene, the neuraminidase gene, into the Hong Kong virus is sufficient to result in a strain that plaques in MDBK cells [19]. In this system, the neuraminidase appears to be important in conferring the ability to plaque in the MDBK cell system, but this in no way proves that it is the only factor responsible for plaquing in MDBK cells. Replacement of a non-neuraminidase gene of Hong Kong virus with a gene derived from yet another strain may also lead to a reassortant that plaques in MDBK cells. Examples abound which suggest that virulence genes in influenza viruses are difficult to define. For instance, neurovirulence involving reassortants derived from WSN virus appears to be influenced by the derivation of the NA, M, and NS genes [20]. On the other hand, reassortant viruses derived from A/FPV/Rostock/34 and A/PR/8/34 virus show a different neurovirulence pattern [16]. When this pair is used in the experiment, neurovirulence always appears to be determined by the derivation of the PB1 gene and the HA gene. Thus, as more and more studies are being undertaken, it may become clear that the importance of particular RNA segments in determining virulence depends upon the genetic makeup of the parental viruses used in the analysis.

In the reovirus system, reassortant analysis appears to give less ambiguous results with respect to the specific genes involved in viral pathogenesis. For example, the $\sigma 1$ protein appears to be associated with cell and tissue tropism and the $\sigma 3$ protein has been found to inhibit host-cell protein and RNA synthesis and to play a role in persistent infection [5]. However, further experiments will be necessary to confirm these findings for all reovirus strains. Particularly, many biological properties may be influenced by mutations in more than one gene and different reovirus strains may have different genetic backgrounds. As a consequence, in different strains a particular virulence pattern may be associated with different genes. Thus, the assignment of specific viral functions to a specific gene(s) may be complicated.

Applications of Genetic Reassortment

Changes introduced by reassortment have been used to construct strains with specific properties. For example, low yielding virus strains in which many or all genes coding for nonsurface proteins have been replaced by A/PR/8/34 virus RNAs show increased growth capacity in embryonated chicken eggs. These high-yielding reassortments, which maintain the surface proteins of the other parent, can then be used as vaccine strains [21]. Their use as killed virus vaccines significantly reduces the cost of manufacturing. Although the molecular basis for attenuation of influenza viruses remains a mystery, reassortment of genes can be used to arrive at live virus vaccine strains of influenza virus. Laboratory derived temperature sensitive (ts) and cold adapted mutants can be used as donors in a reassortment experiment to make wild type viruses less virulent [22]. Such a transfer of genes from attenuated strains to "fresh" virus isolates is an interesting application of the principle of reassortment for the purpose of manufacturing viruses with novel properties. Another application concerns the introduction of genetic determinants that change the cell tropism of a virus. This approach has been used to extend the host range of rotaviruses to tissue culture cells, greatly enhancing the possibilities for further genetic manipulation and use for vaccines [23]. However, reassortment of genes may also lead to an instability of the tailor-made constructs. For example, live virus vaccine strains that are attenuated may—through reassortment with other strains in nature—acquire wild type virulence or traits that are even more pathogenic than those of the parent.

With the caveats mentioned above, experimentation with new reassortments remains a valuable tool for the genetic analysis of viruses and may be expected to provide additional practical applications in the field of virology. Obviously, if we succeed in applying the newly acquired recombinant DNA techniques to changing the RNA information in the segmented RNA viruses, this will add a new dimension to the already valuable ability to manipulate the genetic properties of viruses through reassortment.

References
1. Vezza A, Cash P, Jahrling P, Eddy G, Bishop DHL (1980) Arenavirus recombination: The formation of recombinants between prototype Pichinde and Pichinde Munchique viruses and evidence that arenavirus S RNA codes for N polypeptide. Virology 106:250–260.

2. Compans, RW, Bishop DHL (eds.) (1983) Double-stranded RNA viruses. Elsevier Biomedical, New York
3. Bishop DHL, Shope RE (1979) Bunyaviridae. *In* Fraenkel-Conrat H, Wagner RR (eds) Comprehensive Virology, Vol 14. Plenum Press, New York, p 1
4. Palese P, Kingsbury DW (eds) (1983) Genetics of influenza viruses. Springer-Verlag, Vienna
5. Fields BN, Green MI (1982) Genetics and molecular mechanisms of viral pathogenesis: Implications for prevention and treatment. Nature 300:19–23
6. Tolskaya EA, Romanova LA, Kolesnikova MS, Agol VI (1983) Intertypic recombination in poliovirus: Genetic and biochemical studies. Virology 124:121–132
7. Stanley J, Gay MR (1983) Nucleotide sequence of cassava latent virus DNA. Nature 301:260–262
8. Burnet FM (1956) Structure of influenza virus. Science 123:1101–1104.
9. Lubeck MD, Palese P, Schulman JL (1979) Nonrandom association of parental genes in influenza A virus recombinants. Virology 95:269–274
10. Ramig RF (1983) Factors that affect genetic interaction during mixed infection with temperature-sensitive mutants of simian rotavirus SA11. Virology 127:91–99
11. Webster RG, Laver WG, Air GM, Schild GC (1982) Molecular mechanisms of variation in influenza viruses. Nature 296:115–121
12. Young JF, Palese P (1979) Evolution of human influenza A viruses in nature: Recombination contributes to genetic variation of H1N1 strains. Proc Natl Acad Sci USA 76:6547–6551
13. Desselberger, U., Nakajima, K., Alfino, P., Pedersen FS, Haseltine WA, Hannoun C, Palese P (1978) Biochemical evidence that "new" influenza virus strains in nature may arise by recombination (reassortment). Proc Natl Acad Sci USA 75: 3341–3345.
14. Ushijima H, Clerx-Van Haaster CM, Bishop DHL (1981) Analysis of Patois group bunyaviruses: Evidence of naturally occurring recombinant bunyaviruses and existence of immune precipitable and nonprecipitable nonvirion proteins induced in bunyavirus-infected cells. Virology 110:318–332
15. Chanock SJ, Wenske EA, Fields BN (1983) Human rotaviruses and genome RNA. J Infect Dis 148:49–50
16. Bonin J, Scholtissek C (1983) Mouse neurotropic recombinants of influenza A viruses. Arch Virol 75:255–268
17. Palese P (1977) The genes of influenza virus. Cell 10:1–10
18. Mustoe RA, Ramig RF, Sharpe AH, Fields BN (1978) Genetics of reovirus: Identification of the dsRNA segments encoding the polypeptides of the μ and δ size classes. Virology 89:594–604
19. Schulman JL, Palese P (1977) Virulence factors of influenza viruses. WSN virus neuraminidase is required for productive infection of MDBK cells. J Virol 24:170–176
20. Sugiura A, Ueda M (1980) Neurovirulence of influenza virus in mice. I. Neurovirulence of recombinants between virulent and avirulent strains. Virology 101:440–449
21. Kilbourne ED (1969) Future influenza vaccines and the use of genetic recombinants. Bull WHO 41:643–645
22. Murphy BR, Markoff LJ, Chanock RM, Spring SS, Maassab HF, Kendal AP,

Cox NJ, Levine MM, Douglas RG, Betts RF, Couch RB, Cate TR (1980) Genetic approaches to attenuation of influenza A viruses for man. Phil Trans R Soc (London) B288, 401–415
23. Greenberg HB, Kalica AR, Wyatt RG, Jones RW, Kapikian AZ, Chanock RM (1981) Rescue of noncultivatable human rotavirus by gene reassortment during mixed infection with ts mutants of a cultivatable bovine rotavirus. Proc Natl Acad Sci USA 78:420–424

CHAPTER 22

Immune Selection of Virus Variants

JANICE E. CLEMENTS AND OPENDRA NARAYAN*

Virus infections in animals involve complex virus–host interactions in which various schemes of replication of the agent are pitted against a multitude of host-defense mechanisms. The immune system is the most effective of the antiviral host defenses and usually succeeds in eliminating pathogens after brief periods of infection. Failure to cure infection indicates a shift in favor of the virus and this involves either some novel mechanism of viral replication that eludes the immune system or a defect, often virus induced, in the immune system that interferes with antiviral functions. This review examines one mode of viral persistence in which normal replication of the agent can occur in the functionally competent immune environment. The basic mechanism of this process is mutation of the virus genome with attendant alteration in surface antigens, permitting escape from immune control.

Antigenic Drift and Persistence

Antigenic drift or change in the antigenicity of the surface glycoprotein, is a well-established mechanism among parasites for escaping immune elimination [1]. It is the principal means of persistence of protozoans such as trypanosomes, plasmodia, and several species of bacteria. Recent studies suggest that this phenomenon may also be important in persistence of several viruses. The best-known example of viral antigenic drift is among influenza viruses where variants with minor antigenic changes in the hemagglutinin and/or the neuraminidase glycoproteins are selected during sequential epidemics by populations immune to preceding strains of virus [2]. Recent

---

* The Johns Hopkins University, Baltimore, Maryland.

studies show that variant strains of rabies and measles viruses can replicate in the presence of antibody to other strains of the same virus, suggesting antigenic variation [3,4]. The best model for viral drift in a single animal, however, comes from studies of infections with certain nononcogenic retroviruses. The primary examples are equine infectious anemia (EIA) [5] and ovine visna [6,7]. Horses infected with equine infectious anemia virus develop sequential episodes of acute hemolytic crises during persistent infection. The animals produce neutralizing antibody to the virus after the first episode but succeeding disease crises are associated with mutant viruses that are not neutralized by the preexisting antibody. This is reminiscent of the protozoan disease, but the mechanisms of the process are difficult to study because the equine agents replicate poorly in cell culture. Visna virus of sheep also causes persistent infection, although episodic disease does not occur. However, the agent undergoes antigenic drift in the animal during persistent infection similar to EIA in horses. Further, antigenic drift by visna virus could be duplicated in cell culture systems using plaque-purified virus for inoculation and monospecific antibody for selection [6,8]. This provided an ideal model system for studying viral genetic mechanisms of antigenic variation.

Antigenic Drift of Visna Virus

In this review we summarize studies on genetic changes occurring in visna virus during persistent infection in two immunocompetent sheep and correlate these changes with serological responses of these infected animals [9,10]. Further, these variant viruses were compared with visna virus mutants selected in tissue culture in the presence of early immune sera from these two sheep. Two sheep were inoculated with plaque-purified virus and at various intervals during a 3-year period blood from the animals provided serum for measuring neutralizing antibody and white blood cells were cocultivated with sheep fibroblasts for virus isolation [6,7]. These studies showed that the earliest antibodies produced by the sheep neutralized the parental virus (strain 1514) used for inoculation but not viruses isolated later from blood leukocytes (sheep viruses, LV). The antibody responses broadened with time and sera obtained late in the infection neutralized the variant viruses. However, the last serum sample from each sheep could still distinguish the parental and variant viruses, emphasizing the difference in antigenic character of the agents. To test the ability of immune sera from these animals to select variants in tissue culture, we inoculated cell cultures with 10^3 pfu of plaque-purified virus 1514 and maintained the cultures with medium containing early immune sera from the two animals [6,8]. Viruses developing in these cultures (tissue culture variants, AD1-1 and AD4-1) were no longer neutralized by the sera used to select them and were thus antigenic variants like those arising in the animal. These *in-vitro*-derived viruses were compared with those variants obtained from the animals during persistent infection.

Genetic Basis of Antigenic Drift

The molecular basis for viral antigenic drift has been postulated to be point mutational changes in the viral genome, which produce biologically significant changes in the viral coat protein [9]. Visna virus has a single surface glycoprotein, gp135, which elicits neutralizing antibody [11]. The coding region for this protein is unknown, however, by analogy to other retroviruses, it would be expected to be in the 3000 bases at the 3' terminus of the viral genome. Thus, in order to investigate the molecular basis for the altered antigenicity of the naturally occurring variants, as well as the tissue-culture-derived mutants, the genomic RNAs of these viruses were compared by RNase T_1-resistant oligonucleotide fingerprinting [9,10]. As a first step in these studies, the parental virus 1514 and various variant viruses were plaque purified and grown in tissue culture. The viruses were purified by gradient centrifugation and genomic RNA from each was purified. Eighty unique RNAse T_1-resistant oligonucleotides were identified and the locations of these unique oligonucleotides were mapped on the viral RNA. Sequence analysis was performed: (1) to determine that the unique oligonucleotides in the different viruses were the same, and (2) to determine the differences between altered oligonucleotides. These experiments showed that the antigenic variants were very closely related to the parental virus and differed from it by a few oligonucleotide changes localized in the 3'-terminal region of the viral genome. Further nucleotide sequence analysis of the oligonucleotides that were changed in mutant viruses showed that many of the changes in the oligonucleotides were common among the variants, reflecting the same point mutation. This suggested that the genetic changes accumulated in the genomes of the variant viruses and that antigenic drift occurred in a progressive manner. Oligonucleotide changes outside of the 3' region were rare and no one change was observed in more than one virus. This suggested that these mutations were random and had no selective advantage for the virus carrying them. However, neutralizing antibody in the immune animal appeared to have a strong selection for mutational changes in the 3' region of the visna virus genome and these changes were maintained and passed to progeny virus.

The progressive nature of antigenic mutation of visna virus in sheep is illustrated by the changes detected in the viruses isolated from sheep 1. Virus LV1-2, isolated from this sheep, was of parental serotype; however, it was found to differ from the parental virus by two oligonucleotides in the 3' terminus. (Fig. 1 summarizes the genetic changes and serologic characterization of the variants from two sheep.) This change was found in all subsequent antigenic variants from this animal, suggesting that all the variants descended from this virus and that it had a selective advantage over the

Source Virus		Neutralization		Genetic Changes
		early sera	late sera	
1514		+ + + +	+ + + +	
Sheep 1	LV1-2	+ + + +	+ + + +	A B
	LV1-7	−	+ +	A B
	LV1-4	−	+ +	A B
	LV1-3	−	+ +	A B C
	LV1-6	−	+	A B C
	LV1-5	−	+	A B C
	LV1-1	−	+	A B C D E F
Sheep 4	LV4-2	+ + + +	+ + + +	A E F
	LV4-3	−	+ +	A E F
	LV4-1	−	+ +	A C E F
Tissue Culture	AD1-1	−	+ +	A B E
	AD4-1	−	+ +	A B E

Fig. 1. Antigenic variants of visna virus strain 1514, isolated from two persistently infected sheep and from tissue culture. The tissue-culture-derived virus variants AD1-1 and AD4-1 were selected with early immune sera from sheep 1 and 4, respectively. The genetic changes are those located in the 3' terminus of the viral genome.

parental virus. The original serological classification of the variants (Fig. 1) is consistent with the cumulative changes observed in the 3'-terminal region of their genomes.

When variants from a second infected animal (sheep 4) were examined, it was surprising to find that the same oligonucleotide changes were detected and that these changes reflected the same point mutation, as determined by sequence analysis. These common changes in the variants from the two animals suggest that the selective pressure for the phenotypic change in the viral antigenicity requires very specific alterations in the viral genome. Although mutations probably occurred randomly throughout the genome of the virus and independently in the viruses in the two animals, there was a highly specific selective pressure for the survival of particular point mutations. These mutations would be expected to specifically alter the site of interaction of the neutralizing antibody molecule with the viral glycoprotein. This might involve 3 or 4 epitopes on the protein, each containing a small number of amino acids. Thus, the selection of the same point mutations from independently inoculated animals might have been predicted.

A further extension of this observation is that the specificity of the initial antibody response in the animal would determine the minimum number of mutations or antigenic changes necessary for a virus variant to evade neutralization. Using sera from immune animals having narrow, medium, and wide neutralizing specificity (as determined by their ability to neutralize homologous and heterologous strains of visna virus) to select mutants in tissue culture, it was found that the antigenic profile of the variant was indeed determined by the specificity of the antibody used for selection [8]. Although fingerprint analysis has been done on only two of these viruses (AD1-1 and AD4-1), the small number of mutations in these variants (Fig. 1) can be correlated with the narrow spectrum of antibody used for their selection.

It was further observed that the frequency of isolation of variants from infected cell cultures was dependent on the specificity of the sera used [8]. With early sera or sera of narrow specificity, variants could be isolated from every culture. However, with late sera, or sera of broad specificity, variants could be isolated from only 20% of the cultures. Thus, the hypothesis was put forward by Narayan that in the natural infection with visna virus the sheep initially responds with a narrow, clonal range of neutralizing antibody and that this broadens as the animal responds to newly arising variants. This is supported by fingerprint studies of variants from sheep 1 and 4 (Fig. 1). It is interesting to postulate further that this progressive virus mutation may eventually cause the animal to develop broad enough neutralizing antibody to prevent further mutation and perhaps also prevent the progression of disease. This is supported by the low frequency of variant isolation with hyperimmune sera.

This eventual cessation of drift may also occur in horses infected with equine infectious anemia. Infected horses develop only a limited number of episodes of acute disease. The cessation of acute disease may be due to the broad specificity of the antibody response, which limits further virus mutation.

Thus, the molecular mechanism of antigenic drift of visna virus appears to be the accumulation of point mutational changes in a specific region of the viral genome. This region probably corresponds to the genetic coding region for the envelope glycoprotein, the target of neutralizing antibody. These mutations are presumed to alter the interaction of the glycoprotein molecule with the antibody molecule. As mutants evolve and elicit neutralizing antibody of broader specificity, greater numbers of mutations are required to alter the glycoprotein. It seems likely that this would eventually cause sufficient changes in the protein structure to be lethal to the variants. Thus, there would be an end point at which the humoral immunity of the infected animal would prevent further variants from escaping.

Acknowledgments. The work discussed in this review was supported by grants from the National Institutes of Health (NS-16145 and NS-15721) and the National Multiple Sclerosis Society (1232-B2). We thank Linda Kelly for preparation of the manuscript.

References
1. Bloom BR (1979) Games parasites play: how parasites evade immune surveillance. Nature 279:21–26
2. Webster RG, Laver WG, Air GM, Schild GC (1982) Molecular mechanisms of variation in influenza viruses. Nature 296:115–121

3. Wiktor TJ, Koprowski H (1980) Antigenic variants of rabies virus. J Exp Med 152:99–112
4. Ter Muelen V, Loffler S, Carter MJ, Stephenson JR (1981) Antigenic characterization of measles and SSPE virus haemmaglutinin by monoclonal antibodies. J Gen Virol 57:357–364
5. Kono Y, Kobayashi K, Fukunaga Y (1973) Antigenic drift of equine infectious anemia virus in chronically infected horses. Archiv Virusforchung 41:1–10
6. Narayan O, Griffin DE, Chase J (1977) Antigenic shift of visna virus in persistently infected sheep. Science 197:376–378
7. Narayan O, Griffin DE, Clements JE (1978) Virus mutation during "slow infection." Temporal development and characterization of mutants of visna virus recovered from sheep. J Gen Virol 41:343–352
8. Narayan O, Clements JE, Griffin DE, Wolinsky JS (1981) Neutralizing antibody spectrum determines the antigenic profiles of emerging mutants of visna virus. Infect Immun 32:1045–1050
9. Clements JE, Petersen FS, Narayan O, Haseltine WS (1980) Genomic changes associated with antigenic variation of visna virus during persistent infection. Proc Natl Acad Sci USA 77:4454–4458
10. Clements JE, D'Antonio N, Narayan O (1982) Genomic changes associated with antigenic variation of visna virus. II. Common nucleotide changes detected in variants from independent isolations. J Mol Biol 158:415–434
11. Scott JV, Stowring L, Haase AT, Narayan O, Vigne R (1979) Antigenic variation in visna virus. Cell 18:321–327

CHAPTER 23

Antigenic Variants of Viruses and Their Relevance to Clinical Disease

BELLUR S. PRABHAKAR AND ABNER LOUIS NOTKINS*

Certain viruses have the capacity to produce a variety of different diseases in humans. Infection with Coxsackie B4 virus can result in meningoencephalitis, myocarditis, hepatitis, epidemic pleurodynia, pharyngitis, and perhaps some cases of diabetes. It has never been clear, however, whether these different diseases are due to the existence of naturally occurring viral variants or a chance encounter of the virus with a given organ. Antigenic differences among isolates of Coxsackie B4 virus have been difficult to demonstrate with reference hyperimmune serum. Recently, however, a panel of monoclonal antibodies to Coxsackie B4 virus revealed the existence of a number of naturally occurring variants that are antigenically distinct [1]. All of the clinical isolates of Coxsackie B4 virus tested were found to differ from the prototype virus by anywhere from 2 to 12 epitopes. The frequency of mutation per antigenic determinant was calculated to be as high as 10^{-4}. In fact, different antigenic variants were isolated from different organs from the same individual during the course of a single infection. These and other studies indicate that antigenic diversity is a common feature of many viruses [2]. However, systematic studies have not yet been performed to see if these newly described antigenic variants of Coxsackie B4 virus produce different diseases.

There is considerable evidence that viral variants selected by other methods (e.g., plaque morphology) do, in fact, produce different diseases. For example, the YN strain of parainfluenza 3 virus consists of three variants, designated LT, Sc, and M [3]. The M variant produces an acute illness with

---

* National Institutes of Health, Bethesda, Maryland.

atrophy of lymphoid organs and death, whereas the LT and Sc variants produce a chronic illness with hydrocephalus. Two variants of Coxsackie B3 viruses, $CVB3_M$ and $CVB3_O$, have been identified. Inoculation of mice with $CVB3_M$ results in myocarditis, while inoculation with $CVB3_O$ produces minimal or no lesions [4]. The M variant of encephalomyocarditis virus consists of at least two variants, designated B and D. In mice, the D variant, but not the B variant, causes diabetes [5]. Despite the different diseases produced, the variants within each of these groups cannot be distinguished antigenically by hyperimmune sera.

How viral variants cause different diseases is not clear, but changes in the structural polypeptides seem to be important. In the case of avian influenza virus, the susceptibility of the hemagglutinin to proteolytic cleavage and the presence of proteases in the host cells determine the virulence of the virus for chickens. Structural polypeptides from a number of different viral isolates have been studied using polyacrylamide gel electrophoresis. Isolates with a hemagglutinin that can be cleaved by proteases are pathogenic for chickens, whereas those that resist cleavage are nonpathogenic [6,7]. The cleavage of the hemagglutinin enables the virus to infect cells through fusion of the virus with the host-cell membrane. Differences in the pathogenesis of variants of Newcastle disease virus and Sendai virus have also been attributed to differences in the cleavability of structural polypeptides by proteolytic enzymes [6,8].

Differences in the structural polypeptides of reovirus also influence pathogenesis [9]. The capacity of reovirus types 1 and 3 to infect different cells has been attributed to differences in the sigma 1 protein, which is coded for by the *S1* segment of the viral genome. This protein interacts with receptors on the surface of cells. A virus that has the sigma 1 protein of reovirus type 3 infects neurons and causes lethal encephalitis, whereas a virus that has the sigma 1 protein of reovirus type 1 infects ependymal but not neuronal cells causing hydrocephalus in some cases [10]. Analysis of various reoviruses with monoclonal antibodies revealed that a small region on the sigma 1 protein interacts with viral receptors [11]. Using monoclonal antibodies as selecting agents, variants that lack determinants required for the interaction of the virus with neuronal cells have been isolated from neurovirulent pools of reovirus type 3. These variants, when inoculated into mice, show markedly reduced neurovirulence.

Recent studies by Dietzschold *et al.* [12] with rabies virus have provided information at the molecular level on how antigenic alterations influence viral pathogenesis. Antigenic variants of a highly pathogenic strain of rabies virus were selected in the presence of a neutralizing monoclonal antibody. The variants that were resistant to neutralization by this antibody were shown to be avirulent in adult mice. Amino acid sequence data obtained for the glycoproteins of the virulent and avirulent variants showed that substitution of a single amino acid in position 333 altered virulence. Virus with arginine in position 333 of the glycoprotein was virulent, while substitution of this amino acid by either isoleucine or glutamine rendered the virus nonvirulent.

By use of monoclonal antibodies, antigenic variants of many viruses have now been found [2], but, thus far, very few of the variants have been examined for differences in pathogenicity. Many of the antigenic changes will undoubtedly turn out to be irrelevant in terms of pathogenicity. Others will certainly be important and could alter pathogenicity by any one of several mechanisms. For example, the antigenic change could be located at a critical site on a structural polypeptide involved in virus binding to its receptor [13,14]. The variant virus might bind poorly or not at all to receptors that the parental virus uses. Conversely, the expression of new antigenic determinants on the variant virus might allow recognition of cell receptors that are not ordinarily recognized by the parental virus, thus resulting in altered tissue tropism. As already indicated, the tropism of certain antigenic variants of viruses also may be affected by the susceptibility of structural polypeptides to host-cell proteases.

The generation of antigenic variants may influence the course and outcome of viral infections in still other ways [15]. For example, visna virus, which produces a chronic progressive neurological disease in sheep, does so by undergoing a series of antigenic shifts during the course of a single infection [16]. The antigenic variants arise *in vivo,* probably as a result of the selection pressure exerted by the antibody [17]. These variants react very poorly with antibodies generated in response to the original viral inoculum, suggesting that antigenic shift *in vivo* may be a mechanism by which the virus persists in spite of the host's immune response. A similar mechanism is also involved in the appearance of antigenic variants of influenza virus [18]. The new variants that arise *in vivo* show altered expression of antigenic determinants, and most of these changes are due to replacement of a limited number of amino acids at particular sites on the hemagglutinin molecule [19].

Identification of antigenic variants in what were once thought to be homogeneous virus populations may help explain the wide spectrum of clinical diseases seen with certain viral infections. Antigenic variants also may be one of the explanations for vaccine failures. Death as a result of rabies virus infection in animals or individuals previously vaccinated has been attributed to a poor response to the vaccine. However, this may be due to naturally generated variants that differ antigenically from the vaccine strain. In such situations, monoclonal antibodies could prove invaluable in the differential diagnosis of viral infections [20].

In conclusion, the isolation and identification of a large number of antigenic variants and a better understanding of the relationship between viral structure and function is broadening our knowledge of the different diseases produced by viral infections. This information will undoubtedly lead to new subgroupings and perhaps reclassification of many common viruses.

References

1. Prabhakar BS, Haspel MV, McClintock PR, Notkins AL (1982) High frequency of antigenic variants among naturally occurring human Coxsackie B4 virus isolates identified by monoclonal antibodies. Nature 300:374–376
2. Prabhakar BS, Haspel MV, Notkins AL (1984) Monoclonal antibody techniques applied to viruses. *In* Maramorosch K, Koprowski H (eds) Methods in Virology. Academic Press, New York, in press
3. Shibuta H, Adachi A, Kanda T, Matumoto M (1982) Experimental parainfluenza virus infection in mice: Fatal illness with atrophy of thymus and spleen in mice caused by a variant of parainfluenza 3 virus. Infect Immun 35:437–441
4. Gauntt CJ, Trousdale MD, LaBadie DRL, Paque RE, Nealon T (1979) Properties of Coxsackievirus B3 variants which are amyocarditic or myocarditic for mice. J Med Virol 3:207–220
5. Yoon JW, McClintock PR, Onodera T, Notkins AL (1980) Virus-induced diabetes mellitus. XVIII. Inhibition by a non-diabetogenic variant of encephalomyocarditis virus. J Exp Med 152:878–892
6. Choppin PW, Scheid A (1980) The role of viral glycoproteins in adsorption, penetration and pathogenicity of viruses. Rev Inf Dis 2:40–61
7. Bosch FX, Orlich M, Klenk HD, Rott R (1979) The structure of the hemagglutinin, a determinant for the pathogenicity of influenza viruses. Virology 94:197–207
8. Nagai Y, Klenk HD, Rott R (1976) Proteolytic cleavage of the viral glycoproteins and its significance for the virulence of Newcastle disease virus. Virology 72:494–508
9. Weiner HL, Drayna D, Averill DR Jr, Fields BN (1977) Molecular basis of reovirus virulence: Role of the S1 gene. Proc Natl Acad Sci USA 74: 5744–5748
10. Weiner HL, Powers LM, Fields BN (1980) Absolute linkage of virulence and central nervous system cell tropism of reovirus to viral hemagglutinin. J Infect Dis 141:609–616
11. Burstein SJ, Spriggs DR, Fields BN (1982) Evidence for functional domains on the reovirus type 3 hemagglutinin. Virology 117:146–155
12. Dietzschold B, Wunner WH, Wiktor TJ, Lopes AD, Lafon M, Smith CL, Koprowski H (1983) Characterization of an antigenic determinant of the glycoprotein that correlates with pathogenicity of rabies virus. Proc Natl Acad Sci USA 80:70–74
13. Morishima T, McClintock PR, Aulakh GS, Billups LC, Notkins AL (1982) Genomic and receptor attachment differences between mengovirus and encephalomyocarditis virus. Virology 122:461–465
14. Spriggs DR, Bronson RT, Fields BN (1983) Hemagglutinin variants of reovirus type 3 have altered central nervous system tropism. Science 220:505–507
15. Finberg R, Weiner HL, Fields BN, Benacerraf B, Burakoff J (1979) Generation of cytolytic T lymphyocytes after reovirus infection: Role of S1 gene. Proc Natl Acad Sci USA 76:442–446
16. Narayan O, Griffin DE, Chase J (1977) Antigenic shift of visna virus in persistently infected sheep. Science 197:376–378
17. Narayan O, Griffin DE, Silverstein AM (1977) Slow virus infection: Replication and mechanisms of persistence of visna virus in sheep. J Infect Dis 135:800–806
18. Webster RG, Laver WG (1975) Antigenic variation of influenza viruses. *In* Kil-

bourne ED (ed) The Influenza Viruses and Influenza. Academic Press, New York, p 269
19. Laver WG, Gerhard W, Webster RG, Frankel ME, Air GM (1979) Antigenic drift in type A influenza virus: Peptide mapping and antigenic analysis in A/PR/8/34 (HON1) variants selected with monoclonal antibodies. Proc Natl Acad Sci USA 76:1425–1429
20. Wiktor TJ, Flamand A, Koprowski H (1980) Use of monoclonal antibodies in diagnosis of rabies virus infection and differentiation of rabies and rabies-related viruses. J Virol Meth 1:33–46

Virus Persistence

24. Unique Interactions of Retroviruses with Eukaryotic Cells
    S. R. TRONICK AND S. A. AARONSON 165

25. Molecular Biology of Herpes Simplex Virus Latency
    EDOUARD M. CANTIN, ALVARO PUGA, AND
    ABNER LOUIS NOTKINS 172

26. Cellular Oncogenes and the Pathogenesis of Cancer
    ROBERT A. WEINBERG 178

27. Antibody Initiates Virus Persistence: Immune Modulation and
    Measles Virus Infection
    ROBERT S. FUJINAMI AND MICHAEL B. A. OLDSTONE.......... 187

28. Ovarian Infection and Transovarial Transmission of
    Viruses in Insects
    LEON ROSEN.. 194

CHAPTER 24
Unique Interactions of Retroviruses with Eukaryotic Cells

S. R. Tronick and S. A. Aaronson*

The past few years have witnessed a dramatic increase in our understanding of how normal cells become malignant. Such efforts have been aided immeasurably by investigations of oncogenic retroviruses (oncoviruses). These studies have revealed the intimate associations that exist between retroviruses and their host cells. This review focuses on the structure and function of retrovirus genomes with emphasis on how their organization enables retroviruses to persist in vertebrate cell genomes, induce malignant transformation, and influence host-cell gene structure and expression.

Distribution and Classification of Retroviruses

All members of the retrovirus family possess an RNA genome and a reverse transcriptase. Retroviruses are widely distributed among vertebrate species including mammals, birds, and reptiles [1], and recent evidence indicates that they are present in invertebrates [2] as well. Similarities of the retroviral life cycle to that of the cauliflower mosaic virus [3] suggest that the plant kingdom may also host its own versions of retroviruses. Three retrovirus subfamilies have been recognized and include the oncoviruses, the lentiviruses, and the spumaviruses. Because of their ability to induce a wide spectrum of tumors, the oncoviruses have been most intensively studied.

---

* National Institutes of Health, Bethesda, Maryland.

Oncoviruses have been classified on the basis of morphologic criteria into four genera designated as types A, B, C, and D [4]. There are two groups of oncoviruses, readily distinguishable based on their pathogenicity [5]. The chronic viruses often induce tumors but only after a prolonged latent period. When inoculated in susceptible animals, type C viruses most commonly cause leukemias but can induce other tumors and even neurologic diseases. In contrast, type B viruses primarily cause mammary tumors. Chronic viruses replicate in the absence of any apparent transforming effect on known assay cells in tissue culture. They can also establish persistent infections without apparent symptoms.

One subgroup of type C viruses, acute transforming viruses, induce tumors within a very short period of days to weeks. Moreover, these agents induce a wide spectrum of neoplasias, including sarcomas, hematopoietic tumors, and carcinomas. In tissue culture, these viruses generally induce foci of transformation in appropriate assay cells. A large body of evidence has led to the understanding that acute transforming viruses have arisen in nature by recombination of chronic viruses with cellular (c-*onc*) genes [6]. Within the virus, such discrete segments of genetic information (v-*onc*) have been shown to be required for viral transforming functions.

Mode of Replication

Following retrovirus infection, the viral RNA genome is copied by the viral reverse transcriptase into a double-stranded DNA molecule [1]. The steps in the synthesis of retroviral DNA involve a priming reaction utilizing a specific tRNA bound to the 5' end of the viral RNA, elongation of the DNA transcript and "jumping" of the reverse transcriptase to the 3' end of the same or different viral RNA template, removal of the RNA template by the RNase H activity of the reverse transcriptase, and finally generation of the second DNA strand. First linear and then supercoiled circular DNA species are found *in vivo*. Although not proven, it is suspected that the circular DNA species (provirus) undergoes integrative recombination with the host-cell genome. By this mechanism, the retrovirus can establish an intimate and lasting association with the host-cell genome.

An important feature of the proviral genome is the presence of large (300 to 1300 bp) terminal direct repeats, termed LTRs. Included within each LTR are small direct repeats (R) derived from the termini of the viral RNA and stretches of sequences uniquely representing the 5' (U5) and 3' (U3) ends. This results in an LTR structure, U3-R-U5. LTRs contain signals for the initiation, termination, and enhancement of transcription, and also contain at each end small inverted repeats. Thus, the retroviral genome closely resembles transposable elements that have been isolated from bacteria, yeast, and *Drosophila* [7].

Endogenous Retroviruses

The development of continuous mouse cell lines led to the demonstration that viruses closely related to chronic retroviruses could be induced from such cells spontaneously or following treatment with certain chemicals. These findings have been extended so that it is now known that endogenous retroviruses are present within the genomes of a wide variety of vertebrate species, including human, and are transmitted in Mendelian fashion. Moreover, molecular hybridization studies have shown that endogenous viral genomes have been transmitted in some species for millions of years. Endogenous retroviral sequences are generally found in multiple copies in the chromosomes of the host cell. Sequences related to type A, B, C, and D oncoviruses have been detected in the genomes of a wide variety of species. It has been calculated that the total amount of endogenous retroviral sequences carried within the germline of the mouse, for example, may approach 0.5% of the DNA (for reviews on endogenous viruses, see [8–10]).

Mammalian type C viruses share antigenic determinants among several of their gene products (*gag, pol,* and *env*), implying that these viruses have a common progenitor [11]. However, immunological and molecular hybridization analyses have indicated, in addition, that the evolution of present oncovirus groups has involved genetic interactions among their progenitors. Thus, type C and D oncoviruses share *gag* gene-encoded antigenic determinants as do type B and D oncoviruses [12]. Certain mammalian type C and D oncoviruses are antigenically related in their *env* gene-coded gp70 molecule and also possess nucleotide sequence homology within their p15E coding region. Recent studies have also revealed evolutionary relationships among the reverse transcriptases of most oncovirus genera [36].

Interactions between Retroviral and Cellular Genes

Retroviral genomes, as a consequence of their replication by means of reverse transcription, their ability to integrate into many sites in the host chromosome, and their similarity to transposable elements, can have profound effects on cellular genes. The studies of Hayward et al. [13] were the first to show transcriptional activation of a gene by the insertion of retroviral sequences into the cell genome. Analysis of tumors induced in chickens by avian lymphoid leukosis virus (LLV) revealed the presence of increased levels of mRNA related to a cellular oncogene, c-*myc*. These transcripts contained sequences derived from the 5' terminus of the LLV genome that had integrated near c-*myc* coding sequences. Similar results were obtained by others [14] and have led to the promoter insertion model of transformation by chronic oncoviruses. Subsequently, other examples of specific integrations of exogenous retrovirus sequences in tumor cells have been obtained [15,16].

A more recent and equally dramatic example of gene activation by retroviral sequences has been obtained in a system involving an endogenous retroviral genome [17,18]. Mouse myeloma cells were analyzed for DNA sequence rearrangements of cellular oncogenes. In one tumor, sequences closely related to the LTR of type A retroviruses were discovered to have inserted into c-*mos* resulting in its activation as a transforming gene by deletion of some 5' sequences and transcriptional activation of the gene. Whether activation of the *mos* oncogene resulted from transposition of A-type retroviral sequences from their genomic residence or by reinsertion of a replicating A-type particle remains to be resolved.

Type A retroviral sequences have also been implicated in the inactivation of genes. Mutant mouse hybridoma cell lines were found to be defective in the production of kappa light chains [19]. Analysis of the kappa light chain genes revealed the presence of type A viral sequences in the introns. Each of the two mutant genes contained distinguishable type A viral sequences [20]. These results further demonstrate the ability of retroviral genes to appear in different locations in cellular DNA.

There are at least three other notable examples of retroviruses acting as insertional mutagens. Genetic analysis of DBA/2J mice indicated that an endogenous ecotropic type C provirus segregated concordantly with a coat color (dilute) mutation. The lack of DNA sequences of this provirus at this locus in a spontaneous mutant strongly suggests that the dilute mutation was caused by insertion of a provirus [21]. Deliberate infections of mouse germline cells by Moloney leukemia virus were found to cause recessive lethal mutations that resulted in early death of the developing embryos [22]. The proviral genome was found to have integrated into the 5' end of the alpha 1(I) collagen gene resulting in cessation of its transcription. Exogenous infection of RSV-transformed mouse cells by Moloney leukemia virus caused reversion of the cells to the normal phenotype. The integrated RSV provirus in one class of revertants contained inserted viral sequences, whereas another set of revertants contained deletions of both *src* and upstream sequences [23].

The capacity of both exogenous and endogenous retroviruses to integrate stably into different locations in cell DNA and greatly affect gene activity makes them potentially significant in altering normal pathways of development and differentiation. The expression of A-type viral sequences in mouse germ cells and in preimplantation embryos would be consistent with their involvement in these processes [24–26].

Capture of Cellular Oncogenes by Retroviruses

The architecture of the cellular genome may be altered in yet other ways by retroviruses. As discussed earlier, acute transforming viruses contain cellular sequences required for their oncogenic activity. Analysis of these sequences and their normal cellular counterparts (proto-oncogenes) have shown that the v-*onc* sequences do not contain introns, and most likely represent copies of proto-oncogene mRNAs. Thus, one possible mechanism for the formation of the acute viruses would involve reverse transcription of a processed RNA transcript followed by recombination with retroviral sequences. Alternatively, when a retrovirus integrates near a proto-oncogene, an mRNA containing both viral and oncogene sequences could be produced. This molecule could be packaged into a virus particle and undergo reverse transcription and further recombination during subsequent rounds of virus replication.

Although proto-oncogenes are the only cellular genes known to have been captured and transmitted as viruses, it is reasonable to assume that other cellular genes could be transduced and introduced to other cells of the same organism or could be horizontally transmitted to other organisms. In fact, pseudogenes that have structures remarkably similar to mRNA molecules have been identified and have been shown to reside in different locations from their active counterparts [27]. It is not surprising that retroviral A-type LTR sequences reside on both sides of one such pseudogene (mouse α-globin aψ3) [28]. Although this may be a fortuitous association, it is consistent with transduction and reintegration of the processed gene.

Retroviruses and Human Disease

The involvement of retroviruses in human disease has been a matter of speculation. Studies of these viruses have led to the identification of c-*onc* genes in human DNA and in some cases the detection and even elucidation of their mechanisms of activation as oncogenes in human tumors [29,30]. Recent studies have also established the presence of chronic retroviruses either in exogenous or endogenous form in human cells. For example, the first human retrovirus, designated HTLV, has been implicated as the causative agent of certain human leukemias [31,32]. Sequences related to this virus have not been detected in human DNA, indicating that HTLV is an exogenous retrovirus. However, advances in molecular biology have also made it possible to establish that human DNA does contain sequences related to retroviruses [33–35]. It seems reasonable to assume that interactions of the type leading to gene activation and inactivation, the reintegration of processed genes, and other yet to be defined effects of the intimate association of these endogenous viral genetic elements with vertebrates will be applicable to humans as well. Thus, the large fund of knowledge that has been gained from the studies of retroviruses in animal systems by many investigators over a number of years is now being directly applied to studies of human disease.

References
1. Weiss RA, Teich N, Varmus H, Coffin J (eds) (1982) Molecular biology of tumor viruses, 2nd ed, RNA tumor viruses. Cold Spring Harbor Laboratory, New York

2. Shika T, Saigo K (1983) Retrovirus-like particles containing RNA homologous to the transposable element *copia* in *Drosophila melanogaster*. Nature 302:119–124
3. Varmus H (1983) Reverse transcription in plants (?). Nature 304:113–117
4. Schidlovsky G (1977) Structure of RNA tumor viruses. *In* Gallo R (ed) Recent Advances in Cancer Research: Cell Biology, Molecular Biology and Tumor Virology. CRC, Cleveland, Ohio, p 189
5. Gross L (1970) Oncogenic viruses, 2nd ed. Pergamon Press, Oxford
6. Duesberg PH (1983) Retroviral transforming genes in normal cells (?). Nature 304:219–226
7. Temin H (1982) Function of the retrovirus long terminal repeat. Cell 28:3–5
8. Aaronson SA, Stephenson JR (1976) Endogenous type C RNA viruses of mammalian cells. Biochem Biophys Acta 458:323–354
9. Jaenisch R (1983) Endogenous retroviruses. Cell 32:5–6
10. Aaronson S (1983) Unique aspects of the interactions of retroviruses with vertebrate cells: Rhoads CP memorial lecture, Cancer Res 43:1–5
11. Aaronson S, Barbacid M, Hino S, Tronick SR, Krakower J (1978) Common progenitors in the evolution of mammalian retroviruses: Implications in the search for RNA tumor virus expression in man. *In* Bentvelzen P, Hilgers J, Yohn DS (eds) Advances in Comparative Leukemia Research. Elsevier, Amsterdam, p 127
12. Barbacid J, Long LK, Aaronson SA (1980) Major structural proteins of type B, type C, and type D oncoviruses share interspecies antigenic determinants. Proc Natl Acad Sci USA 77:72–76
13. Hayward WS, Neel BG, Astrin SM (1981) Activation of a cellular *onc* gene by promoter insertion in ALV-induced lymphoid leukosis. Nature 290:475–480
14. Payne GS, Bishop JM, Varmus HE (1982) Multiple arrangements of viral DNA and an activated host oncogene in Bursal lymphomas. Nature 295:209–214
15. Fung YK, Lewis WG, Crittenden LB, Kung HJ (1983) Activation of the cellular oncogene c-*erb* B by LTR insertion: Molecular basis for induction of erythroblastosis by avian leukosis virus. Cell 33:357–368
16. Nusse R, Varmus HE (1982) Many tumors induced by the mouse mammary tumor virus contain a provirus integrated in the same region of the host genome. Cell 31:99–109
17. Rechavi G, Givol D, Canaani E (1982) Activation of a cellular oncogene by DNA rearrangement: Possible involvement of an IS-like element. Nature 300:607–611
18. Kuff EL, Feenstra A, Lueders K, Rechavi G, Givol D, Canaani E (1983) Homology between an endogenous viral LTR and sequences inserted in an activated cellular oncogene. Nature 302:547–548
19. Hawley RG, Shulman MJ, Murialdo H, Gibson DM, Hozumi NI (1982) Mutant immunoglobulin genes have repetitive DNA elements inserted into their intervening sequences. Proc Natl Acad Sci USA 79:7425–7429
20. Kuff EL, Feenstra A, Lueders K, Smith L, Hawley R, Hozumi N, Shulman M (1983) Intracisternal A-particle genes as movable elements in the mouse genome. Proc Natl Acad Sci USA 80:1992–1996
21. Jenkins NA, Copeland NG, Taylor BA, Lee BK (1981) Dilute (d) coat colour mutation of DBA/2J mice is associated with the site of integration of an ecotropic MuLV genome. Nature 293:370–374
22. Schnieke A, Harkers K, Jaenisch R (1983) Embryonic lethal mutation in mice

induced by retrovirus insertion into the alpha 1(I) collagen gene. Nature 304:315–320
23. Varmus HE, Quintrell N, Ortiz S (1981) Retroviruses as mutagens: Insertion and excision of a nontransforming provirus alter expression of a resident transforming provirus. Cell 125:23–26
24. Calarco PG, Szöllösi D (1973) Intracisternal A particles in ova and preimplantation stages of the mouse. Nature New Biol 243:91–93
25. Biczysko W, Prenkowski M, Solter D, Koprowski H (1973) Virus particles in early mouse embryos. J Natl Cancer Inst 51:1041–1050
26. Chase DG, Pikò L (1973) Expression of A- and C-type particles in early mouse embryos. J Natl Cancer Inst 51:1971–1975
27. Hollis GF, Hieter PA, McBride OW, Swan D, Leder P (1982) Processed genes: A dispersed human immunoglobulin gene bearing evidence of RNA-type processing. Nature 296:321–325
28. Leuders K, Leder A, Leder P, Kuff E (1982) Association between a transposed α-globin pseudogene and retrovirus-like elements in the BALB/C mouse genome. Nature 295:426–428
29. Cooper GM (1982) Cellular transforming genes. Science 218:801–806
30. Weinberg RA (1982) Fewer and fewer oncogenes. Cell 30:3–4
31. Poiesz BJ, Ruscetti FW, Gazdar AF, Bunn PA, Minna JD, Gallo RC (1980) Detection and isolation of type C retrovirus particles from fresh and cultured lymphocytes of a patient with cutaneous T-cell lymphoma. Proc Natl Acad Sci USA 77:7415–7419
32. Yoshida M, Miyoshi I, Hinuma Y (1982) Isolation and characterization of retrovirus from cell lines of human adult T-cell leukemia and its implication in the disease. Proc Natl Acad Sci USA 79:2031–2035
33. Martin MA, Bryan T, Rasheed S, Khan A (1981) Identification and cloning of endogenous retroviral sequences present in human DNA. Proc Natl Acad Sci USA 78:4892–4896
34. Callahan R, Drohan W, Tronick S, Schlom J (1982) Detection and cloning of human DNA sequences related to the mouse mammary tumor virus genome. Proc Natl Acad Sci USA 79:5503–5507
35. Bonner TI, O'Connell C, Cohen M (1982) Cloned endogenous retroviral sequences from human DNA. Proc Natl Acad Sci USA 79:4709–4713
36. Chiu, I-M, Callahan, R, Tronick SR, Schlom J, Aaronson SA (1984) Major *pol* gene progenitors in the evolution of oncoviruses. Science 223:364–370.

CHAPTER 25
Molecular Biology of Herpes Simplex Virus Latency

EDOUARD M. CANTIN, ALVARO PUGA, AND
ABNER LOUIS NOTKINS*

Herpes simplex virus (HSV) has the ability to produce a latent as well as a productive infection in its host. The virus replicates initially in epithelial cells at the site of infection and then spreads by retrograde axonal transport to the sensory ganglia that innervate the infected dermatomes [1,2]. It establishes a productive infection in these ganglia, but this subsides after a few days. The latent infection then ensues, characterized by the asymptomatic persistence of the viral genome in neurons of the trigeminal and sensory dorsal root ganglia [2–7] and, to a lesser extent, of the autonomic and central nervous systems of humans and experimentally inoculated animals [2,8,9]. For reasons as yet poorly understood, the virus may sporadically reactivate and cause a recurrence at or in the neighborhood of the primary site of infection. The immune system plays a critical role in controlling and eliminating both primary and recurrent infections at epithelial surfaces, as well as perhaps clearing subclinical ganglionic reactivations. However, the demonstration that latency can persist in the absence of neutralizing antibodies [10] suggests that molecular factors operating within the infected neuron may be major regulators of latency. This brief review will describe what is known about the molecular biology of the latent HSV infection, and how this information may already suggest possible regulatory mechanisms.

Detection of the Viral Genome During Latency

Unfortunately for the applicability of biochemical techniques, only a very small fraction of the neurons, perhaps not more than 0.1%, harbor the latent viral genome. However, with the advent of very sensitive nucleic acid hy-

---

* National Institutes of Health, Bethesda, Maryland.

bridization techniques, detection of HSV-specific nucleic acid sequences in a variety of tissues of neural origin has become possible. In trigeminal ganglia of mice, titration of viral DNA sequences by reassociation kinetics indicated that the viral genome was present at an average of 1-2 genome equivalents per cell during the acute infection, and thereafter declined to 0.05-0.10 equivalents per cell during the latent stage [11]. The presence of reactivable viral genomes in 50%-80% of unselected human trigeminal ganglia has long been demonstrated [3], although no attempts have been made to determine the number of copies present. More recently, Brown et al. [12] have been able to detect in approximately 50% of virus-negative human trigeminal tissues the presence of HSV DNA sequences by the novel use of HSV temperature sensitive mutants as genetic probes. Whether these sequences exist as entire or defective genomes is at present undetermined.

HSV DNA sequences have also been detected in brain tissues. In latently infected mice the levels vary for each individual animal analyzed, but, in general, are one order of magnitude lower than in trigeminal ganglia [13]. It was proposed that these viral DNA sequences in brain could represent incomplete genomes, since the virus could not be reactivated by standard explantation techniques [13,14]. However, more recently, Fraser and colleagues have been able to demonstrate by restriction enzyme analyses that the majority, if not all of the viral genome, is present in brain stem tissue of latently infected mice [15]. Information on human brain tissues is more scarce. In one study, Fraser and coworkers have demonstrated the presence of HSV-1 DNA sequences in restriction endonuclease-cleaved human brain DNA [16]. More than 50% of the brain samples from humans free of clinical signs of HSV contained viral DNA fragments that hybridized with a cloned HSV DNA probe in a manner indicative, in most cases, of the presence of the entire viral genome.

Physical State of the Genome During Latency

In what state is the viral genome during latency? One possibility is that the viral DNA is maintained in an episomal form, either linear, as it exists in virions, or circularized through the terminal repetitions (TR). A second possibility is that it exists as an entity integrated into chromosomal DNA. These two possibilities are not mutually exclusive, since both may coexist in the same or different cells, as is the case for Epstein-Barr virus. Evidence exists to indicate that coexistence of both forms also may be the case with HSV.

Evidence for linearity comes from the finding that most HSV-positive human brain DNA samples analyzed show the presence of terminal genomic fragments at molecular sizes identical to those found in virion DNA [14], indicating that they were part of free linear DNA molecules and not circularized or covalently joined to cellular DNA. Evidence for the presence of nonlinear forms comes from restriction enzyme analyses of nervous tissue

DNA from infected mice. In brain stems of latently infected animals, terminal genomic fragments were absent, although they were readily discernible in DNA samples from brain stems of acutely infected mice [15]. This finding strongly suggests that the HSV genome was present in a form distinct from linear virion-derived DNA, perhaps as a circular or concatemeric structure similar to the replicative intermediates that accumulate during a tissue culture infection. In another study [17], in which large quantities of trigeminal ganglion DNA from latently infected mice were enriched for HSV-sequences by RPC-5 chromatography, terminal genomic fragments were demonstrated at an array of molecular sizes, ranging from 1 to 20 kilobase pairs. These results suggest that the latent HSV-genome may be molecularly rearranged or perhaps integrated into the cellular DNA.

These apparent differences regarding the molecular state of the genome may stem from essential differences between latency in the mouse and human systems or between brain and trigeminal ganglion. Alternatively, as indicated earlier, various molecular forms of the viral genome may coexist. Cloning of the latent viral genome by recombinant DNA techniques should eventually resolve these questions.

Gene Expression During Latency

Is the viral genome expressed during latency? Several experiments indicate that a transcriptional block occurs concomitantly with the establishment of the latent state. In one series of experiments [11], viral-specific RNA transcripts were detected at the level of 0.1–0.2 equivalents per cell during the acute phase in mouse trigeminal ganglia, and became undetectable during the latent state. Thus, progression to the latent state seems to be accompanied by an apparent silencing of the viral genome which, however, need not be absolute, since limited transcription would not have been detected in this study. In agreement with this proviso, Galloway and coworkers [18,19] have been able to detect by *in-situ* hybridization HSV-2 specific transcripts in human paravertebral ganglia, but only from a group of genes located within the left hand 30% of the long unique sequences. Interpretation of these findings should, however, be tempered by the recent demonstration of sequence homology between certain regions of the HSV genome and sequences within the uninfected mammalian genome [20,21]. At least one of the regions detected by *in-situ* hybridization corresponds to one of the regions of homology, which is transcribed in uninfected human placental tissue [22]. The nature of these viral transcripts detected must await further analysis, since they could be cellular transcripts activated as a consequence of the establishment of latency or transcripts from other as yet unidentified regions of homology.

Studies on the nature of the transcripts indicate that none of the immediate early (IE) viral genes are transcribed during latency in human ganglia

[18,19], but studies with temperature sensitive mutants of HSV-1 [23] suggest that expression of at least one IE gene is required for the establishment of latency. This gene, IE3, encodes the viral polypeptide Vmw175 (ICP4), which plays a crucial role in viral gene expression. Vmw175 is an autoregulated protein that also controls expression of delayed-early and late viral genes; its expression is continuously required throughout the replicative cycle and is also responsible for the activation or deregulation of cellular stress proteins [24]. Recently, the Vmw175 polypeptide has been detected in trigeminal ganglia of latently infected rabbits, in the apparent absence of expression of other HSV antigens [25]. The immediate paradox as to the detection of the polypeptide in the absence of its corresponding mRNA can be ascribed to differential sensitivities of the methods used as well as differences between latency in humans and rabbits. However, the implication of the work in rabbits would be that expression of the viral genome may be regulated in a different manner during latency, or, alternatively, that a non-functional Vmw175-like polypeptide was being detected. Similar conflicting reports exist regarding expression of the viral thymidine kinase gene ([19,26] and E. Cantin, unpublished).

Summary and Future Directions

The viral genome has been clearly demonstrated in sensory ganglia from humans and mice. Evidence has also accrued that viral sequences are present as well in autonomic and central nervous system tissues. Although the answers are not yet in, the majority of the available evidence indicates that the latent viral genome is not a linear, virion-like DNA molecule. It should be possible to resolve this issue with evidence gathered from cloning experiments. However, unless the answers are absolutely clearcut, an immediate understanding of the latent infection may not be gained from these experiments. If, for instance, episomal and integrated genomes are shown to coexist, it will be difficult to determine their relative role in the maintenance of the latent infection. Nevertheless, since the herpes genome consists of four isomeric forms, all of which may not be biologically active [1], cloning may make it possible to determine whether latency is specifically related to the predominance of one or more of these genomic isomers.

It appears that the latent viral genome is for the most part in a non-transcribed state, although some genes may be expressed. However, little is known as to the effect that the various latency model systems used (e.g., guinea pig, rabbit, mouse) have on the outcome of experiments designed to study gene expression. Different animal models show large differences in the rate of spontaneous reactivation, which would considerably influence not only the extent of transcription but also the nature of the transcripts detected. Cloning of cDNA molecules synthesized on the viral transcripts found will eventually elucidate their nature and tell us whether these genes are continuously expressed during latency and/or are required to maintain the latent state.

From these studies it may be possible to define latency in terms of the interactions occurring between viral and neuron-specific gene products. It is an appealing idea, already gathering momentum, that establishment of latency may be intimately tied to the functional activity of Vmwl75. It is possible that Vmwl75 affects the expression of neuronal factors, such as perhaps the stress proteins, and conversely that neuronal factors affect the expression of Vmwl75. Thus, latency could result from a host-induced silencing of Vmwl75. Conceivably, this interaction may be reversed in response to metabolic changes, both within the infected neuron and at the periphery, which would then serve as stimuli to reactivate the latent genome.

Results from work in these new areas should provide a more complete understanding of the molecular events leading to the establishment, maintenance, and reactivation of the latent infection.

References

1. Nahmias AJ, Dorodle WR, Schinazi RF (eds) (1981) Latency and oncogenesis in the human herpesviruses: An interdisciplinary approach. Elsevier, Amsterdam
2. Wildy P, Field HJ, Nash AA (1980) Classical herpes latency revisited. In Mahy BW, Minson AC, Darby GK (eds) Viral Persistence. Cambridge University Press, Cambridge, England, p 135
3. Baringer JR, Swoveland P (1973) Recovery of herpes simplex from human trigeminal ganglia. N Engl J Med 288:648–650
4. Walz MA, Price RW, Notkins AL (1974) Latent ganglionic infection with herpes simplex virus types 1 and 2: Viral reactivation in vivo after neurectomy. Science 184:1185–1187
5. Canton CA, Kilbourne ED (1952) Activation of latent herpes by trigeminal sensory root section. N Engl J Med 246:172–176
6. Cook ML, Stevens JG (1976) Latent herpetic infections following experimental viraemia. J Gen Virol 31:75–80
7. McLennan JL, Darby G (1980) Herpes simplex virus latency: The cellular location of virus in dorsal root ganglia and the fate of the infected cell following virus activation. J Gen Virol 51:233–243
8. Openshaw H, Sekizawa T, Cantin EM, Puga A, Notkins AL (1981) Latency and reactivation of herpes simplex virus: Animal models. In Hook J, Jordan G (eds) Viral Infections in Oral Medicine. Elsevier/North Holland, Amsterdam, p 79
9. Price RW, Katz BJ, Notkins AL (1975) Latent infection of the peripheral ANS with herpes simplex virus. Nature 257:686–688.
10. Sekizawa T, Openshaw H, Wohlenberg C, Notkins AL (1980) Latency of herpes simplex virus in absence of neutralizing antibody: Mode of reactivation. Science 210:1026–1028
11. Puga A, Rosenthal JD, Openshaw H, Notkins AL (1978) Herpes simplex DNA and mRNA sequences in acutely and chronically infected trigeminal ganglia of mice. Virology 89:102–111
12. Brown SM, Subak-Sharpe JH, Warren KG, Wroblewska Z, Koprowski H (1979)

Detection by complementation of defective or uninducible (herpes simplex type 1) virus genomes latent in human ganglia. Proc Natl Acad Sci USA 76:2364–2368
13. Cabrera CV, Wohlenberg C, Openshaw H, Rey-Mendez M, Puga A, Notkins AL (1980) Herpes simplex virus DNA sequences in the central nervous system of latently infected mice. Nature 288:288–290
14. Kastoukoff L, Long C, Doherty PC, Wroblewska Z, Koprowski H (1981) Isolation of virus from brain after immune suppression of mice with latent herpes simplex. Nature 291:432–433
15. Rock DL, Fraser NW (1983) Detection of HSV-1 genome in central nervous system of latently infected mice. Nature 302:523–525
16. Fraser NW, Lawrence WC, Wroblewska Z, Gilden DH, Koprowski H (1981) Herpes simplex type 1 DNA in human brain tissue. Proc Natl Acad Sci USA 78:6461–6465
17. Puga A, Cantin EM, Wohlenberg C, Openshaw H, Notkins AL (1984) Different sizes of restriction endonuclease fragments from the terminal repetitions of the herpes simplex virus type 1 genome latent in trigeminal ganglia of mice. J Gen Virol 65:437–444
18. Galloway DA, Fenoglio C, Sherchuk M, McDougall JK (1979) Detection of herpes simplex RNA in human sensory ganglia. Virology 95:265–268
19. Galloway DA, Fenoglio CM, McDougall JK (1982) Limited transcription of the herpes simplex virus genome when latent in human sensory ganglia. J Virol 41:686–691
20. Puga A, Cantin EM, Notkins AL (1982) Homology between murine and human cellular DNA sequences and terminal repetition of S component of herpes simplex virus type 1 genome. Cell 31:81–87
21. Peden K, Mounts P, Hayward GS (1982) Homology between mammalian cell DNA sequences and human herpesvirus genomes detected by a hybridization procedure with high-complexity probe. Cell 31:71–80
22. Maitland NJ, Kinross JH, Busuttie G, Ludgate SM, Sarant GE, Jones KW (1981) The detection of DNA tumor virus-specific RNA sequences in abnormal human cervical biopsies by hybridization. J Gen Virol 55:123–137
23. Watson K, Stevens JG, Cook ML, Subak-Sharpe JH (1980) Latency competence of thirteen HSV 1 temperature sensitive mutants. J Gen Virol 49:149–159
24. Notarianni EL, Preston CM (1982) Activation of cellular stress protein genes by herpes simplex virus temperature sensitive mutants which overproduce immediate early polypeptides. Virology 123:113–122
25. Green MT, Courtney RJ, Dunkel E (1981) Detection of an immediate early herpes simplex virus type 1 polypeptide in trigeminal ganglia from latently infected animals. Infect Immun 34:987–992
26. Yamamoto H, Walz MA, Notkins AL (1977) Viral specific thymidine kinase in sensory ganglia of mice infected with herpes simplex virus. Virology 76:866–869

CHAPTER 26
Cellular Oncogenes and the Pathogenesis of Cancer

ROBERT A. WEINBERG*

Studies in two areas of biology have pointed to the existence of cellular genes that exhibit an oncogenic potential after activation. These normal cellular genes, sometimes termed "proto-oncogenes," can become activated by several different types of genetic alterations, some involving the intervention of viral genomes. These proto-oncogenes and derived oncogenes represent centrally important components of the molecular mechanisms that create the cancer phenotype.

Our knowledge of cellular proto-oncogenes stems from two areas of experimental work. The initial demonstration of their existence came from work on Rous sarcoma virus (RSV). A molecular analysis of the viral genome revealed two component parts. The first part was seen to be constituted of a group of viral genes that are necessary for the growth of the virus—functions such as genome replication and virion formation. The second component consisted of only one gene—the oncogene used by the virus to transform infected cells. This viral oncogene, termed *src,* carries all the information required for the synthesis of the pp60src, the protein whose kinase activity is responsible for induction of the transformed phenotype.

Of prime interest was the origin of this RSV-associated *src* gene. Unlike virtually all other viral genes studied until then, this gene was found to have a nonviral origin. Nucleic acid hybridization experiments showed definitely that its origin lay in the genome of the normal cell [1]. It became clear that the virus acquired a cellular gene by a recombination event; transduced this gene as part of its own genome; and then exploited this gene to transform any cells that it happened to infect.

---

* Whitehead Institute for Biomedical Research and Massachusetts Institute of Technology, Cambridge, Massachusetts.

This demonstrated that within the normal cellular genome, one could find genes of latent oncogenic potential. Moreover, retroviruses such as RSV could be exploited to retrieve and reveal these hitherto unknown genes. In subsequent years, at least 16 other cellular proto-oncogenes have been discovered in a similar manner [2]. Each of these genes has been found by virtue of its association, on one or more occasions, with an actively transforming retrovirus. These virus-associated oncogenes are listed in Table 1.

A second experimental strategy has also revealed cellular proto-oncogenes and related oncogenes. DNAs have been extracted from a variety of tumors of largely nonviral origin. These DNAs have then been applied, using gene transfer procedures, to monolayer cultures of NIH3T3 mouse fibroblasts. The subsequent appearance of foci of transformed cells in these monolayers has signaled the presence of transforming sequences in the applied donor DNA (reviewed in [3,4]).

Some of these transforming sequences have been isolated by molecular cloning [5–8]. These cellular transforming sequences have also been termed oncogenes. Use of these isolated oncogene clones in sequence hybridization experiments reveals, once again, that they are closely related to normal cell genes. These normal cellular antecedents are called, as before, proto-oncogenes. In this case, however, the conversion of the proto-oncogene into an

Table 1. Virus-associated oncogenes

Designation	Name of an Associated Virus	Species of Associated Virus
rel	avian reticuloendotheliosis	turkey
src	Rous sarcoma	chicken
myb	avian myeloblastosis	chicken
myc	avian myelocytomatosis	chicken
*erb*A	avian erythroblastosis	chicken
*erb*B	avian erythroblastosis	chicken
fps/fes	Fujinami/feline sarcoma	chicken/cat
yes	Y73 avian sarcoma	chicken
ros	UR2 avian sarcoma	chicken
mos	Moloney murine sarcoma	mouse
H-*ras*	Harvey murine sarcoma	rat
K-*ras*	Kirsten murine sarcoma	rat
abl	Abelson murine leukemia	mouse
fms	McDonough feline sarcoma	cat
sis	Simian sarcoma	woolly monkey
ros	UR2 sarcoma	chicken
fos	FBJ osteosarcoma	mouse
raf	3G11 murine sarcoma	mouse
ski	SKV770 avian transforming	chicken

active oncogene has not depended upon intervention of a viral genome. Instead, a mutational event in a cellular gene has been the cause of the activation.

The two experimental strategies, retrovirus retrieval and gene transfer, have yielded two distinct repertoires of cellular proto-oncogenes. One may ask whether these two groups of genes are mutually exclusive, or whether they have members in common. Recent work has shown that the two groups are in fact overlapping: Genes found by retrovirus retrieval can also be found in the genomes of non-virus-induced tumors by use of gene transfer [9–11]. An example of this is the H-*ras* oncogene, activated from the rat genome by Harvey murine sarcoma virus. The related proto-oncogene in humans has become converted into an active oncogene, detected by gene transfer in the DNA of EJ/T24 bladder carcinoma cell line. Other examples of this overlap are provided in Table 2. These overlaps show that several cellular proto-oncogenes can be activated by two distinct mechanisms: affiliation with a retrovirus (occurring in certain animal species) or somatic mutation (occurring in "spontaneous" human tumors).

The discovery of these oncogenes provokes a number of questions. An initial one concerns the role of the related proto-oncogenes in the normal cellular genome. Why does the genome carry an array of genes that may, if altered, become active agents of pathogenesis? A partial answer to this is already possible.

These proto-oncogenes appears to be highly conserved over long evolutionary time periods. For example, homologues of the above-mentioned H-*ras* and *src* genes can be found in the genomes of organisms as distantly related as *Drosophila* [12]. This means that these proto-oncogenes were already evolved early in metazoan evolution, and have been strongly conserved in the intervening time because they serve vital cellular or organismic functions. Unfortunately, the exact nature of these functions remains quite obscure.

Table 2. Relationships of transfected oncogenes with retrovirus-associated oncogenes

Oncogene	Associated Tumors or Tumor Cell Lines	
H-*ras*	EJ	bladder carcinoma
K-*ras*	LX-1	small cell lung carcinoma
	Calu	lung carcinoma
	A549	lung carcinoma
	A427	lung carcinoma
	SW480	colon carcinoma
	SK-C01	colon carcinoma
	A2233	colon carcinoma
	A1692	bladder carcinoma
	1189	pancreatic carcinoma
	A1604	gall bladder carcinoma

A second and more productive question is also suggested by the discovery of these genes: How do proto-oncogenes become converted into active oncogenes? Here, a number of specific answers are already available.

Activation by Alteration of Encoded Protein

The oncogene of a human bladder carcinoma has been compared with its antecedent proto-oncogene. Molecular clones of the two genes are structurally very similar to one another, yet the clones exhibit dramatic differences in function. The oncogene clone is potently oncogenic while the proto-oncogene clone, under most circumstances, has no effect on cellular phenotype.

Genetic recombination between these two alleles has localized the critical difference between the two genes to a small region of the genes. The subsequently performed sequence analysis revealed that a single point mutation was sufficient to achieve conversion of the proto-oncogene into an oncogene [13–15]. Importantly, this point mutation occurred in a region of the gene encoding the amino acid sequence of the oncogene-associated protein, termed p21. The glycine residue normally present at residue 12 was replaced by a valine. This showed that activation of the gene need not depend upon alteration of regulatory mechanisms controlling levels of gene expression. Instead, simple alteration of the encoded protein was sufficient to activate the gene.

This mode of activation is not limited to non-virus-induced tumors. It may also explain, at least in part, the mechanism by which some retroviruses are able to activate genes that they have acquired from the cellular genome. For example, one can compare the *ras* oncogenes transduced by three different murine sarcoma viruses with their cognate cellular proto-oncogenes. In each case, the amino acid sequence encoded by the oncogene deviates from that of its antecedent at residue 12, indicating a lesion at the same critical site as that involved in activation of the related human bladder carcinoma oncogene. It appears that after acquisition by transducing retrovirus, mutations occur in the acquired cellular genes which impart or augment the oncogenic potential of this gene.

Other examples that follow this paradigm are being discovered. Thus, the oncogene protein encoded by the FBJ murine osteosarcoma virus deviates from its normal antecedent in certain C-terminal amino acid sequences [16]. Alteration of these sequences appears critical to the transforming activity of the viral *fos* oncogene.

Enhancement of Expression by Promoter Replacement

Some proto-oncogenes may assume an oncogenic role when they become expressed at inappropriately high levels. This is most easily achieved experimentally by dissociating the structure-encoding portion of the normal gene from its usual regulator (promoter). In place of the normal promoter, which may allow only low levels of transcription, one may introduce a potent, high-level promoter obtained, for example, from a viral genome. Such an experiment, attempted first with the mouse cellular *mos* proto-oncogene [17], results in creation of an actively oncogenic hybrid gene. Although extended subsequently to a *ras* proto-oncogene [18], it is not obvious that this strategy will suffice to activate all cellular proto-oncogenes.

This experimentally created juxtaposition of viral promoter and cellular gene mimics naturally occurring alliances found in the transducing retroviruses: Without exception, the acquired cellular genes are removed from their normal cellular regulatory elements and placed under viral transcriptional control. Thus, the activation of the *src* gene by RSV may be due in large part to the enormous over-expression ($50\times$) that the gene enjoys once it becomes affiliated with the viral genome [19].

In a previous section we described the activation of the *ras* oncogene by three murine sarcoma viruses, and mentioned the role of altered amino acid sequence in the activation. In addition, each of these virus-associated *ras* oncogenes also became over-expressed after its incorporation into the viral genome, since its transcription became driven by viral transcriptional promoters. Thus, two separate and distinct molecular mechanisms cooperate to activate the acquired cellular genes carried by these murine sarcoma viruses.

Co-transcription of Cellular Genes with Inserted Promoters

We described above two activating mechanisms dependent upon association of viral promoters with cellular protein-encoding sequences—one achieved via experimental fusion of DNA segments, the other via naturally occurring recombinational mechanisms that lead to chimeric viruses. These two mechanisms suggested yet a third mechanism, which derived from another attribute of retroviruses: an ability to integrate viral genomes efficiently into the chromosomes of infected cells.

Retrovirus integration is promiscuous. The viral genome appears able to insert itself into a large, and possibly unlimited, number of sites in the cellular chromosome. Some of these cellular sites may, by coincidence, lie adjacent to proto-oncogenes. Since the integrated viral genome carries potent transcriptional promoters, the integrative event could result in the transcriptional activation of adjacent cellular genes. In this case, transcription may begin from a viral promoter and continue into the adjacent proto-oncogene. Once again, a cellular gene is placed under the control of a viral sequence, but in this case juxtaposition of sequences results from genetic events quite different from those discussed before.

This "promoter insertion" mechanism was described recently in avian leukosis virus-induced lymphomas [20,21], in which the *myc* proto-oncogene is found to be over-transcribed because its transcription is driven by an avian leukosis virus (ALV) promoter. Since this adjacent integration occurs as a random, nontargeted process, it must require a large number of infectious events, as would be expected to occur in a viremic animal. Other proto-oncogene activations mediated by provirus integration have been described [22,23].

Activation via Enhancement

Subtle variants of the promoter/co-transcription mechanism have recently been reported. In these cases, the cellular *myc* proto-oncogene once again becomes associated with a foreign, regulatory sequence, and this appears to lead in turn to activation of the gene. But now the mechanism of activation is more obscure.

In some of the ALV-induced chicken lymphomas, the inserted viral promoter faces away from the adjacent *myc* gene that it activated, i.e., the transcriptional polarity of the promoter prevents direct co-transcription of the linked proto-oncogene [21]. In a series of mouse plasmacytomas and human Burkitt's lymphomas, this *myc* gene appears to become activated following its juxtaposition with another foreign sequence, the immunoglobulin domain. This juxtaposition is achieved by a chromosomal translocation process [24–27]. Here too, the newly associated foreign sequence of immunoglobulin origin faces away from the *myc* gene: Its transcriptional direction is away from the point of fusion with the *myc* gene. In both cases, the *myc* gene cannot depend upon co-transcription with a foreign promoter for its activation.

Instead, one must invoke a novel mechanism, termed enhancement. The enhancer sequences associated with the ALV promoter and perhaps with the immunoglobulin genes are able to activate the transcription of distantly located, linked domains, such as the *myc* genes. While the mechanism is obscure, the result seems very real—deregulation of the *myc* gene with resulting creation of an oncogene.

Role of Viruses in Indirectly Activating Oncogenes

We have described clear and direct mechanisms by which certain retroviruses, notably ALV, are able to activate cellular proto-oncogenes via promoter insertion. Other viruses have etiologic roles in tumors but in these other cases, the connection of the virus with the molecular mechanisms of gene activation is indeed very obscure. Three such cases may be cited.

In the instance of Burkitt's lymphoma, the Epstein–Barr virus has been indicted as an important etiologic agent. Coexisting with the viral genome in tumor cells are characteristic translocations leading to *myc* activation, as described above. It seems unlikely that the virus directly mediates this translocation event. Instead, it would appear that the virus acts as a mitogen, and thereby creates a large, proliferating cohort of target cells. In one of these large number of cells, a rare and improbable genetic alteration may occur spontaneously, leading to oncogene activation.

In the case of the ALV-induced lymphomas of chickens, a second cellular oncogene, distinct from *myc*, also becomes activated. This second oncogene, termed B-*lym*, is unlinked to the *myc* gene and its mechanism of activation appears not to depend on any associated viral sequences [28]. Once again, one may invoke a mechanism by which an initial stimulus, perhaps *myc* activation, leads to a large proliferating clone of cells. In one of these cells, a second, rare event now happens by spontaneous mutation, leading to the activation of the B-*lym* gene.

A third example is to be found in the relationship between the hepatitis B virus (HBV) infection and hepatomas. Although certain hepatomas give evidence of integrated HBV genomes (reviewed in [29]), it is unclear whether these are instrumental in any way in activating cellular proto-oncogenes. One might argue that these integrated genomes are an epiphenomenon, and that the real oncogenic stimulus is quite indirect. For example, it may be that by creating chronic, extensive tissue damage, the virus stimulates carcinogenesis in the same way that other noxious stimuli, such as ethanol, predispose toward cancer. By provoking extensive cellular proliferation, the virus may once again provide conditions that allow a rare somatic mutation to occur, leading in turn to activation of a cellular oncogene. Thus in these various virus-induced tumors, the connection between viral infection and oncogene activation may be a distant one.

The relationship between viruses and cellular oncogenes is thus a complex one. In many animals, retroviruses may activate cellular proto-oncogenes by transduction or adjacent integration. However, in human tumors of suspected or known viral etiology, the role of viruses in creating oncogenes is quite obscure. Because we now have powerful techniques and relevant molecular reagents, these relationships between human viruses and oncogenes should be clarified over the next several years.

References
1. Stehelin D, Varmus HE, Bishop JM, Vogt PK (1976) DNA related to the transforming gene(s) of avian sarcoma viruses is present in normal avian DNA. Nature 260:170–173
2. Coffin JM, Varmus HE, Bishop JM, Essex M, Hardy WD, Martin GS, Rosenberg NE, Scolnick EM, Weinberg RA, Vogt PK (1981) Proposal for naming host cell derived inserts in retrovirus genomes. J Virol 40:953–957
3. Cooper GM (1982) Cellular transforming genes. Science 218:801–806
4. Weinberg RA (1982) Oncogenes of spontaneous and chemically induced tumors. *In* Klein G, Weinhouse S (eds) Advances in Cancer Research, Academic Press, New York, p 149
5. Goldfarb M, Shimizu K, Perucho M, Wigler M (1982) Isolation and preliminary characterization of a human transforming gene from T24 bladder carcinoma cells. Nature 296:404–409

6. Pulciani S, Santos E, Lauver AV, Long LK, Robbins KC, Barbacid M (1982) Oncogenes in human tumor cell lines: Molecular cloning of a transforming gene from human bladder carcinoma cells. Proc Natl Acad Sci USA 79:2845–2849
7. Shih C, Weinberg RA (1982) Isolation of a transforming sequence from a human bladder carcinoma cell line. Cell 29:161–169
8. Goubin G, Goldman DS, Luce J, Neiman PE, Cooper GM (1983) Nature, in press
9. Der CJ, Cooper GM (1983) Altered gene products are associated with activation of cellular ras^k genes in human lung and colon carcinomas. Cell 32:201–208
10. Parada LF, Tabin CJ, Shih C, Weinberg RA (1982) Human EJ bladder carcinoma oncogene is homologue of Harvey sarcoma virus *ras* gene. Nature 297: 474–479
11. Santos E, Tronick S, Aaronson SA, Pulciani S, Barbacid M (1982) T24 human bladder carcinoma oncogene is an activated form of the normal human homologue of BALB and Harvey-MSV transforming genes. Nature 298:343–347
12. Shilo BZ, Weinberg RA (1981) DNA sequences homologous to vertebrate oncogenes are conserved in *Drosophila melanogaster*. Proc Natl Acad Sci USA 78:6789–6792
13. Tabin CJ, Bradley SM, Bargmann CI, Weinberg RA, Papageorge AG, Scolnick EM, Dhar R, Lowy DR, Chang EH (1982) Mechanism of activation of a human oncogene. Nature 300:143–149
14. Reddy EP, Reynolds RK, Santos E, Barbacid M (1982) A point mutation is responsible for the acquisition of transforming properties by the T24 human bladder carcinoma oncogene. Nature 300:149–152
15. Taparowsky E, Suard Y, Fasano O, Shimizu K, Goldfarb M, Wigler M (1982) Activation of the T24 bladder carcinoma transforming gene is linked to a single amino acid change. Nature 300:762–765
16. van Stratten F, Muller R, Curran T, van Beveren C, Verma I (1983) Complete nucleotide sequence of a human c-*onc* gene: Deduced amino acid sequence of the human c-fos protein. Proc Natl Acad Sci USA 80:3183–3187
17. Blair DG, Oskarsson M, Wood TG, McClements WL, Fischinger PJ, Vande Woude GG (1981) Activation of the transforming potential of a normal cell sequence: A molecular model for oncogenesis. Science 212:941–943
18. DeFeo D, Gonda MA, Young HA, Chang EH, Lowy DR, Scolnick EM, Ellis RW (1981) Analysis of two divergent rat genomic clones homologous to the transforming gene of Harvey murine sarcoma virus. Proc Natl Acad Sci USA 78:3328–3332
19. Collett MS, Brugge JS, Erikson RL (1978) Characterization of a normal avian cell protein related to the avian sarcoma virus transforming gene product. Cell 15:1363–1369
20. Hayward WS, Neel BG, Astrin SM (1981) Activation of a cellular *onc* gene by promoter insertion in ALV-induced lymphoid leukosis. Nature 290:475–480
21. Payne GS, Bishop JM, Varmus HE (1982) Multiple arrangements of viral DNA and an activated host oncogene in bursal lymphomas. Nature 295:209–214
22. Fung YKT, Lewis WG, Kung HJ (1983) Cell, in press
23. Nusse R, Varmus JE (1982) Many tumors induced by the mouse mammary tumor virus contain a provirus integrated in the same region of the host genome. Cell 31:99–109
24. Shen-Ong GLC, Keath EJ, Piccoli SP, Cole MD (1982) Novel *myc* oncogene

RNA from abortive immunoglobulin gene recombination in mouse plasmacytomas. Cell 31:443–452
25. Dalla-Favera R, Bregni M, Erikson J, Patterson D, Gallo RC, Croce CM (1982b) Human c-*myc* onc gene is located on the region of chromosome 8 that is translated in Burkitt lymphoma cells. Proc Natl Acad Sci USA 79:7824–7827
26. Taub R, Kirsch I, Morton C, Lenoir G, Swan D, Tronick S, Aaronson S, Leder P (1982) Translocation of the c-*myc* gene into the immunoglobulin heavy chain locus in human Burkitt lymphoma and murine plasmacytoma cells. Proc Natl Acad Sci USA 79:7837–7841
27. Marcu RB, Harris LJ, Stanton LW, Erikson J, Watt R, Croce CM (1983) Transcriptionally active c-*myc* oncogene is contained within NIARD, a DNA sequence associated with chromosome translocations in B-cell neoplasia. Proc Natl Acad Sci USA 80:519–523
28. Cooper GM, Neiman PE (1981) Two distinct candidate transforming genes of lymphoid leukosis virus-induced neoplasms. Nature 292:857–858
29. Tiollais P, Charnay P, Vyas GM (1981) Biology of hepatitis B virus. Science 213:406–410

CHAPTER 27
Antibody Initiates Virus Persistence: Immune Modulation and Measles Virus Infection

ROBERT S. FUJINAMI AND MICHAEL B. A. OLDSTONE*

By antibody-induced antigenic modulation, we mean the removal of antigens from the surface of cells by specific antibody. The result is to render such cells resistant to killing by immune reagents, i.e., antibody and complement or cytotoxic lymphocytes. Stripping of surface antigens to allow cells to withstand immunologic attack was first described in the context of the thymus leukemia (TL) differentiation antigen system. TL antigen could be stripped or modulated *in vivo* and *in vitro* by anti-TL antibody (reviewed in [1]). We have expanded the concept of antibody-induced antigenic modulation as an aid to understanding several persistent infections, with our primary interest in persistent measles virus infection of humans.

Several hypotheses have been proposed to account for viral persistence. One involves the concept of immunologic tolerance; that is, the inability of the immune system to respond to a virus or its antigens. Immune tolerance to a virus was believed to develop when infection took place at birth or *in utero* and was the initial explanation for the basis of persistence with lymphocytic choriomeningitis virus (LCMV) and murine retroviruses (MuLV) (reviewed in [2]). However, when techniques to detect antibodies bound to antigens (viruses) became available and were used, it became clear that such persistently infected animals mounted virus-specific antibody responses. Antibody usually was not free but rather bound to antigen in the state of an immune complex. The formation of virus-induced immune complexes and resultant trapping in tissues and disease is discussed in Chapter 24. Virus-specific cytotoxic T lymphocytes have been reported in the MuLV models but are absent or abnormally low in mice persistently infected with LCMV.

---

* Scripps Clinic and Research Foundation, La Jolla, California.

Further complication arises in the LCMV model in that active suppression of cytotoxic T lymphocytes occurs. In contrast to these conventional viruses, a virus-specific immune response has not yet been detected against presumed viral agents that cause scrapie, kuru, and Creutzfeldt-Jakob disease. Whether these agents are nonimmunogenic or the host is unresponsive is not clear [3].

Viral persistence may be associated with the generation of virus variants with or without immune selection. An example of immune-response selection of variants allowing persistence is discussed in Chapter 22. In this infection, visna virus isolated from sheep may not be neutralized by antiviral antibody from that animal even though the original visna virus used to infect the sheep is neutralized. As the infection progresses, the immune system continuously selects for new virus variants [4]. Thus the virus persists since it continually changes and is never fully cleared from the host.

None of these mechanisms adequately explains persistent measles virus infection. However, the initiation of measles virus persistence by antimeasles virus antibody through antibody-induced antigenic modulation readily fits several patterns characterizing this infection and is the subject of this article. Once virus persistence is established, it may be also regulated, in part, by dysfunction in the production of the matrix protein as suggested by Hall and Choppin [5] and/or regulation of other measles virus proteins as discussed below.

Measles virus infection of humans may result in three different sequelae. The first and usual situation is an acute self-limiting infection in which, as an antiviral immune response is mounted, the virus is cleared from the body. Immunity to repeated infection is life long. The second sequela is a postinfectious encephalomyelitis, generally occurring 5 to 10 days after the acute viral infection. It is unclear whether this expression of measles virus infection results from an unusual measles virus variant, an antiviral immune response against the virus, or an autoimmune response against self antigens of the central nervous system. In any case, virus is hardly ever recovered, thus suggesting that the immune response is most likely causing the disease. The third manifestation is a persistent infection known as subacute sclerosing panencephalitis (SSPE). In this instance, measles virus may persist in the host for years but can be isolated from the central nervous system and lymphoid organs (reviewed in [6]). Measles virus persists even though the infected individual mounts high titers of antimeasles virus antibody and cytotoxic lymphocytes. The question addressed here is: Why isn't the virus eliminated and what role does the immune response play in the process of virus persistence?

We have proposed the following scenario to explain measles virus persistence in the face of a vigorous antiviral immune response (see Fig. 1). Antibody mounted against measles virus binds to viral proteins expressed on the surface of infected cells. This results in the stripping or loss of viral glycoproteins from the cell surface, which can occur in areas of the body where

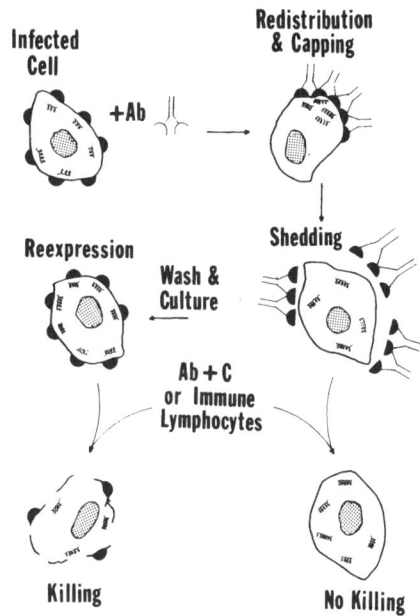

Fig. 1. Addition of virus-specific antibody to infected cells leads to a rearrangement and capping of viral antigens toward one pole of the cell. These antigens are shed from the cell surface as antibody–antigen complexes. At this point, the denuded cells are resistant to lysis by antibody and complement or cytotoxic lymphocytes. When these denuded cells are cultured in the absence of antibody, viral antigens can be reexpressed. At this time, cells reexpress viral antigens on their surfaces and are again susceptible to killing by immunologic reagents.

the complement system is not present, i.e., the central nervous system and thoracic duct fluids. By this means, virus is able to persist and win the race against immune elimination as antibody- and complement-mediated lysis will not occur. Also, viral antigens are presumably removed from the cell surface before lymphoid cells targeted to eliminate the infected cells reach their mark. Further, outside the central nervous system, antibody-induced antigenic modulation may take place in the presence of complement. At critical concentrations of antiviral antibody and only slightly depressed levels of complement, modulation of viral antigens occurs rather than lysis of cells [7]. Therefore, even in fluids containing significant levels of complement, infected cells may resist lysis by immune mechanisms. Fewer molecules of viral antigen are needed to react with antibody for modulation to work than the number of molecules of viral glycoprotein that are needed to bind cytotoxic lymphocytes or antibody and complement for lysis. Thus, quantitatively, the system favors modulation and persistence over immune-mediated lysis.

Several years ago, it was demonstrated that measles virus antigens could be redistributed on the surfaces of infected cells by specific antibody [8]. Further, incubating measles virus-infected cells with specific antibody prevented antibody and complement [9] or immune lymphocytes [10] from killing infected cells. These studies clearly demonstrated that the removal or stripping of the measles virus antigens expressed on the surface of infected cells enabled cells to avoid immune lysis and allowed virus to persist. Since measles virus does not alter cell protein synthesis significantly, infected cells could be maintained for prolonged periods when cultured with antiviral antibodies. Within the cytoplasms of the modulated cells are abnormally positioned nucleocapsids whose picture mirrored those in cells of the nervous system from SSPE patients [11,12]. Hence, antibody initiates measles virus persistence and the cell's phenotype is equivalent to that observed in infected cells from patients with SSPE.

Modulation of the measles virus surface glycoproteins can be reversed. Infected cells cultured in specific antibody for 12 hours lose their surface viral antigens. However, when antibody is removed, measles virus antigens are reexpressed on the cell surface within 24 hours. The reexpression is time-dependent; the longer infected cells are incubated with antimeasles virus antibody, the longer if takes for virus antigens to reappear after the removal of antibody [9]. Measles virus-infected cells cultured with antibody for 6 weeks do not reexpress viral antigens over an additional 6-week observation period [9]. In addition to time-dependency for the expression of viral antigens, the extent of lysis or killing by antiviral antibody and complement or by cytotoxic lymphocytes parallels the amount of antigen on the surface of each cell as well as the number of infected cells expressing antigen. Under modulating conditions, as the number of measles virus-positive cells decreases, the ability of antibody and complement or cytotoxic lymphocytes to kill these infected cells also decreases [9,10].

Measles virus contains six structural polypeptides that are found in purified virions or infected cells. These components are: polymerase, hemagglutinin, phosphoprotein (P protein), nucleocapsid protein, fusion protein, and membrane or matrix protein. Surface labeling infected cells with ^{125}I allows one to follow the turnover of viral hemagglutinin and fusion protein from the cells' surface [13]. When infected surface-labeled cells are cultured in antimeasles virus antibody, initially a rapid loss of the viral glycoprotein occurs, followed by a slower decline. This biphasic loss of the hemagglutinin and fusion proteins seen in the presence of antibody is in contrast to a steady loss of the glycoproteins, observed in the absence of antibody [13].

Besides removing the two viral glycoproteins from the cell's surface, antibody also changes the expression of some intracellular viral polypeptides. When infected cells are incubated with antimeasles virus antibody for 6 hours or more, the amounts of internal viral polypeptides synthesized decrease, as judged by ^{35}S-methionine or ^{32}P-orthophosphate labeling studies. However, there is a preferential alteration in individual viral polypeptides. In addition to the expected decrease in viral glycoproteins, there was marked perturbation of two viral polypeptides found inside the infected cell. These were the P protein and the matrix or membrane (M) protein. The P protein is associated with the nucleocapsid and is probably involved in control of transcription, while the M protein is found under areas of virus budding beneath the viral envelope and is likely important in positioning of the nucleocapsid to virus envelope proteins. Amounts of these two virus proteins were reduced in repeated observation by at least 80% for the P protein and 40% for the M protein. The changes occurred with polyclonal antibody against measles virus or with monoclonal antibodies against measles virus hemagglutinin. Neither polyclonal antibody against the uninfected cell surface determinants nor monoclonal antibody against nucleocapsid or fusion protein induced these changes [14,15].

Antimeasles virus antibody-induced modulation also caused changes in the phosphorylation of proteins. Of the six structural polypeptides, three are phosphorylated: the P protein, nucleocapsid protein, and matrix protein. During modulation, the amount of ^{32}P label found associated with P protein was markedly reduced while, in contrast, the amount of ^{32}P bound to the matrix protein increased even though the total amount of M protein diminished [14,15]. Phosphorylated M protein is primarily found inside the infected cell, whereas the dephosphorylated form is preferentially incorporated into virions. Thus, a shift toward accumulation of the intracellular form of the M protein during modulation is a reflection that less infectious virus is produced. Perhaps the decrease in M protein reported by several workers as occurring in SSPE is a later event of these early effects of antibody-induced modulation.

Critical to accepting the antimeasles virus-antibody modulation hypothesis as inducing virus persistence is the collection of evidence that antibody causes persistence *in vivo*. Such evidence has now come forth not only in animal models but in humans as well. Indeed, prior to the *in vitro* studies, Wear and Rapp demonstrate that persistent measles virus infection developed in weaning hamsters infected with virus at birth and nursed by mothers carrying antibodies to measles virus [16]. Later, Albrecht *et al.* [17] showed that a prolonged measles virus infection occurred in monkeys only when antibody to measles virus was passively transferred. Recently, Rammohan *et al.* [18] produced a chronic and persistent measles virus infection in mice that were passively transferred with monoclonal antibody to measles virus hemagglutinin. Lastly, an unfortunate reconstruction experiment was performed in humans when persistent infection of measles virus developed after a child received passive transfer of measles virus antibody at the time of acute infection [19]. Taken *in toto*, these *in vivo* and complementary *in vitro* observations offer a possible explanation, in part, for the development of SSPE during early childhood years when exposure to measles virus may be coupled with passive transfer of maternal antimeasles virus antibody.

Antibody-induced antigenic modulation has implications for understanding virus infections that occur in antibody excess (as most persistent virus infections in humans do) and for understanding several general biological phenomena and non-viral-related disorders. Concerning measles virus infection, when antibody reduces the amount of hemagglutinin and fusion proteins on the surfaces of infected cells, there is a parallel likelihood that these cells will no longer be recognized and hence, be cleared by the antiviral immune response. Hence, we have a mechanism of viral persistence and escape from immune surveillance. In addition, reduction of fusion protein on the cell surface minimizes or prevents cell–cell fusion, giant cell formation, and hence, death of the cell as a result of measles virus infection. Alterations in P protein probably affect viral RNA synthesis, because this viral polypeptide together with the polymerase and nucleocapsid protein make up the replicative complex of the virus. Decrease in the M protein and/or increase in the phosphorylation of M likely affects viral maturation by altering nucleocapsid recognition and alignment under the plasma membrane. Hence, all these changes could account for the observations seen in SSPE and the SSPE cell in that fewer viral glycoproteins are found on the cell surface, decreased amounts of infectious virus are produced, nucleocapsids are improperly aligned, alterations in the migration and expression of M protein occur and the virus-infected cell persists despite an abundant and determined antiviral antibody and cytotoxic lymphocyte response.

Antibody-induced antigenic modulation is also a viable mechanism by which other viruses can persist and escape immunologic surveillance. Experimental evidence has suggested such a mechanism for several viruses including herpes simplex virus [20] and retrovirus persistent infections including those induced by Gross, Friend, and mammary tumor viruses (reviewed in [21]). Further, several autoimmune illnesses such as myasthenia gravis, some forms of diabetes and thyroiditis may involve antibody modulating specific membrane receptors. Consequently, the search is on to understand the molecular mechanism whereby antibody to a plasma membrane surface determinant alters molecules expressed on the cell surface and generates a transmembranal signal that influences genes or gene expression inside the cell.

Acknowlegments. This is publication no. 3011-IMM from the Department of Immunology, Scripps Clinic and Research Foundation. This research was supported by Grant JF-2009 from the National Multiple Sclerosis Society and US PHS Grants NS-17214, NS-12428, AI-09484, and AI-07007 from the National Institutes of Health. Dr. Fujinami is a Harry Weaver Scholar of the National Multiple Sclerosis Society.

References
1. Stackpole C, Jacobson J (1978) Antigenic modulation. *In* Waters H (ed) The Handbook of Cancer Immunology. Barland STPM Press, New York, p 55
2. Oldstone, MBA (1979) Immune responses, immune tolerance and viruses. *In* Fraenkel-Conrat H, Wagner RR (eds) Comprehensive Virology, Vol 15. Plenum Press, New York, p 1
3. Gajdusek DC (1977) Unconventional viruses and the origin and disappearance of kuru. Science 197:943–960
4. Narayan O, Griffin DE, Chase J (1977) Antigenic shift of visna virus in persistently infected sheep. Science 197:376–378
5. Hall WW, Choppin PW (1979) Evidence for lack of synthesis of the M polypeptide of measles virus in brain cells in subacute sclerosing panencephalitis. Virology 99:443–447
6. ter Meulen V, Katz M, Muller D (1972) Subacute sclerosing panencephalitis: A review. Curr Top Microbiol Immunol 57:1–38

7. Gorman NT, Lachmann PJ (1982) *In vitro* modulation of viral cell surface glycoproteins by antiviral antibody in the presence of complement. Clin Exp Immunol 50:507–514
8. Joseph BS, Oldstone MBA (1974) Antibody-induced redistribution of measles virus antigens on the cell surface. J Immunol 113:1205–1209
9. Joseph BS, Oldstone MBA (1975) Immunologic injury in measles virus infection. II. Suppression of immune injury through antigenic modulation. J Exp Med 142:864–876
10. Oldstone MBA, Tishon A (1978) Immunologic injury in measles virus infection. IV. Antigenic modulation and abrogation of lymphocyte lysis of virus-infected cells. Clin Immunol Immunopathol 9:55–62
11. Iwasaki Y, Koprowski H (1974) Cell to cell transmission of virus in the central nervous system. I. Subacute sclerosing panencephalitis. Lab Invest 31:187–196
12. Lampert PW, Joseph BS, Oldstone MBA (1976) Morphological changes of cells infected with measles or related viruses. *In* Zimmerman HM (ed) Progress in Neuropathology, Vol 3. Grune and Stratton, New York, p 51
13. Fujinami RS, Sissons JGP, Oldstone MBA (1981) Immune reactive measles virus polypeptides on the cell's surface: Turnover and relationship of the glycoproteins to each other and to HLA determinants. J Immunol 127:935–940
14. Fujinami RS, Oldstone MBA (1979) Antiviral antibody reacting on the plasma membrane alters measles virus expression inside the cell. Nature 279:529–530
15. Fujinami RS, Oldstone MBA (1980) Alterations in expression of measles virus polypeptides by antibody: Molecular events in antibody-induced antigenic modulation. J Immunol 125:78–85
16. Wear DJ, Rapp F (1971) Latent measles virus infection of the hamster central nervous system. J Immunol 107:1593–1598
17. Albrecht P, Burnstein T, Klutch MJ, Hicks JT, Ennis FA (1977) Subacute sclerosing panencephalitis: Experimental infection in primates. Science 195:64–66
18. Rammohan KW, McFarland HF, McFarlin DE (1981) Induction of subacute murine measles encephalitis by monoclonal antibody to virus haemagglutinin. Nature 290:588–589
19. Rammohan KW, McFarland HF, McFarlin DE (1982) Subacute sclerosing panencephalitis after passive immunization and natural measles infection: Role of antibody in persistence of measles virus. Neurology 32:390–394
20. Stevens JG, Cook ML (1974) Maintenance of latent herpetic infection: An apparent role for anti-viral IgG. J Immunol 113:1685–1693
21. Oldstone MBA, Fujinami RS, Lampert PW (1980) Membrane and cytoplasmic changes in virus infected cells induced by interactions of antiviral antibody with surface viral antigen. Prog Med Virol 26:45–93

CHAPTER 28
Ovarian Infection and Transovarial Transmission of Viruses in Insects

LEON ROSEN*

In order to survive, viruses must pass from one host to another. Such passage is considered "vertical" when transmission is from a parent organism to its progeny and "horizontal" when transmission occurs under any other circumstances. "Vertical" transmission can also be defined more restrictively as that which takes place via the germ cell line in multicellular organisms as contrasted to that which takes place by other means (e.g., transplacental, peri-, and postnatal). Further, in germ cell transmission, virus can either be integrated into the host genome or be present elsewhere in the cell.

Transmission of viruses via the germ cell line of insects has attracted increasing attention in recent years because of new information about the epidemiology of viruses transmitted to humans and to economically important lower animals and plants by insects. Until recently, the mechanism by which most of these viruses survived periods during which their adult insect hosts were absent or inactive, such as winters or dry seasons, was unknown.

In the case of the large and important group of mosquito-borne viruses of vertebrates, a variety of possible mechanisms had been proposed [1]. These included chronic or recurrent viremias in either homeothermic or poikilothermic vertebrates; reintroduction of virus from other areas by migrating vertebrates (birds or bats); spread by seasonally advancing mosquito–vertebrate–mosquito cycles; survival in hibernating or estivating adult mosquitoes infected by feeding on viremic vertebrates prior to their inactive period; infection of alternative long-lived arthropod hosts (such as bedbugs, mites, or ticks); and transovarial transmission of virus by infected mosquitoes with survival in F1 eggs, larvae, or imagoes during the inactive period.

* University of Hawaii at Manoa, Honolulu, Hawaii.

The concept of transovarial transmission of an infectious agent is not new and, in fact, was demonstrated in the case of the first disease of animals shown to be caused by a microorganism. Pasteur saved the silk industry of France in the mid-19th century by demonstrating that the protozoan etiologic agent of pébrine, a disease of the larva of the silkworm moth, was transmitted from parent to offspring via the egg. The disease was controlled by employing only eggs from female adult moths found free of the protozoan by microscopic examination [2]. Also, the first infectious agent shown to be transmitted to vertebrates by arthropods (the protozoan of Texas cattle fever) was found to be transmitted transovarially in its tick host [3]. In light of this observation, it was postulated that the etiologic agent of yellow fever was transmitted transovarially by mosquitoes before it had even been demonstrated that yellow fever was in fact mosquito borne [4]. In his studies proving mosquito transmission of the virus, Walter Reed failed to find evidence of transovarial transmission. Although positive results were reported by French workers several years later [5], a number of subsequent workers were unable to confirm the work and, until very recently, it was believed to be in error [6]. Similarly, either natural or experimental transovarial transmission of Japanese encephalitis virus by mosquitoes was reported by early Japanese, Russian, and Chinese investigators. Later publications, either by the original authors or others, concluded that transovarial transmission of Japanese encephalitis virus did not occur and that the earlier positive findings could be explained by experimental error [7].

Interest in transovarial transmission of viruses by insects, and especially by mosquitoes, was rekindled in recent years by epidemiologic observations that seemed to rule out other possible means of viral survival. For example, the search for, and discovery of, transovarial transmission by mosquitoes of La Crosse virus (a bunyavirus of the California group) was motivated in part by the fact that the small mammalian hosts of the virus were nonmigratory [8]. Similarly, the rediscovery of transovarial transmission of mosquito-borne flaviviruses was prompted by the apparent absence of other means of survival of yellow fever in gallery forests of West Africa during prolonged dry seasons [9].

It seems clear at this point that transovarial transmission in mosquitoes plays an important role in the survival of viruses of the California group during winters. Very recently, transovarial transmission was demonstrated in nature for another bunyavirus of great economic importance, Rift Valley Fever virus (CL Bailey, 1983, US Army Medical Research Institute of Infectious Diseases, personal communication).

The case for the role of transovarial transmission in the maintenance of flaviviruses in nature is not yet proven. Such transmission has been demonstrated experimentally for such important flaviviruses as dengue [10], yellow fever [6], Japanese encephalitis [7], and St. Louis encephalitis [11] as well as others of lesser importance. Evidence available thus far from nature consists of the recovery of yellow fever and dengue viruses from pools of male mosquitoes captured in forests in West Africa [12] (R Cordellier, 1983, Institute Pasteur de Cote d'Ivoire, personal communication). The only known way in which such male mosquitoes could have been infected is by the transovarial route, but further data are required before it can be concluded that vertical transmission plays an essential role in the survival of the viruses.

Transovarial transmission of viruses by mosquitoes may be important in the maintenance of the viruses in either, or both, of two ways. First, it may provide a mechanism to allow the virus to bridge a dry or cold season or the temporary absence of nonimmune vertebrate hosts. Or, if a stable association occurs with the oogonial cells of the mosquito, the virus may be able to persist from one mosquito generation to another indefinitely in the absence of vertebrate hosts. Such a virus–oogonial cell association is well documented in the case of sigma virus (a rhabdovirus) and the fruitfly, *Drosophila melanogaster* [13] and may occur with certain bunyaviruses and mosquitoes [14].

The relationship of sigma virus to *D. melanogaster* has been the object of extensive laboratory investigation and may prove to be an appropriate model for the interrelationships between vertebrate-pathogenic viruses and their insect hosts. Sigma probably has been the most thoroughly studied virus of a multicellular animal from the point of view of the genetics of its interaction with an intact host. A great deal of information is available on both hereditary virus transmission patterns and interactions between virus and host genomes. More recently, attention has been directed toward the occurrence and behavior of sigma virus in natural and experimental populations of *D. melanogaster*. These studies are motivated in part by the belief that intracellular, extrachromosomal, self-replicating units transmitted vertically function as additional genes and play an important role in the evolution of higher organisms [15,16]. For example, it has been demonstrated that maternally transmitted wall-free prokaryotes, called spiroplasmas, are responsible for a sex-ratio trait (death of male embryos) in *Drosophila willistoni* and related species [17]. (Interestingly, these spiroplasmas are in turn infected by phage-like viruses that may modulate the effects of the former [18].) Other examples include maternally transmitted rickettsia-like *Wolbachia* responsible for the "cytoplasmic incompatibility" sometimes observed when mosquitoes of the *Culex pipiens* complex from one geographic locality are crossed with those from another [19], and vertically transmitted mycoplasma-like endosymbionts of *Drosophila paulistorum* and related species responsible for hybrid-male sterility [20]. It is also becoming apparent that vertically transmitted viruses can increase the mutation rate in insect hosts [21].

Another reason for interest in sigma virus, and in transovarial transmission of viruses in insects in general, is the insight that it may provide into the origin of certain vertebrate-pathogenic viruses transmitted by insects. For example, one striking feature of the viruses of the rhabdovirus family is their direct or indirect link to arthropods—despite the wide range of vertebrates and higher plant hosts in which they cause disease. This is true even for rabies virus, which is serologically related to rhabdoviruses that have been isolated from, and replicate in, insects [22], and for rhabdoviruses causing diseases in fish, which have also been shown to replicate in insects [23]. One rhabdovirus of fish has even been shown to be transmitted by a bloodsucking aquatic arthropod (*Argulus foliaceus*) [24]. All rhabdoviruses pathogenic for higher plants replicate in insects in so far as they have been studied. These data support the view that rhabdoviruses originally were viruses of arthropods and were transmitted to vertebrate and plant hosts as a result of the piercing–sucking feeding habits of some of their arthropod hosts. It is even possible that rabies virus evolved in this manner, since rhabdoviruses are also known to be neurotropic in their insect hosts.

In a somewhat different vein, it has been found that resistance of certain insects to acquired infection with microorganisms pathogenic for vertebrates is conveyed from one generation to another exclusively via maternal inheritance [25,26]. While the mechanism of this resistance is not known, it may be a reflection of infection with indigenous microorganisms transmitted in a similar manner.

In so far as they have been investigated, none of the transovarially transmitted viruses in insects have been found to be associated with the insect genome. In view of this, it is surprising that at least two such viruses can be transmitted by male gametes [13,27]. Since at least one of the viruses, sigma, is known to replicate in the cytoplasm, the virus presumably is contained in the very small amount of cytoplasm associated with the dipteran spermatozoon.

References
1. Reeves WC (1974) Overwintering of arboviruses. Prog Med Virol 17:193–220
2. Pasteur L (1870) Etudes sur la maladie des vers a soie (2 Vols). Gauthier-Villars, Paris.
3. Smith T, Kilbourne FL (1893) Investigation into the nature, causation and prevention of Texas or southern cattle fever. US Dept Agric Bur Anim Ind Bull 1:117–301
4. Finlay CJ (1899) Mosquitoes considered as transmitters of yellow fever and malaria. Med Rec 55:737–739
5. Marchoux E, Simond PL (1906) Etudes sur la fievre jaune. Ann Inst Pasteur 20:16–40
6. Aitken THG, Tesh RB, Beaty BJ, Rosen L (1979) Transovarial transmission of yellow fever virus by mosquitoes (*Aedes aegypti*). Am J Trop Med Hyg 28:119–121
7. Rosen L, Tesh RB, Lien JC, Cross JH (1978) Transovarial transmission of Japanese encephalitis virus by mosquitoes. Science 199:909–911
8. Watts DM, Pantuwatana S, DeFoliart GR, Yuill TM, Thompson WH (1973) Transovarial transmission of La Crosse virus (California encephalitis group) in the mosquito, *Aedes triseriatus*. Science 182:1140–1141
9. Cox J, Valade M, Cornet M, Robin Y (1976) Transmission transovarienne d'un *Flavivirus*, le virus Koutango chez *Aedes aegypti* L.C.R. Acad Sci (D) Paris 283:109–110

10. Rosen L, Shroyer DA, Tesh RB, Freier JE, Lien JC (1983) Transovarial transmission of dengue viruses by mosquitoes: *Aedes albopictus* and *Aedes aegypti*. Am J Trop Med Hyg 32:1108–1119
11. Hardy JL, Rosen L, Kramer LD, Presser SB, Shroyer DA, Turell MJ (1980) Effect of rearing temperature on transovarial transmission of St. Louis encephalitis virus in mosquitoes. Am J Trop Med Hyg 29:963–968
12. Cornet M, Robin Y, Heme G, Adam C, Renaudet J, Valade M, Eyraud M (1979) Une poussée épizootique de fièvre jaune selvatique au Sénégal oriental. Isolement du virus de moustiques adultes mâles et femelles. Med Mal Infect 9:63–66
13. Brun G, Plus N (1980) The viruses of *Drosophila*. In Ashburner A, Novitiski E (eds) Genetics and Biology of *Drosophila*, Vol 2D. Academic Press, New York, p 625
14. Tesh RB, Shroyer DA (1980) The mechanism of arbovirus transovarial transmission in mosquitoes: San Angelo virus in *Aedes albopictus*. Am J Trop Med Hyg 29:1394–1404
15. L'Heritier P (1970) *Drosophila* viruses and their role as evolutionary factors. In Dobzhansky T, Hecht MK, Steere WC (eds) Evolutionary Biology, Vol 4. Appleton-Century-Crofts, New York, p 185
16. Cosmides LM, Tooby J (1981) Cytoplasmic inheritance and intragenomic conflict. J Theor Biol 89:83–129
17. Williamson DL, Poulson DF (1979) Sex ratio organisms (Spiroplasmas) of *Drosophila*. In Whitcomb RF, Tully JG (eds) The Mycoplasmas, Vol 3. Academic Press, New York, p 175
18. Williamson DL, Oishi K, Poulson DF (1977) Viruses of *Drosophila* sex ratio spiroplasma. In Maramorosch K (ed) The Atlas of Insect and Plant Viruses. Academic Press, New York, p 465
19. Yen JH (1975) Transovarial transmission of *Rickettsia*-like microorganisms in mosquitoes. Ann NY Acad Sci 266:152–161
20. Ehrman L, Kernaghan RP (1972) Infectious heredity in *Drosophila paulistorum*. In Pathogenic Mycoplasmas. Elsevier, Amsterdam, p 227
21. Golubovsky MD, Plus N (1982) Mutability studies in two *Drosophila melanogaster* isogenic stocks, endemic for C picronavirus and virus-free. Mutation Res 103:29–32
22. Bauer SP, Murphy FA (1975) Relationship of two arthropod-borne rhabdoviruses (Kotonkan and Obodhiang) to the rabies serogroup. Infect Immun 12:1157–1172
23. Bussereau F, de Kinkelin P, Le Berre M (1975) Infectivity of fish rhabdoviruses for *Drosophila melanogaster*. Ann Microbiol 126A:389–395
24. Pfeil-Putzien C (1978) Experimentelle Ubertragung der Fruhjahrsviramie (spring viraemia) der Karpfen durch Karpfenlause (*Argulus foliaceus*). Abl Vet Med B 25:319–323
25. Bras-Herreng F (1981) Etude d'une souche de *Drosophila melanogaster* non permissive au togavirus Sindbis: Caracteristiques de la souche et determinisme genetique du phenomene. Ann Virol 132E:73–89
26. Trpis M, Duhrkopf RE, Parker KL (1981) Non-Mendelian inheritance of mosquito susceptibility to infection with *Brugia malayi* and *Brugia pahangi*. Science 211:1435–1437
27. Rosen L, Shroyer DA (1981) Natural carbon dioxide sensitivity in mosquitoes caused by a hereditary virus. Ann Virol 132E:543–548

Virus-Interaction with the Immune System

29. Virus-Induced Immune Complex Formation and Disease: Definition, Regulation, Importance
    MICHAEL B. A. OLDSTONE 201

30. Virus-Induced Autoimmunity
    ABNER LOUIS NOTKINS, TAKASHI ONODERA, AND
    BELLUR PRABHAKAR 210

31. The Role of Immunospecific Receptors in Retrovirus-Induced Lymphomagenesis
    M. S. MCGRATH AND I. L. WEISSMAN 216

32. Viruses as Regulators of Delayed Hypersensitivity T-Cell and Suppressor T-Cell Function
    A. A. NASH 225

33. Mechanisms and Biological Implications of Virus-Induced Polyclonal B-Cell Activation
    RAFI AHMED AND MICHAEL B. A. OLDSTONE 231

CHAPTER 29

Virus-Induced Immune Complex Formation and Disease: Definition, Regulation, Importance

MICHAEL B. A. OLDSTONE*

Introduction

The unique molecules of a virus, upon interaction with the host's immune system, can elicit a specific antibody response. These antibodies then bind to the inciting virus and/or their antigens in solution or on the plasma membranes of infected cells to form virus–antibody (V–Ab) complexes [1–3]. Once formed, these complexes can become trapped in certain tissues. Tissues notable for entrapping circulating V–Ab complexes are those (1) having an extensive flow of blood, or (2) containing vessels with fenestrated endothelial cell lining. Cells with this configuration make up parts of the kidney (glomeruli), brain (choroid plexus), spleen and lymph nodes, in addition to the circulatory system. In individuals infected by any of a wide variety of DNA or RNA viruses [1–3], the formation of V–Ab complexes probably represents a predominant host mechanism for clearing infection. However, when a virus infection does not clear but remains in force, the continual stimulation by viral antigens results in a growing load of V–Ab complexes. When production exceeds removal, complexes may accumulate at sites where they can then cause lesions and disease.

Viruses can also induce immune complexes by two other distinct mechanisms. In the first instance, the virus can either act as a mitogen to stimulate polyclonal B lymphocyte activation or can provide and/or enhance nonspecific T lymphocyte helper activity, which may activate B lymphocytes previously primed to a wide variety or nonviral or unrelated virus antigens. Antibodies so made may then bind the corresponding antigens to yield immune complexes. The second mechanism is molecular mimicry. Viral antigens

---

* Scripps Clinic and Research Foundation, La Jolla, California.

may share amino acid sequences in common with host (self) antigens so that antiviral antibodies elicited by the viral antigens may bind to self antigens as well as to those of the virus [4,5]. These two mechanisms may account for the abundance of autoantibodies (i.e., antibodies to actin, myosin, intermediate filaments, nuclei, etc.) that accompany most virus infections, as well as the antibodies to unrelated viruses (i.e., antibodies to measles virus appearing in patients with chronic hepatitis B virus infection at levels as high as those in chronic measles virus infection [subacute sclerosing panencephalitis]). Both virus-induced polyclonal B cell activation and virus-induced molecular mimicry are reviewed in Chapter 33 and Chapter 30, respectively. Regardless of the eliciting mechanism, the outcome is formation of immune complexes. The conclusion that multiple etiologic agents, including RNA and DNA viruses and B and T cell mitogens, may produce autoimmune disease and immune complexes, is supported by experiments with New Zealand and related mice, the traditional animal model of autoimmune and immune complex disorders and by observations from patients with autoimmune disorders and/or chronic virus infections [6-9].

Clemens von Pirquet's astute observations in the early 1900s, in which patients receiving heterologous serum as therapy developed acute serum sickness, defined many aspects of antigen–antibody complexes in disease [10]. Dixon and his associates [11] later substantiated von Pirquet's clinical impressions in a series of elegant and quantitative experiments with heterologous protein-induced acute and chronic serum sickness in rabbits. These investigators found soluble antigen–antibody complexes in the circulation and localized antigen and antibody at sites of tissue injury. Extending their observations from the acute lesion to chronic serum sickness, Dixon and colleagues found that daily injections of small amounts of serum protein produced chronic disease in those animals making amounts of antibody insufficient to clear the antigen. Thus, in animals given repeated doses of bovine serum albumin, only approximately one-third developed chronic immune complex disease. These animals made low antibody responses and formed soluble immune complexes in antigen excess. The remaining two-thirds of the animals remained free from chronic immune complex disease despite receiving identical antigenic challenges. The cause was either lack of an antibody response or a heightened immune response that rapidly cleared the antigen. These data suggested that genetic control over the amount and kind of antibody produced might be an important ingredient in immune complex disease. Other components influencing the course of immune complex formation and disease were established as the reticular endothelial system and mesangial cells as vehicles for clearing complexes [12,13] and the complement system [14,15]. The fact that patients with complement deficiencies are prone to accelerated immune complex deposition and disease indicates complement's participation. It is important to recall in this context that viruses, *per se,* are known to alter complement levels and complement components during infection.

Next came the realization that throughout the animal kingdom were many examples of persistent virus infections associated with continuous host immune responses and resultant immune complex formation. Despite an ongoing antiviral immune response by the host, the interaction of virus with antibody and complement in the circulation may result in the formation of immune complexes while failing to neutralize virus. Interestingly, the infectious virus may circulate bound to host IgG and complement as demonstrated by experiments in which immunochemical precipitation of IgG and C3 (the third component of complement) from the sera of mice persistently infected with lymphocytic choriomeningitis virus (LCMV) essentially removed the infectivity from their sera, whereas precipitation of IgM, IgA, or serum albumin did not [9]. Infectious V–Ab circulating complexes were first noted by Notkins during persistent lactic dehydrogenase virus infection of mice [16]. Complexes of this type are now documented widely in other persistent virus infections, i.e., retroviruses, Aleutian disease of mink, etc. [17,18]. Immune complexes containing infectious virus while in the circulation can bind preferentially to cells with receptors for either Fc or C3. One result is continuation of the infectious process and another is the entrance of viruses into macrophages, B lymphocytes, polymorphonuclear leukocytes, etc., because these cells bear Fc and/or C3 receptors.

Regardless of the virus or nonviral protein-inducing immune complex formation, the resultant morphologic picture is characteristic in all cases of immune complex deposition (Fig. 1). Immunofluorescence microscopy of the glomerular tissue suspected of undergoing immunologic injury shows an accumulation of host immunoglobulin, various complement components and viral antigen(s) in irregular, granular deposits throughout capillary walls and mesangia. Electron microscopy reveals lumpy deposits of electron-dense material along the outer aspects of the basement membrane and in the mesangia. Specific antiviral antibody in deposited complexes is identified and quantitated by eluting tissue-bound immunoglobulin, which is then recovered and absorbed with various virus and tissue antigens. Although such studies are difficult and have limitations, they currently provide the only means of directly identifying and measuring the antibody present in immunologically injured tissue. Specific antibody in the circulation makes up only a small fraction of the total present ($<1\%$); the amount of antibody trapped and specifically concentrated in tissues is at least 10- and usually 50-fold greater than that in the circulation.

Host and Viral Genes Determine the Degree of Immune Complex Formation and Deposition

Host Genes

The events (virus infection, immunization) leading to V–Ab complex formation and disease are relatively clear, but factors modulating the degree of immune complex formation and deposition in tissues are only now being

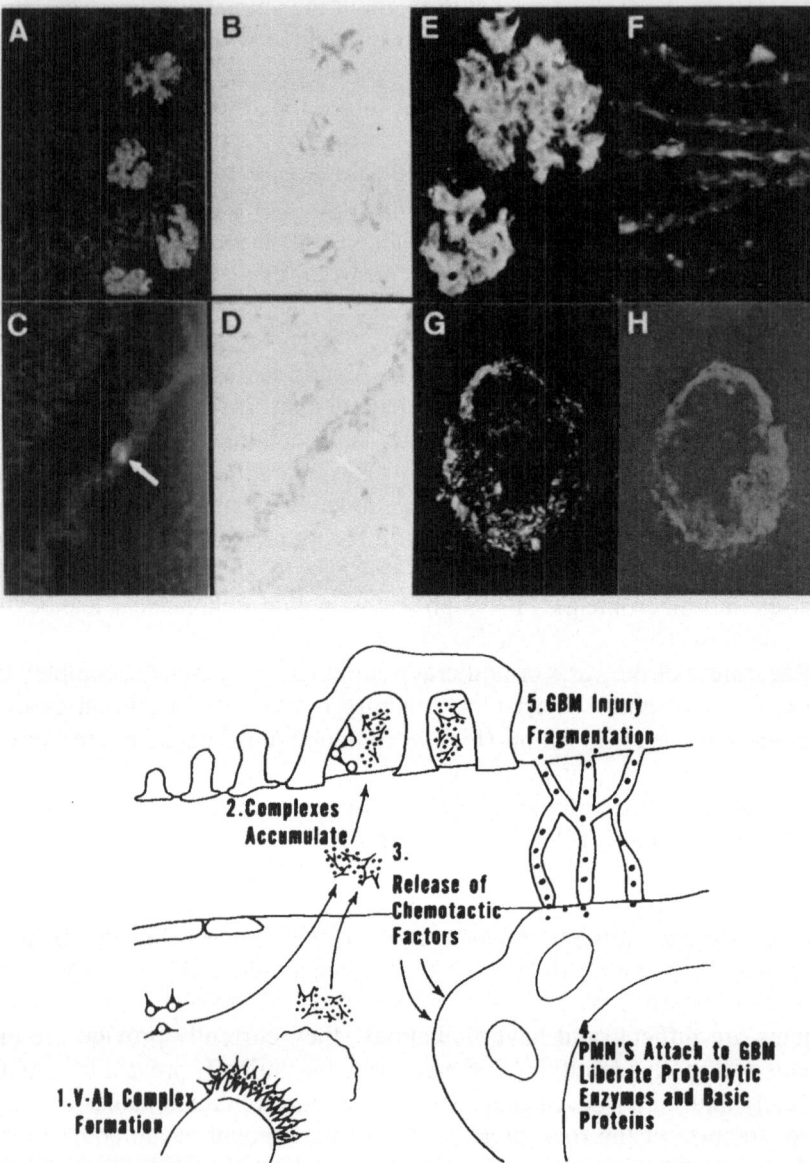

Fig. 1. Principles of virus-induced immune complex formation, deposition, and disease. Antigen elicits a specific antibody that binds to that antigen. The resultant immune complex may become trapped in several tissues, the main ones being renal glomerulus in the kidney, choroid plexus in the central nervous system, or arteries. Panels A–D show trapping of experimentally induced complexes following the injection of ferritin. A and B demonstrate host immunoglobulin (antibody to ferritin), and ferritin trapped in renal glomeruli, respectively. C and D show host immunoglobulin and ferritin deposited as an immune complex in the choroid plexus. (Pictures courtesy of Peter Lampert. See [28] for experimental details.) E–H demonstrate trapped virus–antibody immune complexes during persistent virus infection. E: deposition of

understood. Most of our information has come from the study of LCMV persistent infection of mice, which has served as the prototype model.

Although mice of several strains persistently infected with LCMV mount continuous antibody responses to three structural polypeptides of the virion [19], the amounts of antibodies made vary among the strains [20]. Further, the heightened specific antibody responses of certain inbred strains of mice infected with LCMV correlated directly with their unusually high levels of circulating complement-binding immune complexes detectable by the Clq assay. For example, persistently infected mice of the SWR/J strain make 50-fold more anti-LCMV antibody to the glycoproteins and nucleoproteins of LCMV and contain 7-fold higher levels of circulating complement binding immune complexes than persistently infected BALB mice [20]. Similarly SWR/J mice show heavier deposits of V–Ab complexes in their glomeruli and choroid plexuses than do persistently infected BALB mice, although both strains carry the same load of infectious virus.

Antibody responses of a variety of antigens are controlled by genes mapping in the I region (Ir) of the H-2 complex in mice and HLA-D in humans [21,22]. *Ir* genes are dominant. Studies of two inbred mouse strains, one high (SWR/J) and one low (BALB) immune responders to LCMV, showed high levels of complement-binding immune complexes in the F1 hybrid offspring of the two strains. Back-crosses of these F1 hybrids to both high and low responder parents resulted in a high level of complexes in all the progeny from high responders and comparable levels in approximately one-half of the offspring from the low responders. H-2 typing indicated that these V–Ab complexes were controlled in part by a gene or genes in the H-2 complex that were dominant [20]. Support for *Ir* gene(s) control was further provided by experiments proving that LCMV persistently infected BALB/Kae ($h\text{-}2K^kI^kD^k$) mice and the recombinant inbred stains B10.A ($K^kI^kD^d$) and A.TL ($K^sI^kD^d$) possessed large amounts of circulating V–Ab complexes, but the B10.A (*5r*) ($K^bI^bD^d$) and BALB/St or BALB/cby ($K^dI^dD^d$) strains bore small amounts of these complexes. Additional investigation revealed that other genes, in addition to *Ir* genes, play a role in the V–Ab complex response.

Viral Genes

LCMV is a negative strand RNA virus long considered the prototype of the arenavirus group. Although classified as a single virus, evidence abounds from multiple investigators that LCMV strains in common usage for re-

---

antibody to lymphocytic choriomeningitis virus (LCMV) in the renal glomerulus; F: deposition of antibody to LCMV in the choroid plexus. G: deposition of hepatitis B virus (HBV) antibody; and H: deposition of HBV antigen in medium-size artery. (Pictures of HBV immune complexes through the courtesy of Adam Nowoslawski; see [24].) I: a cartoon of the formation of immune complexes in the circulation or on the plasma membrane of a virus-infected cell and the trapping of such complexes in fenestrated endothelial cells.

search vary considerably in their biologic activity. Generation of a large number of variants as a common theme for RNA viruses is discussed in Chapter 20. Recently through the use of oligonucleotide fingerprinting, tryptic peptide mapping of viral glycoproteins, and a battery of monoclonal antibodies, a number of LCMV variants have been defined. Further, biologic differences among the LCMV strains have correlated with viral variants and their genomic differences [23]. By oligonucleotide mapping, LCMV Arm-1371 and LCMV E-350 are similar (>95% homology) but each differs markedly (by >50%) from LCMV Pasteur, LCMV-WE, LCMV Traub, and LCMV UBC, which also differ from each other by >50% [23]. SWR/J mice persistently infected with LCMV Arm-1371, LCMV E-350, LCMV-WE, or LCMV Pasteur contained high levels of circulating and trapped immune complexes. Amounts of complexes made in mice infected with Arm-1371 were equivalent to those in E-350-infected mice but greater than in mice infected with WE or Pasteur. However, SWR/J mice persistently infected with LCMV Traub produced low to negligible levels of circulating immune complexes and trapped few complexes in their renal glomeruli.

These results with LCMV provide leads for understanding immune complex formation and disease in man. For example, in persistent hepatitis B virus infection, immune complexes form in many but not all individuals and deposit in few [24,25]. The lessons learned with the LCMV system indicate that relevant antibody responses and immune complex formation can vary widely among inbred strains of mice and with different variants of the same virus. Further, although the majority of high responder mice (>95%) within an inbred strain make high levels of V–Ab complexes, there is considerable variability in the amounts made. Thus, fluctuations of immune complexes in human disorders like hepatitis B virus infection probably relate to the genetic composition of the host and perhaps a unique virus variant.

Biologic and Medical Importance

The mechanisms resulting in clearance of virus or virus-infected cells are identical to those causing immunopathologic disease. Immunologically mediated tissue injury during a virus infection basically follows one or both of two pathways. In the first pathway, components of the immune system act either against specific viral antigens or against neoantigens expressed on the surfaces of cells, eventually destroying those cells. The second pathway by which immune constituents cause damage occurs after deposition of immune complexes. In this instance, antibodies combined with their corresponding antigens in the blood or other body fluids form complexes that may become trapped by basement membranes where they cause injury, usually by activating inflammatory mediators. Many of the symptoms and signs of acute virus infections probably result from immunologic injury. For example, the generalized and fluctuating changes in mental status and irritability of some infected patients may relate to shifts in electrolytes and water (increased permeability) in the central nervous system initiated by immune complex deposits in the choroid plexus. Immune complex vasculitis can lead to arthritis, myalgias, and skin rash; immune complex glomerulonephritis may eventuate in acute glomerular injury.

Based on the development of nephritides and arteritides in animals with either experimentally induced immune complex disease or spontaneous disease in the wild, it seems reasonable to think that humans with similar lesions have undergone multiple immunologic assaults via like mechanisms. Indeed, it is clear from immunofluorescence microscopy of kidneys from glomerlonephritic patients that the vast majority (80%–90%) have immune complex deposits.

Viruses may injure kidney tissues by three additional routes. The first includes activities of the virus itself, of toxic products made by the virus or of enzymes (i.e., lysozymes) released during virus infection, all of which can destroy infected cells. The second, less direct route involves molecular mimicry. Owing to the immunologic identity of viral antigens and structural antigens of the kidney, immune responses made against the virus can include formation of antibody that cross-reacts with kidney tissues and initiates cell destruction. A third and also indirect pathway is through the production of interferon. Gresser *et al.* [26,27] showed that viruses may induce the formation of interferon, which can have a toxic effect on rapidly dividing renal glomerular cells. Additionally if antibodies to interferon form and combine with the interferon, the resultant immune complexes add to the total load of V–Ab complexes. Hence, the virus may limit effective clearance of immune complexes by disabling cells whose function it is to clear the complexes (i.e., direct injury or secondarily through interferon activity) while simultaneously inducing antiviral antibodies that form V–Ab complexes.

Finally, because persistent viral infection is frequently associated with the manifestations of the immune complex formation and deposition, a longstanding immune complex syndrome in an individual with disease of unknown etiology may suggest underlying viral causation. Perhaps related is evidence of immune complexes in many, but not all, patients with cancer, demyelinating disorders like multiple sclerosis, and degenerative diseases like amyotrophic lateral sclerosis. For example, immune complexes have been found in humans and mice with neuroblastoma tumors, and antibodies specific to neuroblastoma tumor antigens have been uncovered in these individuals. However, until the identity of an antigen and/or antibody in any immune complex is clear, one cannot be sure that a virus is responsible. Nevertheless, this association is thought-provoking, particularly in the context of much-studied diseases for which no general etiologic agent has yet been identified.

Acknowledgments. This is publication no. 3113-IMM from the Department of Immunology, Scripps Clinic and Research Foundation. This research was supported by US PHS Grants NS-12428, AI-09484, and AI-07007.

References

1. Oldstone MBA (1975) Virus neutralization and virus-induced immune complex disease: Virus–antibody union resulting in immunoprotection or immunologic injury—Two different sides of the same coin. In Melnick JL (ed), Progress in Medical Virology, Vol 19. S. Karger, Basel, p 84
2. Oldstone MBA, Dixon FJ (1975) Immune complex disease associated with viral infections. In Notkins AL (ed), Viral Immunology and Immunopathology. Academic Press, New York, p 341
3. Oldstone MBA, Lampert P, Perrin L (1976) Formation of virus-antiviral antibody immune complexes. In Kluthe R, Vogt A, Batsford SR (eds), Glomerulonephritis. International Conference in Pathogenesis, Pathology and Treatment. Georg Thieme Publishers, Stuttgart, p 12
4. Fujinami RS, Oldstone MBA, Wroblewska Z, Frankel ME, Koprowski H (1983) Molecular mimicry in virus infection: Crossreaction of measles virus phosphoprotein or of herpes simplex virus protein with human intermediate filaments. Proc Natl Acad Sci USA 80:2346–2350
5. Dales S, Fujinami RS, Oldstone MBA (1983) Infection with vaccinia favours the selection of hybridomas synthesizing autoantibodies against intermediate filaments, among them one cross reacting with the virus hemagglutinin. J Immunol, in press
6. Tonetti G, Oldstone MBA, Dixon FJ (1970) The effect of induced chronic viral infections on the immunological diseases of New Zealand mice. J Exp Med 132:89–109
7. Oldstone MBA (1972) Virus induced autoimmune disease: Viruses in the production and prevention of autoimmune disease. In Membranes and Viruses in Immunopathology, Academic Press, New York, p 469
8. Theofilopoulos AN, Dixon FJ (1982) Autoimmune diseases: Immunopathology and etiopathogenesis. Am J Pathol 108:321–365
9. Oldstone MBA, Dixon FJ (1969) Pathogenesis of chronic disease associated with persistent lymphocytic choriomeningitis viral infection. I. Relationship of antibody production to disease in neonatally infected mice. J Exp Med 129:483–505
10. von Pirquet P, Schick B (1905) Die Serum krankheit. Dueticke F, Leipzig and Wein
11. Dixon, FJ (1963) The role of antigen–antibody complexes in disease. Harvey Lect Ser 5821–5852
12. Wilson CB, Dixon FJ (1971) Quantitation of acute and chronic serum sickness in the rabbit. J Exp Med 134:7_s–18_s
13. Haakenstad AO, Case JB, Mannik M (1975) Effect of cortisone on the disappearance kinetics and tissue localization of soluble immune complexes. J Immunol 114:1153–1160
14. Agnello V, de Bracco M, Kunkel H (1972) Hereditary C2 deficiency with some manifestations of systemic lupus erythematosis. J Immunol 108:837–840
15. Peters DK, Williams DG (1974) Complement and mesangeal capillary glomeronephritis: Role of complement deficiency in the pathogenesis of nephritis. Nephron 13:189–197
16. Notkins A, Marh S, Scheele C, Groffman J (1966) Infectious virus–antibody complexes in the blood of chronically infected mice. J Exp Med 124:81–97
17. Porter D, Larson A (1967) Elution disease of mink: Infectious virus–antibody complexes in the serum. Proc Soc Exp Biol Med 126:680–682

18. McGuire T, Crawford T, Henson JB (1972) Equine infectious anemia: Detection of infectious virus complexes in the serum. Immunol Commun 1:543–551
19. Oldstone MBA, Buchmeier MJ, Doyle MV, Tishon A (1980) Virus induced immune complex disease: Specific antiviral antibody and C1q binding material in the circulation during persistent lymphocytic choriomeningitis virus infection. J Immunol 124:831–838
20. Oldstone MBA, Tishon A, Buchmeier MJ (1983) Virus induced immune complex disease: Genetic control of C1q binding complexes in the circulation of mice persistently infected with lymphocytic choriomeningitis virus. J Immunol 130:912–918
21. Kline J (1975) Biology of the mouse histocompatibility-2 complex. Springer-Verlag, New York
22. Benaceraff B, McDevitt H (1972) Histocompatibility linked immune response genes. Science 175:273–279
23. Dutko FJ, Oldstone MBA (1983) Geonomic and biologic variation among commonly used lymphocytic choriomeningitis virus strains. J Gen Virol, in press
24. Brzosko W, Krawaczynski K, Nazarewicz T, Morzytka M, Nowoslawski A (1974) Glomeronephritis associated with hepatitis B surface antigen immune complexes in children. Lancet Vol 2 2477:477–482
25. Trepo C, Zuckerman AJ, Bird RC, Prince AM (1974) The role of circulating hepatitis B antigen–antibody complexes in the pathogenesis of vascular and hepatic manifestations in polyarteritis. J Clin Pathol 27:863–868
26. Gresser I, Morel-Maroger L, Veirous P, Riviere Y, Guillon JC (1978) Anti-interferon globulin inhibits the development of glomerulonephritis in mice infected at birth with lymphocytic choriomeningitis virus. Proc Natl Acad Sci USA 75:3413–3416
27. Riviere Y, Gresser I, Guillon JC, Bandu MT, Ronco P, Morel-Maroger L, Veirous P (1980) Severity of lymphocytic choriomeningitis virus disease in different strains of suckling mice correlates with increasing amounts of endogenous interferon. J Exp Med 152:633–640
28. Lampert P, Garret R, Lampert A (1977) Ferritin immune complex deposits in the choroid plexus. Acta Neuropathol 38:83–86

CHAPTER 30
Virus-Induced Autoimmunity

ABNER LOUIS NOTKINS, TAKASHI ONODERA,
AND BELLUR PRABHAKAR*

Many important human diseases of undetermined etiology have an autoimmune component. In some diseases, the autoimmune component is very broad, involving a number of different organs and tissue antigens. For example, in the case of systemic lupus erythematosus, autoantibodies are found that react with DNA, RNA, cytoplasmic proteins, lymphocytes, and erythrocytes. Similarly, in patients with polyendocrinopathy, autoantibodies that react with the pancreas, thyroid, pituitary, and gastric mucosa have been detected. In contrast, in diseases such as myasthenia gravis, the autoimmune component is far more restricted and predominantly directed against the acetylcholine receptor.

There are many theories as to what triggers the production of autoantibodies ranging from genetically programmed immunological abnormalities to environmental insults. That viruses can sometimes be involved we know from several lines of evidence. In experimental animals, autoantibodies have been found in the sera following a variety of viral infections such as reo, vaccinia, Sendai, rinderpest, Aleutian disease, and retroviruses, to mention just a few [1–7]. Autoantibodies to erythrocytes, nuclei, and intermediate filaments are the ones most frequently observed. Usually, these autoantibodies are of low titer and short lived. Recently, considerable information has been obtained about virus-induced autoimmunity from studies with reovirus. In newborn mice, reovirus type 1 triggers an autoimmune polyendocrine disease characterized by a mild and transient form of diabetes and retardation in growth [1]. The virus infects several organs including cells in the anterior pituitary and the pancreatic islets of Langerhans. In the sera of reovirus-infected mice, autoantibodies that react with tissue sections (e.g.,

---

* National Institutes of Health, Bethesda, Maryland.

anterior pituitary, pancreatic islets, gastric mucosa) from normal, uninfected mice have been detected. Immunosuppression, at the time of infection, prevents the appearance of autoantibodies and the development of the clinical syndrome [8].

In humans, autoantibodies have also been detected after viral infections. Autoantibodies cytotoxic to normal lymphocytes have been observed after infection with rubella, measles, and Epstein–Barr virus (EBV). Autoantibodies reactive with nuclei, smooth muscle, and renal tubules are found more frequently in patients with influenza and recurrent herpes simplex virus infection than in normal controls [9–11]. Other examples include antibody to myelin basic protein in patients with subacute sclerosing panencephalitis [12]; antibody to platelets in patients with rubella, some of whom have thrombocytopenia purpura [13]; rheumatoid factor in patients with congenital cytomegalovirus and hepatitis [14,15]; and antibody to intermediate filaments in patients with several viral infections [16]. As in mice, these antibodies are usually transient and of low titer. The literature contains a number of other reports of an association between viral infections and the development of autoantibodies, but perhaps the most thoroughly studied is EBV; antibodies to cells and tissue antigens (e.g., intermediate filaments, smooth muscle, immunoglobulin, lymphocytes) have been found in the sera of patients infected with this virus [17].

What are the mechanisms by which a virus infection can trigger the production of autoantibodies? There are a number of formal possibilities, but we will restrict our comments to four ideas that are currently receiving considerable attention. The first is that the virus makes the infected cell or some component of it foreign to the host [7,18]. This may occur if the virus: incorporates host antigens into its envelope; inserts, exposes, or modifies host antigens on the cell surface; or leads to the release of sequestered antigens, perhaps in a precursor form, which the host's immune system does not ordinarily encounter. Although this target cell hypothesis has been proposed many times, and in all probability accounts for certain cases of virus-induced antoimmunity, rigorous proof with the exclusion of alternate mechanisms has been hard to obtain.

A second possibility is that the virus triggers autoimmunity by acting on the immunoregulatory system. The virus may stimulate or destroy subpopulations of lymphocytes (e.g., helper or suppressor T cells) or macrophages, and this in turn may lead to the generation of autoantibodies. Immunological abnormalities are known to follow a number of viral infections [19]. One of the best examples comes from studies with EBV. There are receptors for this virus on the surface of B, but not T, lymphocytes. *In vitro*, the virus infects and transforms human B lymphocytes with limited life span into lymphoblastoid cell lines that proliferate indefinitely without need of foreign stimulus. These immortalized cells are capable of secreting immunoglobulins [20], and recent studies show that some of the immunoglobulins, which are predominantly of the IgM class, react with antigens in normal cells. In our laboratory, autoantibodies that react with cells in the thyroid, pituitary, stomach, and pancreas have been prepared by EBV transformation. In fact, this technique is being used in a number of laboratories to prepare human monoclonal antibodies. *In vivo*, EBV is thought similarly to induce lymphoproliferation and the production of antibodies that react with autoantigens. The life span of these EBV-transformed B lymphocytes, however, is usually short because they are destroyed by the host's T-cell response.

A third mechanism by which a virus could trigger an autoimmune response is through what is sometimes referred to as "molecular mimicry." The idea is that antibodies raised against certain viral antigens may cross react with normal host-cell antigens. Although this is an old idea, definitive proof has been difficult to obtain. Monoclonal antibodies are now beginning to provide that proof. For example, a monoclonal antibody to a phosphoprotein of measles virus and another to a herpes virus protein have been shown to react with intermediate filaments of normal host cells, probably vimentin [21]. These antibodies are thought to bind to different antigenic determinants on the vimentin molecule [21]. In other experiments, monoclonal antibodies to SV40 T-antigen were found to react with proteins in normal host cells [22]; a monoclonal antibody to a nonstructural protein (P74) of Japanese encephalitis virus bound to a nuclear antigen in uninfected cells [23]; and a monoclonal antibody to a 19,000-dalton structural protein of human T-cell leukemia virus reacted with normal human thymic epithelium [24]. In our laboratory, 65 Coxsackie B4 virus-neutralizing monoclonal antibodies were tested for reactivity with normal host cells by indirect immunofluorescence. One of these antibodies brilliantly stained mouse myocardium, but not other organs. This antibody is particularly interesting because Coxsackie B4 virus can produce a myocarditis, and it has been speculated that autoimmunity may sometimes be involved in the pathogenesis of this disease [25]. Whether cross-reacting antibodies actually play a role in the pathogenesis of this and other diseases, however, is still not known and may depend on the nature of the antigens (e.g., surface versus cytoplasmic) and the titer of the autoantibodies in the circulation.

A fourth mechanism by which a virus could trigger an autoimmune response is by eliciting anti-idiotypic antibodies. For example, antibodies made to the idiotype of a monoclonal antihormone antibody have been shown to react not only with the antihormone antibody, but also with receptors for the hormone on the cell surface [26]. Presumably the anti-idiotype recognizes a common determinant or structural configuration shared by both the receptor and the antihormone antibody. Recently, Nepom and colleagues [27] made similar observations with reovirus. They found that an anti-idiotypic antibody that reacted with a monoclonal antibody directed against reovirus type 3 hemagglutinin also reacted with receptors for reovirus on the surface of lymphocytes and nerve cells. Similarly, experiments in our laboratory (unpublished data) show that an anti-idiotype made against a monoclonal neutralizing antibody to Coxsackie B4 virus reacts by immunofluorescence with antigens on the surface of several different cell lines. Anti-idiotypic antibodies are thus an appealing mechanism to explain how viruses may sometimes trigger autoimmunity.

Hybridoma technology is making it possible to study virus-induced autoimmunity in still other ways. For example, hybridomas prepared from the spleens of reovirus-infected mice have yielded a large number of monoclonal autoantibodies [28,29]. These antibodies react with some of the same organs as sera from reovirus-infected mice. In addition, monoclonal autoantibodies that react with several other organs have been generated by this technique. The inherent capacity of the animals' B lymphocytes to make these autoantibodies, however, existed prior to the viral infection. We know this because we have succeeded in making some very similar types of autoantibody-producing hybridomas from the spleens of normal uninfected animals ([29] and unpublished data). Thus, these autoantibodies are part of the animals' normal B-lymphocyte repertoire. It appears that *in vivo*, the virus infection, by one or more of the mechanisms discussed, triggers B lymphocytes to secrete these autoantibodies into the circulation.

The studies with monoclonal autoantibodies are revealing other new insights into autoimmunity. By immunofluorescence, many of the monoclonal autoantibodies generated in the reovirus system were found to react with cells in more than one organ [29]. Similarly, human monoclonal autoantibodies, generated from patients with insulin-dependent diabetes and other autoimmune abnormalities, were found to react with cells in multiple organs (e.g., pituitary, thyroid, islets, gastric mucosa) [30]. These multiple organ-reactive antibodies appear to be recognizing either the same molecule present in more than one organ and/or a common antigenic determinant on different molecules in multiple organs. An antibody elicited against a common antigenic determinant in one organ would react with all organs containing that determinant. Multiple organ-reactive antibodies thus may be a partial explanation for the multiple organ autoimmunity observed in some of the human autoimmune diseases [29,30].

In conclusion, autoantibodies are found in sera of animals and patients after certain viral infections. Whether these autoantibodies actually play an important role in the pathogenesis of disease or are largely an epiphenomenon is not clear. Moreover, little is known about the frequency of autoimmune phenomena after known viral infections and the relationship to HLA type. Studies on cell-mediated autoimmune responses [7,31] during and after viral infections are also in their infancy. Although there are many unanswered questions, the development of hybridoma technology, new methods for studying immunoregulation, and recombinant DNA techniques should make it possible to study in greater depth the role of known and still unknown viruses in autoimmune diseases.

References
1. Onodera T, Toniolo A, Ray UR, Jenson AB, Knazek RA, Notkins AL (1981) Virus-induced diabetes mellitus. XX. Polyendocrinopathy and autoimmunity. J Exp Med 153:1457–1473

2. Dales S, Fujinami RS, Oldstone MBA (1983) Infection with vaccinia favors the selection of hybridomas synthesizing autoantibodies against intermediate filaments, one of them cross-reacting with virus hemagglutinin. J Immunol 131: 1546–1553
3. Kay MMB (1979) Parainfluenza infection of aged mice results in autoimmune disease. Clin Immunol Immunopathol 12:301–315
4. Fukuda A, Yamanouchi K (1981) Comparison of autoimmunity induction with virulent and attenuated rinderpest virus in rabbits. Japan J Med Sci Biol 34:149–159
5. Hahn, EC, Kenyon AJ (1980) Anti-deoxyribonucleic acid antibody associated with persistent infection of mink with Aleutian disease virus. Infect Immun 29:452–458
6. Varet, B, Cannat, A, Gisselbrecht, S (1977) Genetic control of antinuclear antibodies in mice infected with Rauscher leukemia virus. Cancer Res 37:1115–1118
7. Hirsch, MS, Proffitt, MR (1975) Autoimmunity in viral infection. In Notkins, AL (ed) Viral Immunology and Immunopathology. Academic Press, New York, p 419
8. Onodera T, Ray UR, Melez KA, Suzuki H, Toniolo A, Notkins AL (1982) Virus-induced diabetes mellitus: Autoimmunity and polyendocrine disease prevented by immunosuppression. Nature 297:66–68
9. Mottironi VD, Tereoski PI (1970) Lymphocytotoxins in disease. I. Infectious mononucleosis, rubella and measles. In Tereoski PI (ed) Histocompatibility Testing. Munskeard, Copenhagen, p 301
10. Loza-Tulimowska M, Semkow R, Michalak T, Nowoslawski A (1976) Autoantibodies in sera of influenza patients. Acta Virol 20:202–207
11. Wutzler, P, Farber I, Ulbricht A (1978) Autoantibodies in herpes virus infections. Acta Virol 22:342
12. Panitch HS, Hooper CJ, Johnson KP (1980) CSF antibody to myelin basic protein. Arch Neurol 37:206–209
13. Kahane S, Dvilanksy, A, Estok L, Nathan I, Zolotov Z, Sarov I (1981) Detection of anti-platelet antibodies in patients with idiopathic thrombocytopenic purpura (ITP) and in patients with rubella and herpes group viral infections. Clin Exp Immunol 44:49–56
14. Stagno S, Pass RF, Reynolds DW, Moore MA, Nahmias AJ, Alford CA (1980) Comparative study of diagnostic procedures for congenital cytomegalovirus infection. Pediatrics 65:251–257
15. Markensen JA, Daniels CA, Notkins AL, Hoofnagle JH, Gerety J, Barker LF (1975) The interaction of rheumatoid factor with hepatitis B surface antigen–antibody complex. Clin Exp Immunol 19:209–217
16. Toh BH, Yildiz A, Sotelo J, Osung O, Holborow EJ, Kanakoudi F, Small JV (1979) Viral infections and IgM autoantibodies to cytoplasmic intermediate filaments. Clin Exp Immunol 37:76–82
17. Linder E, Kurki P, Anderson LC (1979) Autoantibody to "intermediate filaments" in infectious mononucleosis. Clin Immunol Immunopathol 14:411–417
18. Eaten MD (1980) Autoimmunity induced by syngeneic membranes carrying irreversibly adsorbed paramyxovirus. Infect Immun 27:855–861
19. Woodruff JF, Woodruff JJ (1975) The effect of viral infections on the function of the immune system. In Notkins AL (ed) Viral Immunology and Immunopathology. Academic Press, New York, p 393

20. Rosen A, Gergely P, Jondal M, Klein G, Britton S (1977) Polyclonal Ig production after Epstein-Barr virus infection of human lymphocytes in vitro. Nature 267:52-54
21. Fujinami RS, Oldstone MBA, Wroblewska Z, Frankel ME, Koprowski H (1983) Molecular mimicry in virus infection: Crossreaction of measles virus phosphoprotein or of herpes simplex virus protein with human intermediate filaments. Proc Natl Acad Sci USA 80:2346-2350
22. Crawford L, Leppard K, Lane D, Harlow E (1982) Cellular proteins reactive with monoclonal antibodies directed against simian virus 40 T-antigen. J Virol 42:612-620
23. Gould EA, Chanas AC, Buckley A, Clegg CS (1983) Monoclonal immunoglobulin M antibody to Japanese encephalitis virus that can react with a nuclear antigen in mammalian cells. Infect Immun 41:774-779
24. Haynes BF, Robert-Guroff M, Metzgar RS, Franchini G, Kalyanaraman VS, Palker TJ, Gallo RC (1983) Monoclonal antibody against human T cell leukemia virus p19 defines a human thymic epithelial antigen acquired during ontogeny. J Exp Med 157:907-920
25. Woodruff JF (1980) Viral myocarditis: A review. Am J Pathol 101:426-484
26. Sege K, Peterson PA (1978) Use of anti-idiotypic antibodies as cell surface receptor probes. Proc Natl Acad Sci USA 75:2443-2447
27. Nepom JT, Weiner HL, Dichter MA, Tardieu M, Spriggs DR, Gramm CF, Powers ML, Fields BN, Greene MI (1982) Identification of a hemagglutinin-specific idiotype associated with reovirus recognition shared by lymphoid and neural cells. J Exp Med 155:155-167
28. Haspel MV, Onodera T, Prabhakar BS, Horita M, Suzuki H, Notkins AL (1983) Virus-induced autoimmunity: Monoclonal antibodies that react with endocrine tissues. Science 220:304-306
29. Haspel MV, Onodera T, Prabhakar BS, McClintock PR, Essani K, Ray UR, Yagihashi S, Notkins AL (1983) Multiple organ-reactive monoclonal autoantibodies. Nature 303:73-76
30. Satoh J, Prabhakar BS, Haspel MV, Ginsberg-Fellner F, Notkins AL (1983) Human monoclonal autoantibodies that react with multiple endocrine organs. N Engl J Med 309:217-22
31. Watanabe R, Wege H, ter Meulen V (1983) Adoptive transfer of EAE-like lesions from rats with coronavirus-induced demyelinating encephalomyelitis. Nature 305:150-153

CHAPTER 31
The Role of Immunospecific Receptors in Retrovirus-Induced Lymphomagenesis

M. S. McGrath and I. L. Weissman*

It has become increasingly apparent that an understanding of the mechanisms by which oncogenic retroviruses cause neoplastic transformation is likely to lead to an understanding of carcinogenesis in general. Because of rapid advances in the cloning and molecular genetic analysis of these retroviruses, it is now possible to begin to pinpoint the retroviral genes that are involved directly in neoplastic transformation. In those systems wherein neoplastic transformation occurs with defined target cells *in vitro,* progress in the identification of viral oncogenes has been rapid; less rapid is the identification of oncogenes with retroviruses that only transform target cells *in vivo.* The retroviruses that mediate neoplastic transformation in experimental systems fall into two main subcategories. Those are: (1) rapidly transforming retroviruses, which carry cellular-derived oncogenes and cause neoplastic transformation of susceptible cells, *in vitro* or *in vivo,* immediately after infection; and (2) slowly transforming retroviruses, which upon *in-vivo* infection or activation in susceptible animals mainly cause immunologic neoplasms (primarily B- and T-cell lymphomas), only after a prolonged latency period. Although the molecular mechanisms involved in cellular transformation by the class of rapidly transforming viruses have not been completely elucidated, evidence to date shows that rapidly transforming retroviruses carry cellularly derived genes which, upon successful infection, become activated and mediate cellular transformation. Examples include activation of the *src* gene in sarcomagenesis [1], the *erb* gene in erythroblastosis [2], and the *myc* gene in myelocytomatosis [3]. In contrast, there have been no oncogenes discovered in the slowly transforming retrovirus genomes (i.e.,

---

* Stanford University Medical Center, Stanford, California.

murine leukemia virus [MuLV], which primarily causes T-cell lymphomas; avian leukosis virus [ALV], which is associated with avian B-cell lymphomas; or the recently discovered human T-cell leukemia virus [HTLV], which has been associated with a subpopulation of human T-cell neoplasms) [4].

Attempts to elucidate the mechanism by which slowly transforming retroviruses cause immunologic neoplasms have yielded findings consistent with at least three independent models for oncogenesis: (1) *The promoter insertion model.* The discovery of a common integration region in avian bursal lymphoma cells for ALV, with associated activation of the cellular *myc* gene, provided evidence for the promoter insertion hypothesis [5]. This hypothesis states that integration of a retrovirus in the region of a cellular "oncogene" would increase the expression of that oncogene by virtue of the retrovirus-encoded promoter region. The observation that from 70% to 85% of avian bursal lymphomas express elevated levels of the cellular *myc* gene [6,7] prompted many investigators to search for elevated oncogene expression in other tumors. To date there has been mixed evidence that elevation of a cellular oncogene can cause neoplastic transformation; in fact the expression of many oncogenes is elevated during normal embryogenesis [8]. (2) *The DNA transfecting oncogene model.* Cooper, Lane, and Nieman have discovered a DNA sequence (called *B-lym*, personal communication) which, when transfected from avian B cell lymphoma DNA into NIH 3T3 fibroblasts, causes an increased frequency of transformation [9]. This approach suggests that there may be other genes involved in control of neoplastic transformation than those designated as "oncogenes," as the bursal lymphoma transforming DNA in the NIH 3T3 assay has not been shown to be homologous to known oncogenes. (3) *The receptor-mediated leukemogenesis model.* A number of years ago we proposed the receptor-mediated leukemogenesis hypothesis wherein slowly transforming retroviruses would interact with immunospecific and/or growth factor receptors on the surface of B- and T-cell neoplasms to cause continued rounds of mitogenesis (Fig. 1) [10,11]. In this view, the gene responsible for mediating these continued rounds of proliferation would be the envelope gene expressed on the surface of retroviruses and retrovirus-infected cells. The subsequent sections of this review detail the observations we have made in attempting to test this hypothesis.

Murine leukemia virus-induced T-cell lymphomas bear highly specific cell surface receptors that recognize MuLV. To test whether MuLV-induced T-cell lymphomas bound associated MuLVs, we utilized the fluorescence-activated cell sorter (FACS). Retrovirus preparations were purified from supernatants of murine T-cell lymphomas, fluoresceinated, and tested for binding to a series of normal and neoplastic T cells. Moloney virus, radiation leukemia virus (RadLV), and AKR spontaneous lymphoma produced retroviruses (AKR-SL) bound specifically to the lymphomas with which they were associated. This virus binding showed saturation kinetics, was blocked with unlabeled retrovirus, and the polypeptide profile of material bound to

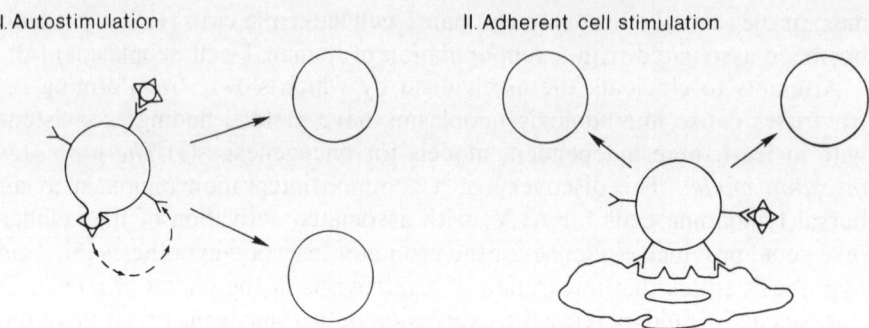

Fig. 1. Receptor-mediated leukemogenesis.

lymphoma cells was representative of intact leukemia virus [12]. In attempts to define the fine specificity of MuLV-binding receptors on T-cell lymphomas, we found that thymotropic leukemogenic RadLV bound to RadLV-induced lymphomas, whereas nonleukemogenic ecotropic and xenotropic retroviruses isolated from the same strain of mouse did not show similar binding abilities [13]. The observation was similar in AKR thymic lymphomas where a hierarchy of virus-binding affinity was clearly demonstrated, which showed AKR SL > MCF-247 > AKR ecotropic virus > AKR xenotropic virus. (For each lymphoma, a distinct AKR-SL virus with a distinct *env* region can be found; MCF247 also has the property of infecting xenogeneic cells.) Further specificity of virus binding was delineated by FACS two-color analysis. RadLV-induced and AKR-spontaneous T-cell lymphomas bound specifically their own retrovirus in an assay that utilized fluoresceinated RadLV with rhodaminated AKR virus in a simultaneous incubation [14]. Therefore, it appeared that MuLVs bound specifically to the T-cell lymphomas with which they were associated, and that this binding was specific for leukemogenic, and not nonleukemogenic retroviruses.

MuLV receptors are markers for irreversible transformation in vivo [10,14]. To test whether the expression of murine leukemia virus receptors on T-cell populations was a marker for transformation, populations of thymocytes from AKR animals of differing ages were analyzed for virus-binding ability. Only 3% of thymocytes present in a newborn thymus bound fluoresceinated AKR virus above background. This binding was completely inhibitable by unfluoresceinated virus. Although the thymocytes from 2–6-month-old AKR mice became infected by endogenous nononcogenic ecotropic AKR MuLV, the subpopulation of MCF-247 virus-receptor-expressing cells remained low (<2%). After approximately 6 months (when AKR animals normally develop spontaneous T-cell lymphomas) three separate phenotypes of thymocyte populations were noted. (1) Frank lymphoma, where the animal had obvious metastatic spread of disease; all thymocytes expressed viral antigen, and all expressed murine leukemia virus receptors. Upon transfer of 10^6 cells into susceptible animals, all recipients died of a donor-derived leukemia within 4 weeks. (2) A phenotypically normal-appearing thymus, which contained approximately 95% viral antigen-expressing cells and <2% MuLV-binding cells. Upon transfer into susceptible animals, no animal developed a donor-derived leukemia within a 4-month observation period. (3) A hypoplastic thymus with 95% of thymocytes expressing murine leukemia virus antigen, but subpopulations of thymocytes were virus-receptor positive (15%–85% in separate animals). FACS separation demonstrated that only animals receiving receptor-positive, virus antigen-positive cells died of donor-derived leukemias in a 30- to 40-day observation period, whereas receptor-negative thymocytes from the same thymus were 100-fold less efficient at leukemia transfer. Therefore, it appears that MuLV-associated T-cell neoplasms bear specific receptors for the leukemogenic retrovirus that they produce and that those receptors are a marker for irreversible cellular transformation.

Lymphoma cell surface receptors for murine leukemia viruses can mediate infection. In contrast to MuLV-induced T-cell lymphomas, X-irradiation induces T-cell lymphomas in C57Bl or C57L mice, which do not initially make an intact retrovirus. This class of lymphomas bears a generalized cell surface receptor for MuLVs, as both leukemogenic and nonleukemogenic ecotropic retroviruses bind to these cells equally well. Xenotropic viruses do not bind to these cells [11,13]. The retroviruses that bind to this class of lymphomas also infect them, and the virus produced by the infected cells has the same polypeptide profile as the input virus [11]. This scheme was used initially to clone the lymphotropic radiation leukemia virus, which replicates poorly *in vitro* in fibroblasts [15]. To analyze whether virus binding may lead to virus infection, we obtained MuLV-VSV pseudotypes that expressed MuLV glycoprotein coats and VSV cores, and, in association with Naomi Rosenberg, Peter Besmer, and David Baltimore, studied their infectivity. In this case, production of VSV required binding and penetration via the MuLV envelope glycoprotein. The non-virus-producing radiation-induced lymphomas were highly infectible by MuLV(VSV) pseudotypes, whereas virus-producing AKR lymphomas (which demonstrated a lesser degree of virus binding) were less infectible by the pseudotype virions [11]. This was in direct contrast to the infectibility of fibroblasts, which demonstrated a complete lack of VSV pseudotype penetration into cells that had been preinfected with MuLV. These observations suggest that MuLVs may interact with receptors on fibroblasts that are different in kind or availability as compared with those on thymic lymphomas.

Monoclonal antibodies that block MuLV binding to T-cell lymphomas also lead to lymphoma cell growth arrest [16–18]. To analyze the role of virus in lymphoma cell proliferation, we prepared a series of monoclonal antibodies to an AKR T-cell lymphoma, KKT-2. Several hybridomas producing KKT-2-binding monoclonal antibodies were found by FACS analysis, a subset of which blocked MCF-247 binding to KKT-2 cells. To determine whether blockade of virus binding would affect KKT-2 cell proliferation, each of the monoclonal antibodies was placed *in vitro* with KKT-2 cells. Several antibodies were discovered that both blocked virus binding and mediated a cell cycle G-1 growth arrest of KKT-2 cells as well as S49 cells. Three of four monoclonal antibodies that blocked virus binding to KKT-2 cells were directed against the T-cell-associated Thy-1 antigen. Because Thy-1 nonexpressing mutants were available (courtesy of R. Hyman) that retained virus receptors, we were able to conclude that the virus did not bind to the Thy-1 molecule. Therefore, the observed growth arrest most likely occurred secondary to steric hindrance of the true virus receptor. Lymphoma growth arrest (and subsequent death) was not merely a side-effect of the inhibiting antibody, as preincubation of KKT-2 cells with live, or UV-inactivated KKT-2SL virus protected KKT-2 cells from antibody-mediated growth arrest, whereas other closely related retroviruses (including MCF-247) did not. Adenylate cyclase and protein kinase minus mutants of the BALB/c S49 T-cell lymphoma line were analyzed in a similar fashion for antibody-mediated growth arrest. No effect on antibody-mediated growth arrest was noted in either of these mutant cell lines. We have demonstrated, therefore, that continued cell surface binding of MuLV was required to maintain several T-lymphoma cell lines in a proliferative state, and that blockade of virus binding with anti-Thy-1 antibodies, but not antibodies directed to H-2, I-A, T-200, Lyt-1, 2, or cell surface viral proteins mediated this growth arrest.

Murine B-cell lymphomas bind associated MuLVs via cell surface IgM [19] (McGrath, in preparation). Because one of the postulates of the receptor-mediated leukemogenesis hypothesis was that B- or T-cell lymphomas should utilize cell surface antigen-specific receptors to bind retroviruses, we decided to test whether B-cell lymphoma receptor immunoglobulins might be retrovirus-binding molecules. BCL_1 is a spontaneous BALB/c B-cell lymphoma that expresses cell surface IgM, and is spleen dependent for *in vivo* growth. When placed *in vitro,* suspensions of BCL_1 splenic lymphoma cells form clusters around surface-adherent cells.

Careful analysis of the lymphoma cells and the adherent cells revealed that the surface-adherent cells had many characteristics of antigen-presenting cells (phagocytic, I–A expressing) and produced high levels of a retrovirus, whereas BCL_1 leukemia cells did not. FACS analysis utilizing fluoresceinated BCL_1 stromal cell-produced virions (BCL_1-V) demonstrated that BCL_1 cells bound very high levels of this virus, whereas only approximately 2% of normal spleen cells bound BCL_1-V to a similar degree. This binding showed saturation characteristics and was blockable with unlabeled virus. Although the binding was most specific for BCL_1-V, many ecotropic retroviruses interacted with the BCL_1 cells above background, with the exception of murine xenotropic virus, which bound poorly. To test whether cell surface immunoglobulin was involved in mediating this virus binding, we obtained a hybridoma between BCL_1 and the mouse nonsecretor myeloma NS1 from Mike Knapp and Samuel Strober. This hybridoma produces large amounts of BCL_1-IgM. BCL_1-IgM was affinity purified, exposed to 3% SDS, and its 19S IgM was isolated on a sucrose gradient. This IgM preparation was tested in a plate binding assay in comparison with two closely related myeloma proteins (HPC-76, MOPC-104E) for binding of internally radiolabeled BCL_1-V. Only BCL-IgM bound the BCL_1-V above background, and this binding could be prevented with monoclonal anti-mu and anti-BCL_1 idiotypic antibodies. The polypeptide pattern of virus bound to BCL_1-IgM was identical to that of the input BCL_1-V preparation. Therefore, it appeared that BCL_1 lymphoma cells bound an associated retrovirus via cell surface IgM. To test whether this was an isolated finding, or was also a property of other murine B-cell lymphomas, a series of B-cell neoplasms was screened for virus binding and blocking of that binding with monoclonal anti-mu antibodies. To date, of 12 B-cell lymphomas tested for binding of BCL_1-V, 11 have been shown to bind this virus; in each case tested the virus binding has been inhibitable with monoclonal anti-mu. These lymphomas include: WEHI-231, WEHI-279, CH-1, BAL-17, 2PK3, NBL, CH-5, CH-7, T-69, A20. Because there appears to be cross-specificity of retrovirus binding by these B-cell lymphomas, it would be of interest to determine whether common immunoglobulin variable regions are expressed by these cells. Investigations are currently in progress to determine whether viruses binding B-cell lymphomas bear similar idiotypic determinants on their cell surface immunoglobulins.

Avian leukosis virus binds to the surface of ALV-induced bursal lymphoma cells via cell surface IgM (McGrath, in preparation). To test whether retrovirus interaction with the surface of neoplastic lymphocytes was a property of other tumor systems, we tested a series of avian bursal lymphomas for binding of fluoresceinated ALV. Analyses of six *in vitro* bursal lymphomas demonstrated fluoresceinated-ALV binding to the surface of bursal lymphoma cells in direct proportion to the level of surface IgM. Cell lines that had lost the ability to produce surface IgM did not bind ALV. The cell line SC2L, which produces a high level of cell surface IgM, bound leukemogenic ALV preparations strongly and bound nonleukemogenic RAV-0 much less efficiently. This binding showed saturation characteristics, and the peptide profile of virus bound was that of the input virus preparation. Monoclonal antibodies detecting immunoglobulin determinants (obtained from M. Cooper, Birmingham, Alabama) blocked virus binding in direct proportion to the level of antibody bound per cell. Affinity-purified (with monoclonal anti-light chain antibodies) SC2L IgM bound radiolabeled ALV in a plate binding assay. A radioimmunoassay utilizing radiolabeled ALV polypeptides showed that the one major ALV peptide bound to the SC2L IgM preparation was an 85,000-molecular-weight protein. Investigations are currently under way to determine whether this 85,000-molecular-weight protein is in fact the major external glycoprotein of ALV,GP85.

To determine whether ALV-receptor expression was a marker for particular subclasses of bursal cells during bursal lymphomagenesis (as previously described in murine T-cell leukemogenesis), we analyzed a series of bursal cell preparations from uninfected, ALV-infected, and tumor-bearing chickens. Two to 4% of normal bursal cells were noted to bind fluoresceinated ALV above background. This binding was blockable with unlabeled ALV. Bursal cell preparations from animals with productive ALV infections but without overt lymphomas showed the same level of virus binding (2%–4%). Bursal cell preparations from three animals bearing metastatic B-cell lymphomas were analyzed for virus binding and more than 75% of cells within these lymphoma populations bound fluoresceinated ALV above background. This binding was blockable with unlabeled ALV, as well as monoclonal antibody directed against the chicken IgM light chain. Therefore, as previously shown in murine T- and B-cell lymphomagenesis, ALV-receptor expression appeared to be a marker for lymphoma cells, but not cells that predominate in the preleukemic period. Because all ALV-induced lymphomas express cell surface IgM, and all lymphomas tested to date bind ALV, ALV binding to bursal lymphomas appears to be as strongly correlated with cellular transformation, as does the elevated level of c-*myc* and the appearance of B-*lym* as noted by other investigators. The relationships between ALV-binding IgM, the appearance of a transfectable B-*lym*, and elevated levels of intracellular c-*myc* are currently unknown.

Human T-cell leukemia virus binds to the surface of HTLV-associated T-cell neoplasms (McGrath, in preparation). In a preliminary set of experiments intended to test whether observations made in mouse and avian tumor systems would be translatable to humans, we fluoresceinated HTLV produced by HUT-102 cells and tested them for binding to a series of human T-cell neoplasms. HTLV was noted to bind to the cells that produced the virus, a cutaneous T-cell lymphoma (designated HUT-102) as well as to several other HTLV-producing and nonproducing cell lines. This binding showed saturation kinetics, and was blockable with unlabeled HTLV. Normal peripheral blood lymphocytes, *in-vitro* human T-cell lines, murine, and gibbon T-cell lymphomas did not bind HTLV above background. Investigations are currently proceeding to determine whether the binding of HTLV will be a marker for transformation of human T cells (as in murine T-cell lymphomagenesis) and to determine the role that this virus–cell surface interaction plays in human leukemogenesis.

In summary, we have described an approach to lymphoid lymphomagenesis that views the lymphoma cells (B and T) as extensions of the normal immune system. During lymphomagenesis, we have proposed that a retrovirus envelope gene product stimulates the responding B or T lymphocyte in an antigen-specific fashion, either by self expression of the antigen (murine T cell) or by accessory-cell production and presentation of that envelope gene product (murine B, and avian B-cell lymphomas) to stimulate antigen-specific neoplastic proliferation. If this indeed is the case (that retrovirus envelope gene products may stimulate antigen-specific proliferation), an extension of that hypothesis should include the possibility that retrovirus envelope gene products may stimulate other subpopulations of immune cells to affect other specific aspects of the immune system. One might therefore predict that a retrovirus interaction with a suppressor cell population could induce immunosuppression. In a like fashion, overstimulation of particular helper T cells may induce an overactivation of immunoglobulin production. At this time it is clear that retrovirus interaction with neoplastic lymphoid populations in several species of animals correlates with the transformed phenotype. The relationship between the expression of receptors for retrovirus and intracellular changes noted in neoplastic cell populations (elevation of oncogenes, chromosomal translocations, and aneuploidy) is currently unknown, and will be the object of intense investigation in the future.

References

1. Hanafusa H (1977) Cell transformation by RNA tumor viruses. *In* Fraenkel-Conrad H, Wagner R (eds) Comprehensive Virology. Plenum Press, New York, p 401
2. Lae M, Neil MC, Vogt PK (1979) Avian erythroblastosis virus: Transformation specific sequences form a contiguous segment of 3.25 kb located in the middle of the 6kb genome. Virology 97:366
3. Sheiness D, Bishop JM (1979) DNA and RNA from uninfected vertebrate cells contain nucleotide sequences related to the putative transforming gene of avian myelocytomatosis virus. J Virol 31:514
4. Gallo RC, Wong-Staal F (1982) Retroviruses as etiologic agents of some animal and human leukemias and lymphomas as tools for evaluating the molecular mechanism of leukemogenesis. Blood 60:545
5. Hayward W, Neel BG, Astrin S (1981) Activation of a cellular *onc* gene by promoter insertion in ALV-induced lymphoid leukosis. Nature 290:475
6. Neel BG, Hayward WS, Robinson HL, Fang J, Astrin S (1981) Avian leukosis virus-induced tumors have common proviral integration sites and synthesize discrete new RNA's: Oncogenesis by promoter insertion. Cell 23:323
7. Payne C, Courtreidge S, Crittenden L, Fadly A, Bishop JM, Varmus H (1981) Analysis of avian leukosis virus DNA and RNA in bursal tumors: Viral gene expression is not required for maintenance of the tumor state. Cell 23:311
8. Bishop JM (1983) Cancer genes come of age. Cell 32:1018
9. Cooper GM, Nieman PE (1981) Two distinct candidate transforming genes by lymphoid leukosis virus-induced neoplasms. Nature 292:857
10. McGrath MS, Weissman IL (1978) A receptor mediated model of viral leukemogenesis: Hypothesis and experiments. *In* Cold Spring Harbor Symposium: Normal and Neoplastic Hematopoietic Cell Differentiation. Cold Spring Harbor Laboratory, New York, p 577
11. McGrath MS, Pillemer E, Kooistra D, Weissman IL (1980) The role of MuLV receptors on T lymphoma cells in lymphoma cell proliferation. Contemp Top Immunobiol 11:156
12. McGrath M, Weissman IL, Baird S, Raschke W, Decleve A, Lieberman M, Kaplan HS (1978) Each T-cell lymphoma induced by a particular murine leuke-

mia virus bears surface receptors specific for that virus. *In* Lerner R, Bergsma D (eds) The Molecular Basis of Cell–Cell Interaction. The National Foundation—March of Dimes Birth Defects: Original Article Series, XIV. Alan R. Liss, New York, p 349
13. McGrath MS, Lieberman M, Decleve A, Kaplan HS, Weissman IL (1978) The specificity of cell surface virus receptors on RadLV and radiation-induced thymic lymphomas. J Virol 28:819
14. McGrath MS, Weissman IL (1979) AKR leukemogenesis: Identification and biological significance of thymic lymphoma receptors for AKR retroviruses. Cell 17:65
15. Lieberman M, Decleve A, Ihle J, Kaplan HS (1979) Rescue of a thymotropic, leukemogenic C-type virus from cultured non-producer lymphoma cells of strain C57Bl/Ka mice. Virology 97:12
16. McGrath MS, Pillemer E, Kooistra D, Jacobs S, Jerabek L, Weissman IL (1980) T lymphoma retroviral receptors and control of T-lymphoma cell proliferation. Cold Spring Harbor Symposium on Quantitative Biology 44:1297
17. McGrath MS, Pillemer E, Weissman IL (1980) Murine leukemogenesis: Monoclonal antibodies to T cell determinants which arrest T lymphoma cell proliferation. Nature 285:259
18. McGrath MS, Jerabek L, Pillemer E, Steinberg RA, Weissman IL (1981) Receptor mediated murine leukemogenesis: Monoclonal antibody induced lymphoma cell growth arrest. *In* Neth R (ed) Modern Trends in Human Leukemia IV. p 360
19. McGrath MS, Jerabek L, Weissman IL (1982) Receptor mediated leukemogenesis: Retrovirus receptors on B and T lymphoma share idiotypic determinants. *In* Baum S, Ledney GD, Therfelder S (eds) Experimental Hematology Today. p 93

CHAPTER 32
Viruses as Regulators of Delayed Hypersensitivity T-Cell and Suppressor T-Cell Function

A. A. NASH*

There are a number of strategies employed by viruses to ensure their survival in a particular host or population. Nearly all such strategies involve outwitting the host's immune defenses either directly or indirectly. Of these virus mechanisms, those that directly affect the balance between helper and suppressor T-cell responses are considered here. Viruses can potentially create this imbalance in a number of ways: For instance specific T cells can become infected, thereby compromising specific regulatory functions, or antigen-presenting cells can become infected thereby interfering with antigen presentation and consequently the induction of specific immune responses. These mechanisms may be highly relevant in the case of certain lymphotropic viruses, such as measles, or macrophage "tropic" viruses such as lactic dehydrogenase virus, both of which are also capable of persisting in their respective target cells. Despite the likely importance of such mechanisms in disturbing specific and nonspecific immune reactivity, attention here is focused on the importance of the route of entry of viruses into the host and the state of a particular virus, i.e., whether infective or not, as determining whether helper/delayed hypersensitivity T-cell or suppressor T-cell responses predominate *in vivo*. In particular, the importance of these T-cell subsets in protecting the host from viral disease, as well as as inducers of immune-mediated pathology is considered. However before this, a knowledge of the induction of delayed hypersensitivity (DH) and suppressor T-cell responses to viruses is presented.

---
* University of Cambridge, Cambridge, England.

Induction of Delayed Hypersensitivity to Viruses: MHC Restriction and Membrane Phenotypes

The induction and expression of DH to viruses depends in particular on the nature of the virus, whether infective or not, and the route of entry into the host. Influenza and Sendai virus infections of mice, infective virus injected subcutaneously (s.c.) or intranasally (i.n.) induces both Lyt 1 + 2−, H-2 IA* region-restricted and Lyt 1 − 2+, H-2 K,D region-restricted DH responses [1]. However if ultraviolet-inactivated influenza or a fusion negative (F−) mutant of Sendai is injected s.c., only Lyt 1 + 2− IA region-restricted T cells are induced. In reovirus infections a similar situation exists, namely that infective virus injected s.c. or intravenously (i.v.) induces a H-2 D or H-2 K/IA region-restricted DH response [2]; ultraviolet-inactivated reovirus injected i.v. induces a state of tolerance to DH. In contrast to these observations, infection of mice with lymphocytic choriomeningitis virus (LCMV) results in a DH response which is H-2 K,D region-restricted, with no apparent role for H-2 IA-restricted T cells in DH responses [3]. On the other hand, DH to infective herpes simplex virus (HSV) injected s.c. is mediated by Lyt 1 + 2− T cells that are H-2 IA region restricted. In this instance, there is no apparent role for Lyt 1 − 2+ H-2 K,D-restricted T cells. Unlike reovirus, both infective and ultraviolet-inactivated HSV injected i.v. leads to the induction of tolerance to HSV-induced DH [4].

What do all these observations tell one about the induction of helper/DH responses to viruses in general? First, that a generalization from one virus to another is not really possible. This is in part due to the ability of different viruses/viral antigens to become associated with the various products of the MHC system. As has been shown, influenza, Sendai, and reovirus (possibly) associate with both class I (i.e., K,D) and class II (i.e., I region) MHC antigens, LCMV with class I and HSV with class II. This association is to some extent determined by the site of virus growth (infection), i.e., HSV in the epidermis and influenza in the lung and possibly blood stream. Indeed it would seem that i.v. routes favor a class I MHC-restricted DH response, while s.c. infections favor predominantly class II MHC responses. It is also likely that the "infectivity" of a particular virus preparation is important; for example where high particle (noninfectious virions)-to-infectivity ratios prevail, as with HSV, then perhaps class II-restricted responses predominate, the reverse being the case for low particle-to-infectivity ratios.

Induction of Suppressor T Cells by Viruses

To illustrate the induction of virus-specific suppressor T cells, the murine model of HSV infection is considered in detail. In studies with this model, it

---

* Lyt 1 + 2−, H-2 IA region restriction characterizes classic T helper cells.

was found that infective (apathogenic mutants or with some mouse strains wild type) virus injected i.v. led to a subsequent inhibition of the induction of DH [4]. Antiherpes antibodies and cytotoxic T-cell (CTL) responses were coinduced, implying a form of split tolerance [5]. The suppression is transferable to normal recipients by splenic T cells as early as day 7 post-infection, and up to at least day 200. In other words, the suppressive response appeared to be life long. Whether or not this is true for other virus-induced suppressor mechanisms remains to be determined. The early suppressor T-cell response is mediated solely by Lyt 1 + 2−, IJ+† cells, but by day 28 both Lyt 1 + 2− IJ+ and Lyt 1 − 2+, IJ+ cells are able to transfer the suppression [6]. Of particular interest is the herpes type specificity of the suppressor T-cell response, i.e., type I herpes induces suppressors that suppress the DH response to type I, but not to type 2 herpes. This was to some extent unexpected in view of the broad cross-reactivity of DH, antibody, and CTL responses for both virus types. It would, therefore, appear that although suppressor T cells recognize predominantly type-specific epitopes, they are capable of suppressing a DH response that "sees" both type-specific and type-common determinants. The target cell in this system is presumably the Lyt 1 + 2−, IA region-restricted T cell or its precursor.

The induction of virus-specific suppressor T cells reactive against DH has been observed in a number of other animal models. For example, high doses of ultraviolet-inactivated reovirus injected i.v. or introduced orally leads to the induction of T-suppressor cells, which inhibit DH induction [7,8]. In a murine model of influenza virus infection, suppressor T cells develop as a result of infective virus administered intranasally. These cells suppress a H-2 IA-restricted DH response, but do not inhibit helper T cells cooperating with B cells to produce anti-influenza antibody [9]. Interestingly, inactivated influenza does not induce suppressor T cells of the type described above, but instead produces suppressor cells that inhibit the K, D-restricted DH response [1]. A similar situation exists for F− Sendai virus. As pointed out by these authors, an apparent symmetry exists between the regulation of class I and class II MHC-restricted DH responses. This type of symmetry to some extent exists in HSV infections of mice. Animals injected s.c. with infective virus sensitize for strong H-2 IA-restricted DH responses, but H-2 K,D-restricted cytotoxic T-cell responses appear to be suppressed *in vivo* [10].

It can be concluded that the i.v. route leads predominantly to suppressor T-cell responses, which are active in suppressing IA-restricted DH responses. Such a response might well be to the advantage of the host, particularly by inhibiting cell-mediated hypersensitivity reactions that could result in immunopathology.

---

† IJ is a marker that characterizes suppressor T cells.

The Role of DH-T Cells and Suppressor T Cells in Protection Against Viruses and in Virus-Induced Immunopathology

In considering the role of helper/DH cells in protection or immunopathology, the site of virus replication, the extent of virus growth and spread, and the subsequent extent of infiltration into the appropriate tissues are important variables. There is good evidence for a protective role for DH in herpes simplex virus infections, where the virus replicates predominantly in epidermal cells or mucocutaneous surfaces. In the mouse ear model, the rapid elimination of infective virus from the ear of an acutely infected animal correlates with the appearance of DH, and this protection can be transferred to infected recipients by Lyt 1 + 2− cells but not Lyt 1 − 2+ cells [7]. The actual mechanism of antiviral activity is not known, but probably involves macrophage activation by lymphokines. Similar findings have been made in ectromelia infections where there appears to be an absolute requirement for immune T cells and monocyte/macrophages in regressing infectious foci in the liver [11]. Hence in these tissues, there appears to be an important role for DH reactions in protection. In other tissues, e.g., heart, lung, and CNS, provided the response is not excessive, one would expect similar protective mechanisms to operate. However, excessive, uncontrolled cell-mediated hypersensitivity reactions could lead to tissue damage, which in some situations might prove fatal.

An example of virus-induced immunopathology is that produced by helper/DH cells in the lungs of influenza-infected mice [12]. Although this model is not typical of influenza infection in man, it does show how sensitization of helper/DH cells to an noninfective preparation of influenza can lead to subsequent immunopathology when such animals are challenged intranasally with live influenza. The damage in this instance is mediated by Lyt 1 + 2− IA region-restricted T cells and involves active recruitment of monocytes/macrophages into the infected lung: In other words, a typical DH response is produced. These observations may go some way to explain the phenomenon encountered when some patients exposed to infectious respiratory syncytial virus or measles virus, having previously received inactivated virus vaccines, suffered severe illness that was attributed to immunopathology [13]. The problem here clearly resides in a failure to induce adequate protective or regulatory responses, notably suppressor T cells, which would serve to limit the extent of DH-cell activation and consequently cellular infiltration. In the murine influenza model, it is known that T suppressor cells, which regulate the induction of DH responses, are not induced following the injection of inactivated virus [9]. The idea that immunopathology is induced in this model because of a failure to induce T suppressor cells was investigated by F.Y. Liew (personal communication). He found that the adoptive transfer of suppressor T cells to mice injected with inactivated influenza prevented lung consolidation and death when these animals were subsequently challenged intranasally with live influenza.

Thus a role for suppressor cells as important regulators of cell-mediated hypersensitivity responses has been demonstrated. However, preferential induction of suppressors could work against the host, particularly in situations where a DH response is beneficial, e.g., as with cutaneous herpes infections. Although in the murine model of herpes simplex virus infection, the induction of suppressor T cells does not predispose the animal to a severe primary infection (in fact, it is known that a primary DH response is induced in this model but subsequent responses are inhibited), the role of suppressor cells in interfering with immunosurveillance against recurrent virus episodes is still a possibility [14]. In humans, T cells undergoing proliferation or producing macrophage inhibition factor (MIF) in response to specific antigens are depressed in patients undergoing herpes recurrences. Furthermore analysis of the T-cell population in such patients shows an increased proportion of OKT 8+ cells (suppressor T-cell phenotype); *in vitro* OKT 8+ T cells from patients with recrudescences do appear to inhibit specifically the proliferative response of lymphocytes from convalescent patients to herpes antigens [15].

Conclusions

Clearly, the balance between helper and suppressor responses can be affected by viruses in a number of ways. The nature of viruses, in particular with respect to their target tissues and to the preferential association with MHC products of infective or noninfective viruses, is central to whether recovery from infection follows or pathogenesis ensues. On the whole, it is concluded that suppressor T cells are beneficial to the host in regulating the intensity of certain immune reactions, but could be detrimental in suppressing protective responses, for situations where only one mechanism is central to recovery (for the majority of viruses this appears unlikely). Other lessons that can be learned here involve the use of vaccines. Clearly, live, attenuated vaccines are more likely to elicit a balanced T-cell response and hence initiate broader antiviral defenses. On the other hand, inactivated vaccines and possibly peptide vaccines need to be used carefully with knowledge of the type of responses they elicit. If T-cell memory is central to long-lived immune responses against viruses, then regimes designed to induce the right balance, i.e., not to oversensitize for one particular mechanism, are crucial.

References

1. Ada GL, Leung KN, Ertl H (1981) An analysis of effector T cell generation and function in mice exposed to influenza or Sendai viruses. Immunol Rev 58:5–24
2. Weiner, HL, Greene MI, Fields BN (1980) Delayed hypersensitivity in mice infected with reovirus. I. Identification of host and viral gene products responsible for the immune response. J Immunol 125:278–282
3. Zinkernagel RM (1976) H-2 restriction of virus-specific T cell mediated effector functions in vivo. II. Adoptive transfer of DTH to murine lymphocytic chorio-

meningitis virus is restricted by the K and D regions of H-2. J Exp Med 144:776–787
4. Nash AA, Gell PGH, Wildy P (1981) Tolerance and immunity in mice infected with herpes simplex virus: Simultaneous induction of protective immunity and tolerance to delayed-type hypersensitivity. Immunology 4:153–159
5. Nash AA, Ashford NPN (1982) Split T cell tolerance in herpes simplex virus-infected mice and its implication for anti-viral immunity. Immunology 45:761–767
6. Nash AA, Gell PGH (1983) Membrane phenotype of murine effector and suppressor T cells involved in delayed hypersensitivity and protective immunity to herpes simplex virus. Cell Immunol 75:348–355
7. Greene MI, Weiner HL (1980) Delayed hypersensitivity in the mouse infected with reovirus. II. Induction of tolerance and suppressor T cells to viral specific gene products. J Immunol 125:283–287
8. Rubin D, Weiner HL, Fields BN, Greene MI (1981) Immunologic tolerance after oral administration of reovirus: Requirement for two viral gene products for tolerance induction. J Immunol 127:1697–1701
9. Liew FY, Russell SM (1980) Delayed-type hypersensitivity to influenza viruses: Induction of antigen specific suppressor T cells for delayed type hypersensitivity to haemagglutinin during influenza virus infection in mice. J Exp Med 151:799–814
10. Pfizenmaier K, Jung H, Starzinski-Powitz A, Rollinghoff M, Wagner H (1977) The role of T cells in anti-herpes simplex virus immunity. I. Induction of antigen-specific cytotoxic T lymphocytes. J Immunol 119:939–944
11. Blanden RV (1971) Mechanisms of recovery from a generalized viral infection: Mousepox. III. Regression of infectious foci. J Exp Med 133:1090–1104
12. Leung KN, Ada GL (1982) Different functions of subsets of effector T cells in murine influenza infection. Cell Immunol 67:312–324
13. Isacson P (1968) Delayed dermal hypersensitivity after viral immunization. *In* Pollard M (ed) Perspectives in Virology. Academic Press, New York, p 141
14. Nash AA, Gell PGH (1981) The delayed hypersensitivity T cell and its interaction with other T cells. Immunol Today 2:162–165
15. Sheridan JF, Donnenberg AD, Aurelian L, Elpern DJ (1982) Immunity to herpes simplex virus type 2. IV. Impaired lymphokine production during recrudescence correlates with an imbalance in T lymphocyte subsets. J Immunol 129:326–331

CHAPTER 33
Mechanisms and Biological Implications of Virus-Induced Polyclonal B-Cell Activation

RAFI AHMED AND MICHAEL B. A. OLDSTONE*

A remarkable feature of the immune system is its ability to make a specific response to a variety of antigens. The concept of specificity is such a dominant theme in immunology that one tends to overlook the fact that immunization with certain antigens not only results in the formation of specific antibody, but also leads to the nonspecific activation of B cells, as reflected by an increased number of cells secreting immunoglobulin (Ig) and elevated levels of Ig in the serum [1–3]. Characteristic of most viral infections are increases in both virus-specific and nonspecific antibodies. Some of these nonspecific antibodies are autoantibodies reactive with host components. This polyclonal activation of B cells, including autoantibody production, can develop during both acute and chronic infections by a variety of DNA and RNA viruses in experimental animals and in humans [4]. In this article, we examine the mechanisms by which viruses can induce polyclonal B-cell activation and discuss its biological significance.

Mechanism of Virus-Induced Polyclonal B-Cell Activation

Viruses as Direct B-Cell Mitogens
A large number of substances such as lipopolysaccharide (LPS), dextran sulfate, pokeweed mitogen, purified protein derivative of tuberculin (PPD) etc., are direct B-cell mitogens and polyclonal B-cell activators (PBAs) [5]. Such substances directly activate B cells to proliferate and secrete antibody by interacting with receptors other than Ig receptors. Consequently, this activation is not immunologically specific but polyclonal and nonspecific.

---
* Scripps Clinic and Research Foundation, La Jolla, California.

Although the origins of the various PBAs are quite diverse, they have certain structural similarities. PBAs are usually high molecular weight polymers (of polysaccharides or proteins) with repeating antigenic determinants. Viruses resemble the known PBAs, in that, inherent in the structure of many viruses is a geometrical array of repeating epitopes. Thus, viruses satisfy the structural requisites for acting as B-cell mitogens.

The best-characterized viral PBA is Epstein–Barr virus (EBV) [6]. EBV is a strong polyclonal activator *in vitro* of human B cells, which it infects and transforms. Transformation and subsequent immortalization of the B cells by EBV are closely associated with the polyclonal activation, since virus inactivated by ultraviolet light can bind to the B cell but does not induce proliferation or antibody secretion. Thus, the mechanism of B-cell activation by EBV differs from that of the classical PBAs, such as LPS and pokeweed, i.e., binding *per se* to receptors on B cells does not cause activation. The ability of EBV to act as a PBA *in vitro* is of considerable biological significance *in vivo*, because EBV is the etiological agent of infectious mononucleosis, a disease in which there is a pronounced increase of antibody with diverse specificities, including autoantibodies. Hence, part of the disease may be autoimmune in origin and reflect the generation of autoantibodies that outlast the initial viral infection.

Several other viruses including influenza virus, herpes simplex virus (HSV), vesicular stomatitis virus (VSV), adenovirus, African swine fever virus, and sindbis virus have been shown to act as B-cell mitogens *in vitro* [7–12]. Except for HSV, infectivity by these viruses is not required for B-cell activation, suggesting that the mitogenicity is due to a structural protein of the virion. Subsequent studies have shown that isolated glycoproteins of influenza virus, VSV, and sindbis virus, and the fiber protein of adenovirus, are mitogenic for B cells [13,14]. These potentially important results indicate that certain viral proteins, in particular glycoproteins, may function directly as polyclonal B-cell activators.

In studies examining the mitogenic potential of viruses *in vitro*, one must show conclusively that the mitogenic activity is, indeed, due to the virus and not some contaminant. We stress this point because the currently improved methods for detecting mycoplasma have made clear that a large number of routinely used tissue culture lines are contaminated with mycoplasma, several species of which are extremely strong PBAs [15]. In fact, one report describing that HSV is a PBA was subsequently modified after the investigators discovered that their HSV stock was contaminated with mycoplasma [16]. In addition, we found that certain stocks of lymphocytic choriomeningitis virus (LCMV) and gradient-purified virus from these stocks were PBAs, whereas some other LCMV stocks had no mitogenic activity. After screening for mycoplasma, we found that all stocks (including the "purified" LCMV) with polyclonal B-cell activity were contaminated with mycoplasma, whereas LCMV stocks without mitogenic activity were mycoplasma-free. Thus, the potential hazards of mycoplasma contamination when examining virus–lymphocyte interactions *in vitro* cannot be overemphasized.

Polyclonal B-Cell Activation Due to Inactivation of Suppressor T Cells

Although Gershon and Kondo's [17] conclusion that suppressor T cells exist met with initial resistance, it is now generally accepted that suppressor T cells provide a regulatory function in antibody production. Both antigen-specific and nonspecific suppressor T cells have been described. Since removal or inactivation of suppressor T cells results in increased antibody production, a possible mechanism by which viruses can induce polyclonal B-cell activation is to infect and subsequently impair the function of suppressor T cells. Such a mechanism was postulated to explain VSV-induced augmentation of the antibody response to sheep red blood cells (SRBC) *in vitro* [18]. For a virus to increase antibody production by this mechanism, it should selectively infect and inactivate only suppressor T cells. No such tropism has been shown for VSV, which infects all actively replicating T cells; however, certain viruses may show unique tropisms for particular subpopulations of T cells. Identifiable surface antigens are expressed exclusively on functionally distinct subsets of T cells, and these antigens may act as receptors for viral attachment.

Polyclonal B-Cell Activation Mediated by Virus-Specific Helper T Cells

In studying the antibody response during acute viral infection, we found that intraperitoneal injection of LCMV into adult BALB/c WEHI (+/+) mice induced an LCMV specific antibody response as well as a polyclonal antibody response (Ahmed *et al.*, in preparation). The total number of Ig-secreting cells increased tenfold in the spleens of these mice 7 days after infection. The LCMV-infected mice also developed two- to threefold increases in the number of trinitrophenyl and fluorescein antibody plaque-forming cells (PFC). In addition, LCMV infection of mice primed with SRBC increased the number of anti-SRBC PFC, indicating the activation of bystander B cells *in vivo*.

The following experimental evidence strongly suggests that the polyclonal B-cell activation seen during acute LCMV infection is mediated by LCMV-specific helper T cells: (1) LCMV is not a direct B-cell mitogen. (2) The kinetics with which LCMV-specific T helper cells proliferate in the lymph nodes and spleens of infected mice closely parallel the rate of increase in numbers of Ig-secreting cells. B-cell activation peaks a couple of days after the peak activity of LCMV-specific T-proliferating cells, and then the number of Ig-secreting cells rapidly declines as the LCMV-specific T-helper activity disappears. (3) In striking contrast to the polyclonal B-cell activation seen during acute LCMV infection of BALB/c WEHI (+/+) mice, no polyclonal B-cell activation occurs in congenitally athymic nude BALB/c WEHI (nu/nu) mice after LCMV infection. These nude mice made considerable amounts of LCMV-specific antibody but showed no increase in the total number of Ig-secreting cells. As expected, no proliferation of LCMV-specific T cells was detectable in the infected nude mice.

We propose that the polyclonal activation of B cells commonly observed during viral infection is mediated by soluble nonspecific B-cell growth factors and differentiation factors (BCGF and BCDF) released by virus-specific helper T cells. We think that this mechanism may account for much of the polyclonal B-cell activation that accompanies viral infections for the following reasons: (1) Most viruses induce a strong T-cell response resulting in activation of virus-specific helper T cells [19]. (2) According to recent studies of human and murine B-cell activation, antigen-specific T cells release nonspecific factors that can polyclonally activate B cells to proliferate and secrete antibody [20–23]. The nonspecific BCGF and BCDF released by antigen-specific T helper cells have minimal or no effect on resting B cells. However, these factors can activate B cells that were previously primed (i.e., cells that have received the first signal from a specific antigen). At any given time, an individual has a substantial number of "primed" B cells that are sensitive to the nonspecific lymphokines released by activated T cells. Thus, a virally infected host that undergoes activation of virus-specific helper T cells subsequently develops polyclonal activation of bystander B cells. This proposed scheme of virus-induced polyclonal B-cell activation is shown in Fig. 1.

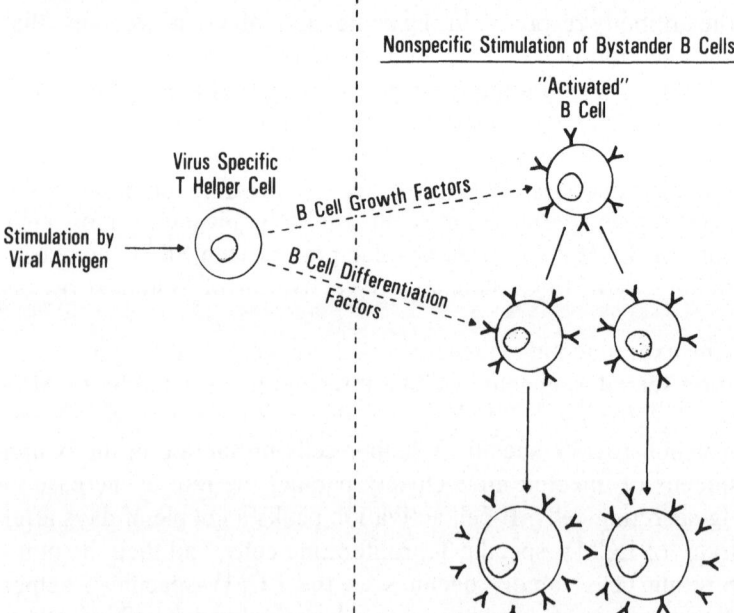

Fig. 1. Polyclonal B-cell activation mediated by virus-specific helper T cells. Viral infections result in activation of virus-specific T helper cells, which release B-cell growth and differentiation factors. These soluble lymphokines can nonspecifically activate "primed" bystander B cells.

Biological Significance of Virus-Induced Polyclonal B-Cell Activation

A major implication of virus-induced polyclonal B-cell activation is the potential for immune response disease to follow. For example, antibodies reacting against self are a common feature of many viral infections [4]. During acute viral infections the autoimmune response is transient, but in persistent viral infections, there is continuous production of autoantibodies with the potential for harmful autoreactivity against cells and tissues bearing recognizable self antigens. In addition, these autoantibodies when reacting with antigen in fluids form antigen–antibody complexes. Such immune complexes may become trapped in the renal glomeruli, arteries, or choroid plexus and lead to immune complex disease. The mechanism(s) and manifestations of immune complex disease are addressed in Chapter 29.

A well-established fact is that normal individuals contain self-reactive lymphocytes but maintain tolerance by immune regulation of these cells. Chiller et al. [24] have shown that, for the induction of tolerance, T and B cells each require different concentration of antigen, and that escape from a tolerant state occurs much faster at the B-cell level than at the T-cell level. On the basis of these studies, it has been proposed that unresponsiveness to self is maintained by tolerance at the helper T-cell level. When unresponsiveness exists only in autoreactive T cells but competent autoreactive B cells are present, the unresponsive state may terminate if the self B cells become activated by nonspecific BCGF and BCDF released by virus-specific helper T cells. With the availability of cloned virus-specific helper T cells and well-defined models of immunologic tolerance, this hypothesis is testable.

Recent studies employing monoclonal antibodies have shown that certain viral proteins share common antigenic determinants with normal host constituents [25]. This mimicking of host antigen by microbial antigens is referred to as molecular mimicry (see Chapter 30). If a viral antigen is similar to normal host antigen, the immune response to this antigen should be weak since the host is tolerant to the "self" antigen. Thus, molecular mimicry *alone* may not result in autoimmunity. Nevertheless, antibody response against these "self" determinants is made following infection with viruses that have cross-reacting proteins. A mechanism to explain the antibody response against such "self" determinants is outlined in Fig. 2. In this model, tolerance to the "self" determinant is broken by help from virus-specific T helper cells, thus bypassing the requirement for autoreactive T cells. The role that virus-specific T helper cells play in breaking tolerance to shared "self" determinants would be quite different from their role in the activation of bystander B cells described in Fig. 1. The activation of "primed" bystander B cells is mediated primarily by the action of the soluble lymphokines BCGF and BCDF. On the other hand, the production of antibody against shared "self" determinants involves a hapten-carrier-like

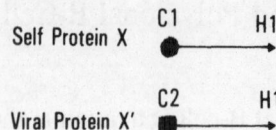

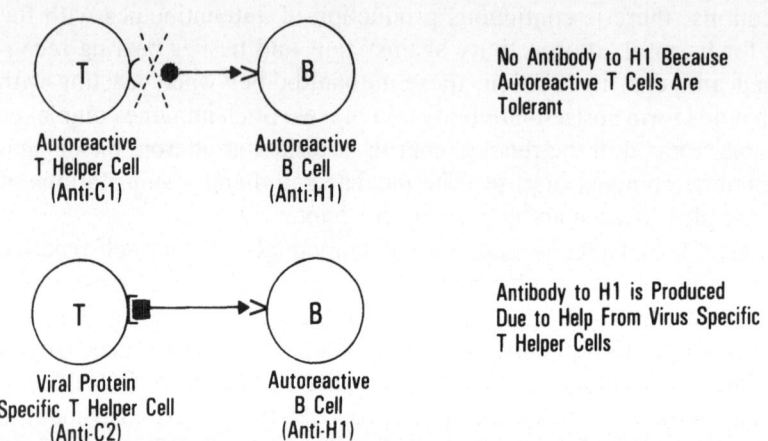

Fig. 2. Breaking tolerance to shared "self" determinants by help from virus-specific helper T cells. Self protein X has the carrier determinant C1 and hapten determinant H1. Viral protein X' shows cross-reactivity with the self protein and has the same hapten determinant H1, but a different carrier determinant C2. Antibody against the shared "self" H1 epitope is produced by help from virus-specific T helper cells reactive with the carrier determinant C2.

effect in addition to the action of BCGF and BCDF. This may be an important distinction, and may possibly influence the magnitude of the antibody response as well as the quality (avidity) of the antibody produced. As a general rule, effective T–B cell cooperative interactions and subsequent antibody production are elicited when the hapten-carrier determinants exist on the same molecule.

A potential benefit to the host of virus-induced polyclonal B-cell activation may be the maintenance of immunologic memory in the absence of specific antigenic stimulation. We know that immunity to certain microbial agents can persist for long periods, apparently without reexposure to the agent [19]. Stimulation of bystander memory B cells by activated virus-specific helper T cells may be one of the factors responsible for maintaining long-term immunity.

Acknowledgments. This is publication no. 2998-IMM from the Department of Immunology, Scripps Clinic and Research Foundation. This research was supported by US PHS Grants AI-09484, AI-07007, and NS-12428.

References

1. Boyd WC, Bernard H (1937) Quantitative changes in antibody and globulin fractions in sera of rabbits injected with several antigens. J Immunol 33:111–122
2. Moticka EJ (1974) The non-specific stimulation of immunoglobulin secretion following specific stimulation of the immune system. Immunology 27:401–412
3. Rosenberg YJ, Chiller JM (1979) Ability of antigen-specific helper cells to effect a class-restricted increase in total Ig-secreting cells in spleens after immunization with the antigen. J Exp Med 150:517–530
4. Hirsh RS, Proffitt MR (1975) Autoimmunity in viral infections. In Notkins AL (ed) Viral Immunology and Immunopathology. Academic Press, New York, p 419
5. Waldmann TA, Broder S (1982) Polyclonal B cell activators in the study of the regulation of immunoglobulin synthesis in the human system. Adv Immunol 32:1–63
6. Bird AG, Britton S (1979) A new approach to the study of human B lymphocyte function using an indirect plaque assay and a direct B cell activator. Immunol Rev 45:41–67
7. Butchko GM, Armstrong RB, Martin WJ, Ennis FA (1978) Influenza A viruses of the H2N2 subtype are lymphocyte mitogens. Nature 271:66–67
8. Goodman-Snitkoff GW, McSharry JJ (1980) Activation of mouse lymphocytes by vesicular stomatitis virus. J Virol 35:757–765
9. Mochizuki D, Hedrick S, Watson J, Kingsbury DT (1977) The interaction of herpes simplex virus with murine lymphocyte. I. Mitogenic properties of herpes simplex virus. J Exp Med 146:1500–1510
10. Gibson M, Tiensiwakul P, Khoobyarian N (1982) Adenovirus fiber protein (FP) functions as a mitogen and an adjuvant. Cell Immunol 73:397–403.
11. Goodman-Snitkoff G, McSharry JJ (1982) Mitogenic activity of sindbis virus and its isolated glycoproteins. Infect Immun 38:1242–1248
12. Wardley RC (1982) Effect of African swine fever on lymphocyte mitogenesis. Immunology 46:215–220
13. Armstrong RB, Butchko GM, Kiley SC, Phelan MA, Ennis FA (1981) Mitogenicity of influenza hemagglutinin glycoproteins and influenza viruses bearing H2 hemagglutinin. Infect Immun 34:140–143
14. Goodman-Snitkoff G, Mannino RJ, McSharry JJ (1981) The glycoprotein isolated from vesicular stomatitis virus is mitogenic for mouse B lymphocytes. J Exp Med 13:1489–1502
15. Stanbridge EJ, Weiss RL (1978) Mycoplasma capping on lymphocytes. Nature 276:583–587
16. Mochizuki D, Watson J, Kingsbury DT (1981) Comparative mitogenic effects of herpes simplex virus and mycoplasma on murine lymphocytes. J Gen Virol 54:185–190
17. Gershon RK, Kondo K (1970) Cell interactions in the induction of tolerance: The role of thymic lymphocytes. Immunology 18:723–737
18. Minato N, Katsura Y (1978) Virus-replicating T cells in the immune response of mice. III. Role of vesicular stomatitis virus-replicating T cells in the antibody response. J Exp Med 148:850–861
19. Mims CA (1982) The pathogenesis of infectious disease. Academic Press, London

20. Dutton RW (1975) Separate signals for the initiation of proliferation and differentiation in the B cell response to antigen. Transplant Rev 23:66–77
21. Altman A, Katz DH (1982) The biology of monoclonal lymphokines secreted by T cell lines and hybridomas. Adv Immunol 33:73–166
22. Muraguchi A, Butler JL, Kehre JH, Fauci AS (1983) Differential sensitivity of human B cell subsets to activation signals delivered by anti-μ antibody and proliferative signals delivered by a monoclonal B cell growth factor. J Exp Med 157:530–546
23. Howard M, Paul WE (1983) Regulation of B cell growth and differentiation by soluble factors. Ann Rev Immunol 1:307–333
24. Chiller JM, Habicht GS, Weigle WO (1971) Kinetic differences in unresponsiveness of thymus and bone marrow cells. Science 171:813–815
25. Fujinami RS, Oldstone MBA, Wroblewska Z, Frankel ME, Koprowski H (1983) Molecular mimicry in virus infection: Cross-reaction of measles virus phosphoprotein or of herpes simplex virus protein with human intermediate filaments. Proc Natl Acad Sci USA 80:2346–2350

Evolving Concepts in Viral Pathogenesis Illustrated by Selected Animal Models

34. Virus-Induced Diabetes Mellitus
    ABNER LOUIS NOTKINS AND JI-WON YOON 241

35. Herpesvirus-Induced Atherosclerosis
    CATHERINE G. FABRICANT 248

36. Retrovirus-Induced Arthritis
    TRAVIS C. MCGUIRE....................................... 254

37. Virus-Induced Demyelination
    PETER W. LAMPERT AND MOSES RODRIGUEZ.................. 260

38. Virus Can Alter Functions Without Causing Cell Pathology: Disorder Functions Leads to Imbalance of Homeostasis and Disease
    MICHAEL B. A. OLDSTONE 269

CHAPTER 34
Virus-Induced Diabetes Mellitus

ABNER LOUIS NOTKINS AND JI-WON YOON*

Insulin-dependent diabetes mellitus (IDDM), also known as juvenile diabetes, results from destruction of the insulin-producing beta cells in the pancreas [1]. Both viruses and autoimmunity have been suggested as possible causes. The evidence for viruses comes largely from experiments in animals, but several studies in humans point to viruses as an occasional trigger of this disease.

In experimental animals, viruses can induce diabetes by several different mechanisms. First, the virus may infect and directly destroy beta cells. The most thoroughly studied model is encephalomyocarditis (EMC) virus in mice. Destruction of beta cells by this virus leads to hypoinsulinemia, and this results in diabetes, which is characterized by hyperglycemia, glycosuria, polydipsia, and polyphagia. The development of EMC-induced diabetes is dependent on two important factors: the genetic background of the host, and the genetic makeup of the virus. Only certain inbred strains of mice are susceptible to EMC-induced diabetes [2]. By choosing appropriate susceptible (e.g., SJL) and resistant (e.g., C57B1/6) strains, it was found that susceptibility is inherited as an autosomal recessive trait, and that a single locus plays a major role in controlling susceptibility. The reason that susceptible strains, as compared to resistant strains, develop diabetes is because the virus infects and destroys more of their beta cells. In fact, between 10 and 100 times more virus is found in the islets of susceptible mice [3]. Why certain strains of mice are more susceptible is not known, but differences in the number or type of viral receptors (Chapter 14, this volume) on the surface of beta cells or differences in the sensitivity of beta cells to the antiviral actions of interferon have been put forth as possible explanations.

---

* National Institutes of Health, Bethesda, Maryland.

The genetic makeup of the virus is also important. The original virus pool that we used contained two variants: one, designated D, which is highly diabetogenic, and the other, designated B, which is nondiabetogenic [4]. When these two variants are mixed together and inoculated into mice, the nondiabetogenic B variant markedly inhibits the development of diabetes by the D variant. This appears to be due to the greater interferon-inducing capacity of the B variant, which protects beta cells from infection by the D variant.

The separation of the EMC virus pool into the B and D variants has made it possible to ask a number of questions about the pathogenesis of virus-induced diabetes. First at the molecular level, what makes these two viruses different? Thus far, it has not been possible to distinguish them by hyperimmune sera or by nucleic acid hybridization [5]. However, RNA fingerprints revealed that at least one oligonucleotide (20 to 25 nucleotides long) is missing from the B variant as compared to the D variant.

The separation of these variants also has made it possible to look for some of the long-term complications of diabetes. In the absence of the B variant, the D variant produces much more severe and prolonged diabetes [4]. Animals infected with the D variant and that are diabetic for 4 to 6 months show some of the same long-term complications as humans with diabetes mellitus. These include diffuse and nodular type of glomerulosclerosis, thickening of the capillary basement membrane, and early retinal changes [6]. In addition, the isolation of these variants has made it possible to develop a vaccine. Since the B and D variants are antigenically very similar, immunization of mice with the nondiabetogenic B variant completely prevents them from developing diabetes when exposed months later to the diabetogenic D variant [7]. Thus, in mice, EMC virus causes both the acute and long-term complications of diabetes, and the disease can be prevented by use of a live, attenuated vaccine.

Other viruses also can induce diabetes in mice. Members of the Coxsackie B virus group typically infect and destroy the acinar tissue of the pancreas while sparing the islets of Langerhans. If the viruses are passaged a number of times in beta cell cultures, the resulting viruses show a shift in tropism from acinar cells to beta cells and cause a mild and transient type of diabetes [8,9]. Coxsackievirus-induced diabetes is also dependent on the genetic background of the host and only certain inbred strains of mice develop glucose abnormalities [8]. Moreover, recent studies show that there are many antigenic variants of Coxsackie B4 virus. Some of these variants differ from each other by more than a dozen epitopes. It has been suggested that these antigenic changes may influence the tissue tropism of the virus, and this could result in different diseases [10].

A second mechanism by which viruses can trigger diabetes is through cumulative environmental insults. In the animal model, the severity of diabetes is dependent upon the degree of beta cell damage. In some cases, the virus may destroy too few beta cells to result in clinical diabetes. If, however, the animal is first treated with a subdiabetogenic dose of a beta cell toxin such as streptozotocin, and then infected with a virus that normally produces little or no diabetes, the cumulative insults result in diabetes [11]. By this technique, strains of mice that are ordinarily resistant to EMC-induced diabetes have been made diabetic, and viruses that do not generally cause diabetes have made animals hyperglycemic. These studies suggest that animals and patients with a depleted beta cell reserve, for whatever reason, are more likely to develop diabetes when exposed to a beta tropic virus.

A third mechanism by which a virus might cause diabetes is through the establishment of an infection that shuts off the "luxury functions" of a cell [12]. Recently Oldstone and colleagues (personal communication) showed that lymphocytic choriomeningitis (LCM) virus produces a persistent infection in mouse beta cells. Viral antigens were found in beta cells months after inoculation of the virus, insulin levels in sera were decreased, blood glucose levels were elevated, and glucose tolerance tests were abnormal. Since the regenerative capacity of beta cells is thought to be poor, a slow or persistent infection might also result in a gradual reduction in the number of functioning beta cells not dissimilar to the loss of neurons in animals infected with slow viruses [13].

Viruses can produce other alterations in luxury functions. Hamsters infected with Venezuelan encephalitis (VE) virus showed viral antigens in their beta cells during the first week of the infection and glucose abnormalities for almost three weeks. Although there is no evidence that this virus produces a persistent infection, decreased plasma insulin levels were apparent for almost three months. Moreover, morphologically intact islets isolated from these animals showed a marked decrease in glucose-stimulated insulin release for as long as 8 months after infection [14]. The cause of these virus-induced changes is not clear, but there seems to be an abnormality in the generation of cyclic AMP. Thus, the LCM and VE systems are examples of islets that are morphologically normal, but are functionally abnormal.

A fourth mechanism by which a virus might cause diabetes is through the triggering of an autoimmune response. Until very recently, it was generally thought that IDDM had an acute or subacute onset. This was one of the prime reasons for thinking that acute lytic viral infections might be involved in the etiology of this disease. Over the last few years, evidence has accumulated showing both a strong HLA (e.g., DR3 and DR4) association and an autoimmune component in patients with IDDM [15]. Autoantibodies that react with pancreatic islets have been found in many newly diagnosed patients [16]. Moreover, recent studies have shown that the appearance of islet cell antibodies may actually precede the development of clinical diabetes, in at least some patients, by many months or even years [17]. Thus, the emphasis is beginning to switch from viruses that produce acute-onset diabetes to viruses that may initiate a slower-onset disease or an autoimmune process. Evidence that viruses can trigger an autoimmune process comes from studies in mice with reovirus type 1. This virus induces an autoimmune polyendocrine disease characterized by a mild and transient form of diabetes and retarded growth [18]. Autoantibodies are found in the sera that react with several endocrine organs including the islets of Langerhans. Immunosuppression prevents the development of this disease [19]. Other viruses are also known to trigger the production of autoantibodies, and the mechanisms by which they do this are discussed elsewhere (Chapter 30, this volume).

Although there is now substantial evidence that viruses can induce diabetes in animals, the evidence that viruses cause diabetes in humans is more circumstantial. First, there are case reports, going back well over 100 years, showing a temporal relationship between the onset of certain viral infections (e.g., mumps, Coxsackie) and the subsequent development of diabetes. However, since these viral infections are particularly common in children, and since approximately one in 1000 children comes down with diabetes, the relationship between these two events may simply be fortuitous. Second, some, but not other, seroepidemiologic studies have shown a higher titer and frequency of antibodies to Coxsackie B viruses, especially B4, in newly diagnosed diabetics as compared with controls [20,21]. Because the frequency of coxsackievirus infections varies with the season and locale, a definitive relationship between the infection and diabetes has been difficult to establish. Third, some support for the seroepidemiologic data comes from case reports in which infectious virus was isolated or viral antigens identified in tissues. In one of the cases, a ten-year-old boy died of what appeared to be an acute-onset diabetes. At autopsy, the histopathology of the pancreas looked very much like that of virus-infected animals. A Coxsackie B4 virus was isolated from the child's pancreas. A rise in antibody titer to the virus was demonstrated, and when the virus was inoculated into appropriate strains of mice, they came down with diabetes [22]. In a second case of acute-onset diabetes, a Coxsackie B5 was isolated from the stools, a rise in antibody titer was demonstrated, and in certain strains of mice the virus produced abnormal glucose tolerance tests [23]. In a third case, a child came down with diabetes and myocarditis several weeks after surgery for an atrial septic defect. The serum contained a high titer of antibody to Coxsackie B5, and, by immunofluorescence, viral antigens were found in the islets [24]. Since abrupt-onset cases of this type are rare, a fourth approach was to look for beta cell damage in the pancreases of children who had died of well-documented, overwhelming viral infections, but who did not necessarily have diabetes [25]. Beta cell damage and acute inflammatory changes were demonstrated in the pancreases of a number of children with Coxsackie B infections and viral inclusion bodies were found in the islets of children with cytomegalovirus infections. Taken together, these reports suggest that under certain circumstances acute viral infections can lead to beta cell damage and sometimes to IDDM. However, whether the virus infection is capable of producing diabetes in individuals with a normal beta cell reserve or whether the virus triggers the disease only in individuals who already have a depleted beta cell reserve, perhaps as a result of autoimmune destruction, is still not known.

Persistent viral infections may also lead to diabetes in humans. Studies from Australia on 44 patients with congenital rubella showed that close to 20% developed abnormal glucose tolerance tests or diabetes [26]. Recent studies from New York on over 150 patients with congenital rubella showed that approximately 12% developed abnormal glucose tolerance tests or diabetes ([27] and unpublished data). Since the onset of diabetes occurs from many months to years after birth, the mechanism by which the infection leads to diabetes is not clear. It is not known whether the virus damages beta cells, triggers an autoimmune response, or alters glucose regulatory mechanisms.

In conclusion, based on the experiments in animals and the studies in humans, especially with coxsackieviruses and rubella virus, it would appear that at least an occasional case of IDDM can be triggered by a viral infection. Whether viruses are anything more than a minor cause of this disease in humans is not known. If a virus is a major cause, that virus has not yet been isolated or identified. Thus far, most efforts have been directed toward finding a virus that produces an acute lytic infection. Perhaps this has been too restrictive an approach. The new information, largely from studies on islet-cell antibodies, leads us to believe that at least in some individuals there is a long or chronic phase before the clinical symptoms of diabetes become apparent. If this is the case, efforts should be made to look for viruses that produce cumulative insults, or viruses that produce slow or persistent infections that alter luxury functions, or viruses that trigger an autoimmune response. Even if the pathogenesis of IDDM in humans turns out to be primarily autoimmune, the question of what triggers the autoimmune response remains to be answered. Are diseases such as IDDM, systemic lupus erythematosus, and rheumatoid arthritis solely genetic in origin with preprogrammed autoimmune abnormalities, or are the autoimmune abnormalities triggered by an environmental agent such as a virus? Both possibilities seem viable, and must be explored.

References
1. Notkins AL (1979) The causes of diabetes. Scient Am 241:62–73
2. Onodera T, Yoon JW, Brown KS, Notkins AL (1978) Evidence for a single locus controlling susceptibility to virus-induced diabetes mellitus. Nature 274:693–696
3. Yoon JW, Notkins AL (1976) Virus-induced diabetes mellitus. VI. Genetically determined host differences in the replication of encephalomyocarditis virus in pancreatic beta cells. J Exp Med 143:1170–1185
4. Yoon JW, McClintock PR, Onodera T, Notkins AL (1980) Virus-induced diabetes mellitus. XVIII. Inhibition by a non-diabetogenic variant of encephalomyocarditis virus. J Exp Med 152:878–892
5. Ray UR, Aulakh GS, Schubert M, McClintock PR, Yoon JW, Notkins AL (1983) Virus-induced diabetes mellitus: XXV. Differences in the RNA fingerprints of

diabetogenic and non-diabetogenic variants of encephalo myocarditis virus. J Gen Virol 64:947–950

6. Yoon JW, Rodrigues MM, Currier C, Notkins AL (1982) Long-term complications of virus-induced diabetes mellitus in mice. Nature 296:566–569
7. Notkins AL, Yoon JW (1982) Virus-induced diabetes in mice prevented by a live attenuated vaccine. N Engl J Med 306:486
8. Yoon JW, Onodera T, Notkins AL (1978) Virus-induced diabetes mellitus. XV. Beta cell damage and insulin-dependent hyperglycemia in mice infected with coxsackie virus B4. J Exp Med 148:1068–1080
9. Toniolo A, Onodera T, Jordan G, Yoon JW, Notkins AL (1982) Virus-induced diabetes mellitus: Glucose abnormalities produced in mice by the six members of the Coxsackie B virus group. Diabetes 31:496–499
10. Prabhakar B, Haspel MV, McClintock PR, Notkins AL (1982) High frequency of antigenic variants among naturally occurring human Coxsackie B4 virus isolates identified by monoclonal antibodies. Nature 330:374–376
11. Toniolo A, Onodera T, Yoon JW, Notkins AL (1980) Induction of diabetes by cumulative environmental insults from viruses and chemicals. Nature 288:383–385
12. Oldstone, MBA, Sinha YN, Blount P, Tishon A, Rodriguez M, Von Wedel R, Lampert PW (1982) Virus-induced alterations in homeostasis: Alterations in differentiated functions in infected cells in vitro. Science 218:1125–1127
13. Gajdusek DC (1978) Slow infections with unconventional viruses. Harvey Lect Ser 72:283–353
14. Rayfield EJ, Seto Y, Walsh S, McEvoy RC (1981) Virus-induced alterations in insulin release in hamster islets of Langerhans. J Clin Invest 68:1172–1181
15. Nerup J, Lernmark A (1981) Autoimmunity in insulin-dependent diabetes mellitus. Am J Med 70:135–141
16. Dobersen JJ, Scharff JE, Ginsberg-Fellner F, Notkins AL (1980) Cytotoxic autoantibodies to beta cells in the serum of patients with insulin-dependent diabetes mellitus. N Engl J Med 303:1493–1498
17. Gorsuch AN, Spencer KM, Lister S, McNally JM, Dean BM, Bottazzo GF, Cudworth AG (1983) Evidence for a long prediabetic period in Type 1 (insulin-dependent) diabetes mellitus. Lancet ii:1363–1365
18. Onodera T, Toniolo A, Ray UR, Jenson AB, Knazek RA, Notkins AL (1981) Virus-induced diabetes mellitus. XX. Polyendocrinopathy and auto immunity. J Exp Med 153:1457–1473
19. Onodera T, Ray UR, Melez KA, Suzuki H, Toniolo A, Notkins AL (1982) Virus-induced diabetes mellitus: Autoimmunity and polyendocrine disease prevented by immunosuppression. Nature 297:66–68
20. Gamble DR (1980) The epidemiology of insulin-dependent diabetes with particular reference to the relationship of virus infection to its etiology. Epidemiol Rev 2:49–70
21. King ML, Shaikh A, Bidwell D, Voller A, Banatvala JE (1983) Coxsackie B virus specific IgM responses in children with insulin-dependent (juvenile-onset, type 1) diabetes mellitus. Lancet 1:1397–1399
22. Yoon JW, Austin M, Onodera T, Notkins AL (199) Virus-induced diabetes mellitus: Isolation of a virus from the pancreas of a child with diabetic ketoacidosis. N Engl J Med 300:1173–1179
23. Champsaur HR, Bottazzo GF, Bertrams J, Assan R, Bach C (1982) Virologic,

immunologic and genetic factors in insulin-dependent diabetes mellitus. J Pediat 100:15–20
24. Gladisch R, Hofmann W, Waldherr R (1976) Myocarditis and insulitis in coxsackievirus infection. J Kardiol 65:837–849
25. Jenson AB, Rosenberg HS, Notkins AL (1980) Pancreatic islet cell damage in children with fatal viral infections. Lancet ii:354–358
26. Menser MA, Forrest JM, Bransby RD (1978) Rubella infection and diabetes mellitus. Lancet 1:57–60
27. Rubinstein P, Walker ME, Fedun B, Witt ME, Cooper LZ, Ginsberg-Fellner F (1982) The HLA system in congenital rubella patients with and without diabetes. Diabetes 31:1088–1091

CHAPTER 35
Herpesvirus-Induced Atherosclerosis

CATHERINE G. FABRICANT*

Our studies have established that a herpesvirus infection leads to atherosclerosis strikingly similar to the human arterial disease in a normocholesterolemic animal model [1–3]. This finding has introduced two important new concepts in the pathogenesis of atherosclerosis. (1) A herpesvirus may have a primary etiological role in a disease considered for many years to be degenerative or metabolic in origin. (2) Virus-induced alteration of arterial smooth muscle cell lipid metabolism may cause lipid accumulations in atherosclerotic lesions.

We hypothesized a link between herpesvirus infections and the pathogenesis of atherosclerosis in 1973 and 1975 [4,5]. This hypothesis resulted from observations that a variety of cell cultures infected with a feline herpesvirus accumulated considerable quantities of lipid droplets and cholesterol crystals [5,6]. These observations suggested that infections with herpesviruses might induce atherosclerosis by altering cellular lipid metabolism [4,5].

The Paterson and Cottral (1950) report supported our hypothesis and was of particular interest. In this report, these investigators associated neurolymphomatosis (NL) with an increased incidence of arteriosclerosis in chickens [7]. At that time, NL was considered a transmissible neoplastic disease peculiar to chickens. Therefore, Paterson and Cottral concluded that their findings were not relevant to the human arterial disease [7].

The establishment of Marek's disease herpesvirus (MDV) as the etiologic agent of NL in 1967 [8] and our hypothesis suggested that these findings and conclusion merited reinvestigation. Our initial experiments were based on Paterson's experimental design [7]. However, we used a purified MDV strain of known virulence and specific pathogen-free (SPF) chickens of

---

* New York State College of Veterinary Medicine, Ithaca, New York.

known genetic susceptibility to MDV infection. All groups of MDV-infected and uninfected SPF chickens were initially fed a cholesterol-poor diet. For the last half of the experimental period, one half of the infected and uninfected groups of chickens were placed on a cholesterol-supplemented diet. The corresponding infected and uninfected groups remained on the cholesterol-poor diet [1,3].

Results of these experiments established a direct link between the herpesvirus infection and atherosclerosis [1,3]. MDV-infected chickens in both hypercholesterolemic and normocholesterolemic groups developed atherosclerosis. However, uninfected hypercholesterolemic chickens or normocholesterolemic chickens did not develop this arterial disease. Grossly visible frequency occlusive arterial lesions were found in some virus-infected chickens of both groups. These lesions were found in thoracic aortas, main arterial branches, as well as in medium and large coronary arteries. Microscopic arterial lesions were similarly distributed in both groups of infected chickens. These lesions were proliferative and fatty proliferative with fibrous caps overlying areas of atheromatous change. More arterial lesions in hypercholesterolemic infected chickens appeared to be fatty proliferative suggesting a synergy between virus infection and cholesterol supplementation. However, appreciable lipid accumulations were observed in arterial lesions of the virus-infected normocholesterolemic chickens. The MDV-induced arterial lesions closely resembled human atherosclerosis in character and distribution of lesions [1,3]. These results were reproducible in repeated experiments [2].

Pathogenesis experiments of similar design established the time and sequence of arterial lesion development [2]. Grossly visible arterial lesions were found 3 months after infection. On the other hand, microscopic arterial lesions were found 1 month after infection of normocholesterolemic chickens. These microscopic lesions were found in brachiocephalic arteries and in aortal segments proximal to the heart. By the third month, an appreciable number of infected chickens in this group had arterial lesions distributed as described in the initial experiments. At 7 months, there was a marked increase in the number of chickens with lesions in these arterial beds. The results indicated that practically all the infected normocholesterolemic chickens had one or more of these arterial lesions [2].

In contrast, uninfected hypercholesterolemic SPF chickens at 7 months had no microscopic arterial lesions in arterial segments where MDV-infected normocholesterolemic or hypercholesterolemic chickens had numerous lesions [2]. As in our early experiments, there appeared to be a synergy between virus infection and cholesterol supplementation. However, this effect did not appear consistent at every period after cholesterol supplementation [2].

In the early experiments, viral-specific internal antigens (VIA) were found in arterial walls of a few birds by immunofluorescence microscopy [1,3]. These antigens were first found at 1 month post-infection in the pathogenesis experiments and, thereafter, throughout the time-span of arterial lesion development [2]. The VIA were located in focal areas in medial layers of arteries near lesions. These antigens appeared to be in smooth muscle cells (SMC) of this arterial layer [2].

The absence of MDV-VIA in SMC of arterial lesions was not unexpected and did not exclude infection in these cells. Cells containing VIA are productively infected and would be expected to die [9]. Although arterial injury may initiate arterial lesions [10] the evolution and progression of the atherosclerotic plaque depends upon proliferation of altered SMC. In our *in vitro* studies we observed that MDV-induced proliferation of chicken arterial SMC (CG Fabricant, 1979, unpublished observations). In a recent collaborative study, evidence of MDV genome was found in arterial lesions of our infected chickens by *in-situ* DNA hybridization confirming that SMC in these lesions were virally infected (EP Benditt, 1982, personal communication, MDV-DNA probes generously provided by N Ross, Houghton Research Poultry Station, England).

Injury to endothelial cells lining the lumen of arteries has been implicated in atherogenesis [10]. However, in MDV-induced atherosclerosis these cells are not obviously injured. Whether more subtle viral injury occurs in this arterial layer, depends upon establishing the identity and role of the few cells found adherent in some areas of the endothelium in arteries of infected chickens (Minick and Salisbury, 1982, unpublished observations).

Arterial injury mediated by immunologic means has also been implicated in atherogenesis [11]. However, no significant evidence of immune complexes was found in MDV-induced arterial lesions by immunofluorescence microscopy with IgG and C3 complement (J Fabricant *et al.*, 1980, unpublished data [2,3]). Immunologic injury at the cellular level was not excluded by this finding. The effects of such injury in the pathogenesis of virus-induced atherosclerosis are being explored.

As mentioned earlier, significant lipid accumulations were found in arterial lesions of MDV-infected normocholesterolemic chickens [1–3]. These findings were consistent with the observations that led to our hypothesis [4–6] and to this series of experiments. As a result of our *in vivo* observations, experiments were designed to assess the effects of MDV infection on lipid metabolism of cultured chicken arterial SMC [12]. Our results indicated that MDV infection induced quantitative and qualitative differences in lipid accretions in SMC cultures compared with those in uninfected SMC cultures or compared with SMC infected with turkey herpesvirus (HVT), a second avian herpesvirus [12]. Chemical analyses revealed that total lipid content was increased in MDV-infected SMC cultures compared with the uninfected SMC or HVT-infected SMC control cultures. It is of particular interest to note that the lipid increase in MDV-infected cultures was primarily attributable to increases in cholesterol and saturated cholesteryl esters [12]. These are the principal lipids reported to accumulate in human atherosclerotic lesions [13–15]. Preliminary results of chemical analyses of aortas from MDV-infected chickens were similar to the *in vitro* findings [12]. Results of more recent studies [16] indicate that the cholesteryl ester (CE) synthetic activity (ACAT) is almost 7 times greater in MDV-infected SMC cultures than in uninfected SMC. On the other hand, the acid cytosolic CE hydrolytic (ACEH) activity was 3 times greater in the uninfected SMC than in the virus-infected SMC. These findings suggest that increased CE biosynthesis and decreased CE hydrolysis result in the CE accumulations in MDV-infected SMC cultures [16].

Similar mechanisms may contribute to CE accumulations in arterial lesions of herpesvirus-infected chickens. In addition, our results indicate that MDV-induced altered arterial SMC lipid metabolism may have a major role in the pathogenesis of MDV-induced atherosclerosis.

In other experiments, we tested the protective effects of HVT immunization against MDV-induced atherosclerosis. The HVT vaccine has been used successfully to prevent MDV-induced tumors in commercial chicken flocks. Results of these experiments established that immunization with HVT also protects against MDV-induced atherosclerosis [17,18]. These findings established an additional important link associating MDV specifically with the development of atherosclerosis in SPF chickens infected with the virus.

Our findings that MDV infection causes atherosclerosis in pathogen-free normocholesterolemic chickens have far reaching implications. As mentioned earlier, they have introduced new concepts that may be important in the pathogenesis of atherosclerosis. These include: (1) a primary herpesvirus etiology for this disease, and (2) that the herpesvirus infection causes cholesterol and cholesteryl ester accumulations in arterial lesions.

These results and concepts are relevant to other animal models of atherosclerosis research. Animal models in current use for such studies are known to be widely and persistently infected with herpesviruses [19,20]. Our findings suggest that interpretation of experimental results obtained from use of animals (or tissues thereof) not known to be free of infections with these viruses are now subject to question and may be difficult to evaluate.

In addition, our findings may be especially relevant to human atherosclerosis because humans are widely and persistently infected with up to five herpesviruses [21]. These viruses include: herpes simplex type 1 (HSV-1), herpes simplex type 2 (HSV-2), Varicella zoster (VZ), Epstein–Barr virus (EBV), and cytomegalovirus (CMV). One or more of these viruses may be implicated in causing atherosclerosis.

A recent report supports this suggestion as well as the relevance of our findings to human atherosclerosis [22,23]. Evidence of HSV (type undetermined) RNA was found by *in-situ* hybridization with viral probes in 13 of more than 100 human aortal specimens obtained during coronary bypass surgery. As reported, some of the specimens positive for HSV appeared to be in the early stages of atherogenesis. These investigators did not find evidence of EBV or CMV in the specimens examined. They suggested that some atherosclerotic plaques may be initiated and/or the lesion progression enhanced by HSV infection [22,24].

Finally, when a herpesvirus role can be established for human atherosclerosis, our findings may be important for the eventual control of this disease. The demonstrated protective effects of HVT immunization against MDV-induced atherosclerosis suggests that similar procedures or antiviral therapy may prevent this human cardiovascular disease.

Our recent experiments concerned the effect of three human herpesvirus infections on human fetal arterial SMC in culture. Results of these experiments indicated that in these cultures HSV types 1 and 2 infections were lytic; whereas, CMV infections induced cell proliferation and tended to be latent [25]. Melnick et al. recently reported finding evidence of CMV antigens in human arterial SMC cultured from surgical specimens with or without evidence of atherosclerosis [26]. This report and our in vitro results suggest that CMV may have an atherogenic potential.

References
1. Fabricant CG, Fabricant J, Litrenta MM, Minick CR (1978) Virus-induced atherosclerosis. J Exp Med 148:340–345
2. Fabricant CG, Fabricant J, Minick CR, Litrenta MM (1980) Herpesvirus induced atherosclerosis. Cold Spring Harbor Conf. Cell Prolif 7:1251–1258
3. Minick CR, Fabricant CG, Fabricant J, Litrenta MM (1979) Atheroarteriosclerosis induced by infection with a herpesvirus. Am J Pathol 96:673–706
4. Fabricant CG (1975) Herpesvirus induced cholesterol—an added dimension in the pathogenesis, prophylaxis, or therapy of atherosclerosis. Artery 1:361
5. Fabricant CG, Krook L, Gillespie JH (1973) Virus-induced cholesterol crystals. Science 181:566–567
6. Fabricant CG, Gillespie JH (1974) The identification and characterization of a second feline herpesvirus. Infect Immun 9:460–466
7. Paterson JC, Cottral GE (1950) Experimental coronary sclerosis. III. Lymphomatosis as a cause of coronary sclerosis in chickens. Arch Pathol 49:699–707
8. Churchill AE, Biggs PM (1967) Agent of Marek's disease in tissue culture. Nature 215:528–530
9. Roizman B (1972) The biochemical features of herpesvirus-infected cells particularly as they relate to their potential oncogenicity. In Biggs PM, de Thé G, Payne N (eds) Oncogenesis and Herpesvirus. Elsevier, New York, p 1–17
10. Ross R, Glomset MD (1976) The pathogenesis of atherosclerosis. N Engl J Med 295:369–377
11. Minick CR (1976) Immunologic arterial injury in atherogenesis. Ann NY Acad Sci 275:210–227
12. Fabricant CG, Hajjar DJ, Minick CR, Fabricant J (1981) Herpesvirus infection enhances cholesterol and cholesteryl ester accumulation in cultured arterial smooth muscle cells. Am J Pathol 105:176–184
13. Day AJ, Wahlquist ML (1970) Cholesteryl ester and phospholipid composition of normal aortas and atherosclerotic lesions in children. Exp Mol Pathol 13:199–216

14. Portman OW (1970) Arterial composition and metabolism: Esterified fatty acids and cholesterol. Adv Lipid Res 8:41–114
15. Smith EB (1974) The relationship between plasma and tissue lipids in human atherosclerosis. Adv Lipid Res 12:1–49
16. Fabricant CG, Hajjar DJ, Minick CR, Fabricant J (1983) Herpesvirus alters cholesteryl ester metabolism in arterial smooth muscle cells. Fed Proc 43(3):501
17. Fabricant CG, Fabricant J, Minick CR, Litrenta MM (1983) Herpesvirus-induced atherosclerosis in chickens. Fed Proc 42:2476–2479
18. Fabricant J, Fabricant CG, Litrenta MM, Minick CR (1981) Vaccination prevents atherosclerosis induced by Marek's disease herpesvirus. Fed Proc 40:331
19. Barahona H, Melendez LV, Melnick JL (1974) A compendium of herpesviruses isolated from non-human primates. Intervirology 3:175–192
20. McKercher D (1973) Viruses of other vertebrates. *In* Kaplan AS (ed) The Herpesviruses. Academic Press, New York, p 427
21. Nahmias AJ, Dowdle WR, Schinazi RF (eds) (1980) The human herpesviruses. Elsevier, New York
22. Barrett TB, McDougall JK, Benditt EP (1983) Herpesviruses and atherosclerotic lesions in humans. Fed Proc 43(3):501
23. Benditt EP, Benditt JM (1973) Evidence for a monoclonal origin of human atherosclerotic plaques. Proc Natl Acad Sci USA 70:1753–1756
24. Benditt EP, Barrett T, McDougall JK (1983) Viruses in the etiology of atherosclerosis. Proc Natl Acad Sci USA 80:6386–6389
25. Fabricant CG, Schat KA, and Fabricant J. Infections of human smooth muscle cell cultures suggest cytomegalovirus may have a greater atherogenic potential than herpes simplex types 1 or 2. Submitted for publication.
26. Melnick JL, Dreesman GR, McCollum CH, Petrie BL, Burek J, DeBakey ME (1983) Cytomegalovirus antigen within human arterial smooth muscle cells. Lancet ii: 644–647.

CHAPTER 36
Retrovirus-Induced Arthritis

TRAVIS C. MCGUIRE*

Caprine Arthritis-Encephalitis Virus Infection

Initial experiments with caprine arthritis-encephalitis virus (CAEV) demonstrated that leukoencephalomyelitis occurring in young kids of a goat herd could be transmitted with 220-nm filtrates of tissue suspensions [1]. A virus was subsequently isolated by explantation of synovial membrane from an arthritic adult goat from the same herd [2]. This virus replicated in fetal goat synovial cells and caused both arthritis and encephalitis when inoculated into Caesarean-derived, specific pathogen-free goat kids [2]. Evaluation of the physical and biochemical properties of the virus resulted in its classification as a retrovirus [2–4]. Immunologic and morphologic studies indicated similarities among CAE, maedi-visna, and progressive pneumonia viruses of sheep [2,5,6]. The primary immunologic similarity was cross reaction of the major structural protein of the viruses [5,7] even though cross reactions with other viral proteins occurred [6]. Competitive inhibition of hybridization, however, demonstrated that CAEV exhibited less than 20% genome sequence homology with visna and progressive pneumonia viruses and was therefore distinct from these ovine lentiviruses [7].

A survey of serum samples from over 1000 goats in the United States indicated that 81% had antibody to CAEV [8]. Even though the virus persists in infected goats and can be isolated long after infection [2,8,9], clinical disease did not occur in all animals. The percentage of CAEV-infected goats that had signs of arthritis varied among goat herds and was usually much less than 25% [8]. The primary mode of virus transmission is the ingestion of colostrum and milk from infected females by susceptible kids [10]. Other

---

* Washington State University, Pullman, Washington.

ways of transmission occur, including, for instance, horizontal transmission between infected and noninfected weaned goats [10]. The mode of horizontal transmission in these older goats is not known.

Rheumatoid Arthritis-Like Lesions Caused by CAEV

The lesions of adult goats naturally infected with CAEV are associated with synovial-lined structures including joints, tendon sheaths, and bursae [11]. These lesions are typified by synovial cell proliferation, subsynovial mononuclear cell infiltration, fibrin, necrosis, and mineralization [11]. Other postmortem findings in chronic CAEV infection include arteritis, pericarditis, pleuritis, foreign body giant cells, amyloidosis, and nonsuppurative interstitial nephritis [11]. The morphogenesis of CAEV arthritis was determined in a sequential study of experimentally infected goat kids [12]. A mild synovial cell hyperplasia and perivascular mononuclear cell infiltration are the first detectable changes and these progress by 45 days to severe synovial cell hyperplasia, villous hypertrophy, and lymphoid follicle formation in the synovial membrane [12]. Ultrastructural changes in CAEV-infected joints involve both synovial lining cells and connective tissue fibroblasts [13]. Fibroblasts are hypertrophic during the first month of infection and then return to normal. In contrast, the synovial lining cells have hypertrophy, hyperplasia, and necrosis, which continues.

Arthritis in Other Retroviral Infections

An arthritis similar to that described for CAEV infection occurs in some natural ovine progressive pneumonia virus infections and in experimental infections of sheep with this virus [14,15].

In domestic cats, chronic progressive polyarthritis has been etiologically linked to both feline leukemia virus and feline syncytial-forming virus infections [16]. The arthritis was not reproduced in cats inoculated with cell-free synovial tissue from diseased cats or with tissue culture fluid containing both viruses [16]. Lesions in affected cats vary, but include superficial synovial infiltrate of polymorphonuclear neutrophils, and a deep synovial infiltration of lymphocytes and plasma cells. With time there is synovial hypertrophy, panus formation, and some cases have a severely erosive and deforming arthritis [16]. Rheumatoid factor cannot be detected in the serum from cats with chronic progressive arthritis.

One of the mouse models (MRL/1) of systemic lupus erythematosus spontaneously develops features similar to human rheumatoid arthritis, including IgM and IgG rheumatoid factors, and clinical polyarthritis of the hindlegs [17]. Most of these mice have a zenotropic retrovirus [18], but the causal relationship between the retrovirus and the arthritis is unknown.

Mechanisms of CAEV Persistence

The most obvious way for CAEV to persist in the host is by DNA provirus incorporation into the host-cell genome [3,4,19]. Failure by some of the provirus-infected cells to express viral antigens for a period of time would maintain a source of CAEV. Subsequent expression of infectious virus could provide a source of virus to infect susceptible cells perpetuating the infection. Neutralizing antibody, however, should prevent the spread of newly synthesized virus in the persistently infected host. In visna virus infection, virion surface glycoproteins antigenically vary, effectively preventing neutralization [20,21]. In CAEV infection, the question of antigenic variation is being investigated, although the possible importance of the phenomenon is diminished by failure to detect neutralizing antibody in sera from goats with natural infection and from goats hyperimmunized with CAEV [22]. Neutralization of CAEV *in vitro* occurs with hyperimmune rabbit sera [22]. At present, it appears that the lack of neutralizing antibody allows CAEV expression by provirus-infected cells, which explains the presence of infectious virus in the joints of chronically infected goats [23]. Since virus expression is variable over an 18-month period in individual goats [23], it appears that there are additional factors controlling virus expression such as cytotoxic T lymphocytes destroying virus-producing cells and/or cell regulation of provirus expression.

Pathogenesis of CAEV Arthritis

Goat kids inoculated with CAEV have an initial phase of virus replication in the synovial cavity followed by a period in which virus titers in individual goats fluctuate markedly [23]. For instance, 9 of 10 goats had measurable virus 1 month post-infection, 7 of 10 at 5 months, 1 of 10 at 10 months, and 3 of 10 at 12 months. During periods when free virus was undetectable in the joint fluid, cocultivation of synovial fluid cells with susceptible cells revealed the presence of CAEV. Demonstration of viral antigens in cells from the infected joint by immunofluorescence is very difficult and can only be done during the first two weeks after infection [12,23]. Particles that resemble CAEV were present in the synovial lining cells for as long as 45 days, but budding of virus from the surface of these cells was not observed [13]. Available information indicates that macrophages and other cells derived from the bone marrow mononuclear phagocyte system are the primary cells supporting CAEV replication [12,24]. In summary, CAEV persists in the joints of infected goats with low level virus expression occurring in some goats one to several years after infection. Preliminary data indicate that clinical arthritis quantitated by measuring the amount of joint swelling correlates positively with expression of CAEV in the joint [23]. This correlation is remarkable when the number of variables, such as residual joint swelling and involvement of different anatomical structures in causing the enlargement, are considered [23].

During the first month after CAEV infection, lymphocytes and macrophages increase in the synovial fluid until they exceed by tenfold the number in control animals [23]. Concurrently the immunoglobulin concentration in synovial fluid increases dramatically over controls and the increase is comprised almost exclusively of polyclonal IgG_1 [25]. Evidence for the IgG_1 being produced in the joint includes synovial IgG_1 concentrations greater than serum, lower synovial $IgG_2:IgG_1$ ratios than serum, lower synovial albumin : IgG ratios than serum, lower synovial antibody titers than serum antibody titers to systematically administered antigens and the presence of lymphocytes and plasma cells in the inflamed synovial tissues [25]. The synovial fluid IgG_1 reaches a peak by 6 months post-infection and amounts fluctuate but remain elevated for at least 3 years. A significant part of the IgG_1 in the synovial fluid is directed toward viral antigens [26]. Antibodies in synovial fluid precipitate [^{35}S]methionine-labeled CAEV proteins of 125,000, 90,000, 28,000, and 15,000 apparent molecular weights [26]. The 125,000-, 90,000-, and 15,000-molecular-weight proteins precipitated by antibody are surface glycoproteins when examined by external labeling techniques. Synovial fluid antibody titers to gp90,000 exceed those to p28,000, the most abundant viral structural protein, by 100- to 1000-fold [26]. Antibody to p28,000 persists in the serum and provides the basis of a diagnostic test [7,8]. T lymphocytes reactive with CAEV are present by 1 month post-infection and increase for at least 8 months [27], but after 2.5 years, presence of reactivity is variable [28]. No rheumatoid factor is present in either serum or synovial fluid.

In addition to the mechanisms described, which involve specific immune responses to viral antigens, studies with CAEV demonstrate that T lymphocytes of infected goats proliferate more in response to concanavalin A than those of uninfected goats [28]. Also, CAEV induces surface changes and proliferation of T lymphocytes of uninfected goats when concanavalin A is present *in vitro* [29]. These observations suggest that CAEV augments lymphocyte reactivity independent of antigen-sensitive events. Such lymphocyte augmentation could affect both CAEV and non-CAEV lymphocyte responses, enhancing inflammation.

The potential for virus expression in the persistently infected joint, the presence of a vigorous antibody response in the synovial fluid, the presence of lymphocytes in the synovial tissue and fluid, and the positive correlation between clinical disease and virus expression provide an explanation for the arthritis. Initial viral infection stimulates a local and systemic immune response. Failure of the host to remove nonexpressing provirus-infected cells allows subsequent expression of viral antigens that react with specifically sensitized lymphocytes and their products causing inflammation. A possible consequence of this view is that the more viral antigen expressed, the more severe the inflammation. Adding to this specific response is the possibility that lymphocyte responses are augmented by CAEV [29].

The pathogenesis of CAEV arthritis described is similar to antigen-induced arthritis in immunized animals. In this experimental model of arthritis, antigen given intraarticularly to systemically immunized animals causes arthritis [30]. The lesions of antigen-induced arthritis resemble those described for both CAEV and rheumatoid arthritis. Similarities of the morphologic lesions in these conditions probably reflect a common etiology of persistent antigenic stimulus and inflammation in the joint cavity. The antigen in antigen-induced arthritis is defined, while in CAEV at least a part of the antigenic stimulus is viral antigens. In rheumatoid arthritis, a part of the antigenic stimulus is IgG. Whether viral antigens are involved in rheumatoid arthritis is unknown.

References

1. Cork LC, Hadlow WJ, Crawford TB, Gorham JR, Piper RC (1974) Infectious leukoencephalomyelitis of young goats. J Infect Dis 129:134–141
2. Crawford TB, Adams DS, Cheevers WP, Cork LC (1980) Chronic arthritis in goats caused by a retrovirus. Science 207:997–999
3. Clements JE, Narayan O, Cork LC (1980) Biochemical characterization of the virus causing leukoencephalitis and arthritis in goats. J Gen Virol 50:423–427
4. Cheevers WP, Roberson S, Klevjer-Anderson P, Crawford TB (1981) Characterization of caprine arthritis-encephalitis virus: A retrovirus of goats. Arch Virol 67:111–117.
5. Narayan O, Clements JE, Strandberg JD, Cork LC, Griffin DE (1980) Biological characterization of the virus causing leukoencephalitis and arthritis in goats. J Gen Virol 50:69–79
6. Dahlberg JE, Gaskin JM, Perk K (1981) Morphologic and immunological comparison of caprine arthritis encephalitis and ovine progressive pneumonia virus. J Virol 39:914–919
7. Roberson SM, McGuire TC, Klevjer-Anderson P, Gorham JR, Cheevers WP (1982) Caprine arthritis-encephalitis virus is distinct from visna and progressive pneumonia viruses as measured by genome sequence homology. J Virol 44:755–758
8. Crawford TB, Adams DS (1981) Caprine arthritis-encephalitis: Clinical features and presence of antibody in selected goat populations. J Am Vet Med Assoc 178:713–719
9. Cork LC, Narayan O (1980) The pathogenesis of viral leukoencephalitis-arthritis of goats. I. Persistent viral infection with progressive pathologic changes. Lab Invest 42:596–602
10. Adams DS, Klevjer-Anderson P, Carlson JL, McGuire TC, Gorham JR (1983) Transmission and control of caprine-arthritis virus. Am J Vet Res, in press
11. Crawford TB, Adams DS, Sande RD, Gorham JR, Henson JB (1980) The connective tissue component of the caprine arthritis-encephalitis syndrome. Am J Pathol 100:443–454
12. Adams DS, Crawford TB, Klevjer-Anderson P (1980) A pathogenetic study of the early connective tissue lesions of viral caprine arthritis-encephalitis. Am J Pathol 99:257–278
13. Brassfield AL, Adams DS, Crawford TB, McGuire TC (1982) Ultrastructure of arthritis induced by a caprine retrovirus. Arth Rheum 25:930–936
14. Oliver RE, Gorham JR, Parish SF, Hadlow WJ, Narayan O (1981) Ovine progressive pneumonia: Pathologic and virologic studies on the naturally occurring disease. Am J Vet Res 42:1554–1559

15. Oliver RE, Gorham JR, Perryman LE, Spencer GR (1981) Ovine progressive pneumonia: Experimental intrathoracic, intracerebral, and intraarticular infections. Am J Vet Res 42:1560–1564
16. Pedersen NC, Pool RR, O'Brien T (1980) Feline chronic progressive polyarthritis. Am J Vet Res 41:522–535
17. Hang L, Theofilopoulos AN, Dixon FJ (1982) A spontaneous rheumatoid arthritis-like disease in MLR/1 mice. J Exp Med 155:1690–1701
18. Yoshiki T, Mellors RC, Strand M, August JT (1974) The viral envelope glycoprotein of murine leukemia virus and the pathogenesis of immune complex glomerulonephritis of New Zealand mice. J Exp Med 140:1011–1027
19. Klevjer-Anderson P, Cheevers WP (1981) Characterization of the infection of caprine synovial membrane cells by the retrovirus caprine arthritis-encephalitis virus. Virology 110:113–119
20. Narayan O, Griffin DE, Chase J (1977) Antigenic shift of visna virus in persistently infected sheep. Science 197:376–378
21. Scott JV, Haase AT, Narayan O, Vigne R (1979) Antigenic variation in visna virus. Cell 18:321–327
22. Klevjer-Anderson P, McGuire TC (1982) Neutralizing antibody response of rabbits and goats to caprine arthritis-encephalitis virus. Infect Immun 38:455–461
23. Klevjer-Anderson P, Adams DS, Anderson LW, Banks KL, McGuire TC (1984) A sequential study of virus expression in retrovirus induced arthritis of goats. Submitted for publication
24. Klevjer-Anderson P, Anderson LW (1982) Caprine arthritis-encephalitis virus infection of caprine monocytes. J Gen Virol 58:195–198
25. Johnson GC, Adams DS, McGuire TC (1983) Pronounced production of polyclonal IgG_1 in the synovial fluid of goats with caprine arthritis-encephalitis infection. Infect Immun 41:805–815
26. Johnson GC, Barbet AF, Klevjer-Anderson P, McGuire TC (1983) Preferential immune response to virion-surface glycoproteins by caprine arthritis-encephalitis virus. Infect Immun 41:657–665.
27. Adams DS, Crawford TB, Banks KL, McGuire TC, Perryman LE (1980) Immune response of goats persistently infected with caprine arthritis-encephalitis virus. Infect Immun 28:421–427
28. DeMartini JC, Banks KL, Greenlee A, Adams DS, McGuire TC (1983) Augmented T lymphocyte responses and abnormal B lymphocyte numbers in goats chronically infected with the retrovirus causing caprine arthritis-encephalitis. Am J Vet Res 44:2064–2069
29. Jacobs CA (1982) Increased proliferation and alteration of lectin binding receptors of concanavalin A stimulated T lymphocytes on first exposure to a nononcogenic retrovirus. Masters thesis, Washington State University, Pullman, Washington
30. Dumonde DC, Glynn LE (1962) The production of arthritis in rabbits by an immunological reaction to fibrin. Br J Exp Pathol 43:363–383

CHAPTER 37
Virus-Induced Demyelination

PETER W. LAMPERT* AND MOSES RODRIGUEZ†

Demyelination refers to a pathologic process that causes the destruction of myelin sheaths without injury to axons. According to etiology, the demyelinating diseases can be classified into genetic, nutritional, toxic, allergic, or virus-induced disorders. Clues to etiology are found by observing early ultrastructural alterations that may affect either myelin lamellae or the myelin supporting cells, i.e., oligodendrocytes or Schwann cells in the central or peripheral nervous system, respectively. Metabolic changes induced by nutritional deficiencies or intoxications may initially manifest themselves in the most distal extensions of the cytoplasm of the myelin-supporting cells, i.e., at nodes of Ranvier or at the inner glial loops of myelin sheaths (Fig. 1). In other conditions, myelin lamellae are primarily affected resulting in wide lamellar separation as seen in intoxications with hexachlorophene or triethyltin [1]. Early lysis of myelin lamellae without damage to the myelin-supporting cells is seen in allergic neuritis [2] and presumably also occurs in autoimmune encephalomyelitis in which the myelin lamellae rather than the myelin-supporting oligodendrocytes are the target of the allergic reaction. This review is concerned with virus infections of the central nervous system that cause demyelination by different mechanisms such as direct virus-induced lysis of oligodendrocytes, immune-mediated destruction of persistently infected cells, "bystander" demyelination accompanying nonspecific inflammatory reactions, and autoimmune demyelination occurring after virus-induced sensitization to myelin.

---

* University of California-San Diego, La Jolla, California.
† Mayo Clinic and Research Foundation, Rochester, Minnesota.

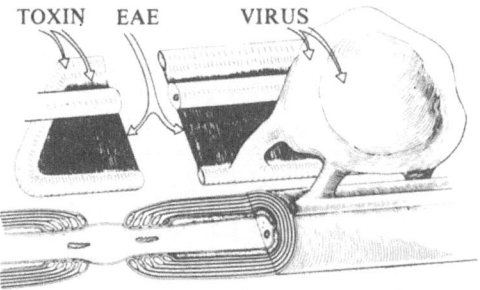

Fig. 1. Oligodendrocyte depicting connections with myelin sheaths, one of which is unwrapped to reveal the continuity of cytoplasm around compact myelin. Early ultrastructural changes provide clues to the etiology of demyelinating diseases. Nutritional deficiencies and intoxications may first affect myelin lamellae or glial cytoplasm in their most distal extensions. Virus-induced demyelination may begin with infection and lysis of oligodendrocytes or by primary injury to myelin lamellae similar to the mechanism of demyelination in experimental autoimmune encephalomyelitis (EAE).

Virus-Induced Lysis of Oligodendrocytes

This mechanism of demyelination is well documented in mice after acute infection with mouse hepatitis virus (HMV) and in humans in progressive multifocal leucoencephalopathy (PML) caused by infection with papova virus.

After inoculation with the JHM strain of MHV, a corona virus, mice develop an acute encephalomyelitis [3]. Depending on virus dosage, inoculation route, age, and strain of mice, the course of the disease can be attenuated with lesions restricted to the white matter [4]. Demyelination most consistently develops after infection of weanling Balb/c mice with a temperature sensitive mutant (*ts8*) [5]. The lesions develop within a week, are randomly scattered throughout the white matter and show no particular relation to vessels. Large plaques of demyelination involving the entire half of the spinal cord are observed in some mice 3 weeks after infection (Fig. 2). Inflammatory cells are rare at the onset of demyelination, but abundant macrophages are seen at later stages. The mortality of infected mice is enhanced by immunosuppressive measures, suggesting that the disease is not mediated by host immune responses but by the direct cytolytic effect of the virus [4].

Electron microscopy confirmed the presence of corona virus in oligodendrocytes (Fig. 2) [6]. The infected cells reveal hypertrophic changes prior to lysis. Their cytoplasm is filled with aggregates of viral protein that bud into cisterns of endoplasmic reticulum. The infected hypertrophic cells frequently show myelin lamellae arising from plasma membranes (Fig. 2), a

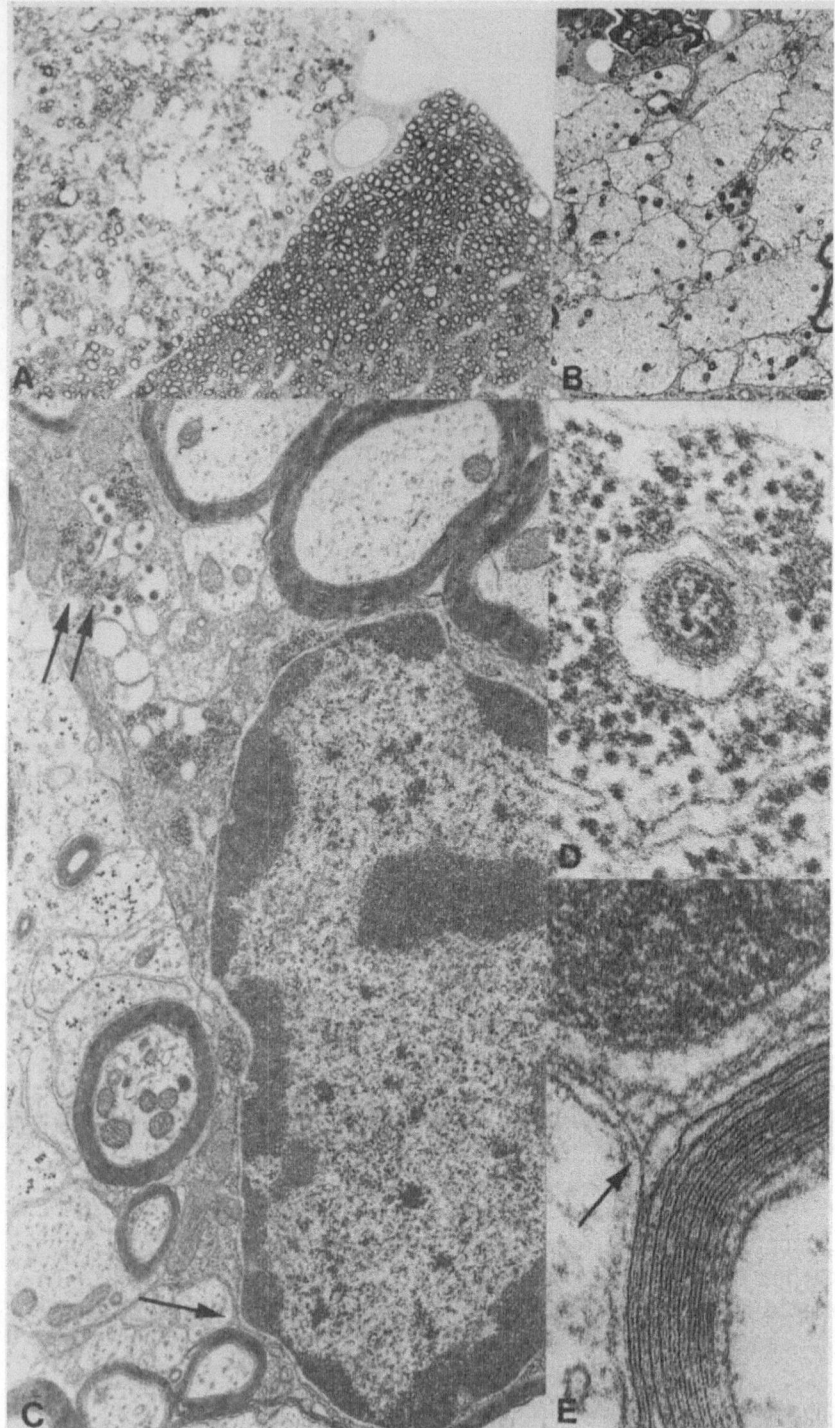

Fig. 2. Demyelination in mouse hepatitis virus encephalomyelitis caused by direct viral cytolysis in oligodendrocytes. A. Large area of demyelination in spinal cord 25 days after infection ×500; B. Completely denuded axons, 21 days after infection

feature difficult to demonstrate in normal oligodendrocytes. After cytolysis, the myelin sheaths related to the necrotic cells are removed by macrophages that strip compact myelin sheaths off their axons. Abundant demyelinated axons interspersed by lipid-laden macrophages are seen 2 to 3 weeks after infection. Characteristically, even mice with large areas of demyelination recover rapidly. Oligodendrocytes proliferate and remyelinate the denuded axons [7]. Remyelination is remarkably effective, resulting in almost complete restoration of myelin sheaths within 2 to 3 months. The recovered mice remain persistently infected at a low titer and small foci of recurring demyelination have been detected in some mice many months after infection [5].

A similar mechanism of demyelination beginning with virus-induced cytolysis of oligodendrocytes is seen in humans afflicted with progressive multifocal leucoencephalopathy [8]. This rare demyelinating disease may develop in debilitated or immunosuppressed patients, particularly in patients suffering from leukemia or lymphoma. Multiple patches of demyelination are found in the brains of patients after a progressive neurologic illness that may last for months. The lesions show no relation to vessels and, except for macrophages, inflammatory infiltrates may be completely lacking. Oligodendrocytes with enlarged nuclei full of papova virus are pathognomonic. Astrocytes with great bizarre, hyperchromatic nuclei that also contain viral antigen are seen in more advanced lesions. It is not clear how this virus infects the brain or what might activate its proliferation in oligodendrocytes. Papova virus isolated from cases of PML has been most frequently identified as JC virus, which commonly infects humans [2].

Immune-Mediated Lysis of Virus-Infected Oligodendrocytes

In contrast to demyelination in PML or MHV, the pathogenesis of other virus-induced demyelinating disorders probably also involves interaction of virus or virus-infected cells with host immune responses. Subacute sclerosing panencephalitis (SSPE) and Theiler's mouse polioencephalomyelitis (TME) will serve as examples.

SSPE is caused by persistent infection with measles virus. It is a slowly progressive disease in children and young adults leading to death within months to a few years. The virus persists in both neurons and glial cells including oligodendrocytes, leading to demyelination. Characteristically, both the spinal fluid and the infected brain contain high titers of antimeasles antibodies and there are perivascular infiltrates of lymphocytes and plasma

---

×5000. C. Oligodendrocyte identified by connections with myelin lamellae (arrow) contains numerous virus particles within cisterns of endoplasmic reticulum (double arrows) ×15,000. D. Corona virus within a cistern of endoplasmic reticulum of an oligodendrocyte ×150,000. E. Enlargement of area marked by arrow in C to show connection of oligodendrocyte with a myelin lamella (arrow) ×100,000.

cells. Ultrastructural studies demonstrated that lysis of infected oligodendrocytes occurs in the presence of antibodies that bind to nucleocapsids of measles virus [9]. Although the persistently infected cells are eventually destroyed by competent host immune responses, one wonders how the virus can persist in the presence of specific antiviral antibodies. Morphologic studies show that persistently infected oligodendrocytes rarely reveal surface expression of viral antigen. Nucleocapsids are found within the nucleus and cytoplasm but no virus buds from the cell surface and there is no alignment of nucleocapsids beneath the plasma membrane. Experimentally, lysis of infected cells is induced by antibody and complement when virus is expressed on the cell surface. Viral antigen can be stripped off the cell membrane by antibody in the absence of complement, rendering the infected cell resistant to immunologic injury [9]. However, after removal of antibody, the cell will again express viral antigen on its surface and become vulnerable to immunologic injury. Survival or lysis of persistently infected cells in SSPE may thus depend on titer and ratio of complement and antibody within the affected brain, which may be subject to changes in the permeability of the blood brain barrier.

TME is a demyelinating disease that develops in mice 1 to 3 months after infection with the DA strain of Theiler's mouse encephalomyelitis virus (TMEV), a picorna virus [10,11]. The virus first affects neurons but a few weeks later perivascular demyelinating lesions develop in the spinal cord (Fig. 3). Demyelination is preceded by perivascular mononuclear cell infiltrates. Patchy perivascular demyelination associated with inflammatory infiltrates is seen for many months in persistently infected mice. Remyelination by oligodendrocytes occurs, but compared with efficient remyelination following MHV infection, the restoration of myelin sheaths is delayed, abortive, or incomplete. In some mice, however, Schwann cells enter the damaged cord via nerve roots and remyelinate axons more effectively [11].

The biphasic course of the disease, as well as the perivascular demyelination occurring within mononuclear cell infiltrates, suggested an autoimmune pathogenesis following host sensitization by the initial virus invasion of the brain. This view is supported by observations indicating that immunosuppressive measures prevented the development of the demyelinating lesions [12]. However, at present no convincing evidence exists to prove that the demyelinating disease can be transferred to uninfected mice much like the transfer of experimental autoimmune encephalomyelitis that is induced by sensitization to myelin. On the other hand, there are reports indicating that virus persists at low titers within the spinal cord and that glial cells (in particular, oligodendrocytes) are persistently infected in this chronic recurrent demyelinating disease (Fig. 3) [13,14]. These observations favor the interpretation of immune-mediated injury to persistently infected oligodendrocytes. Ultrastructural studies confirmed the presence of degenerating oligodendrocytes and the stripping of myelin from axons by invading macrophages.

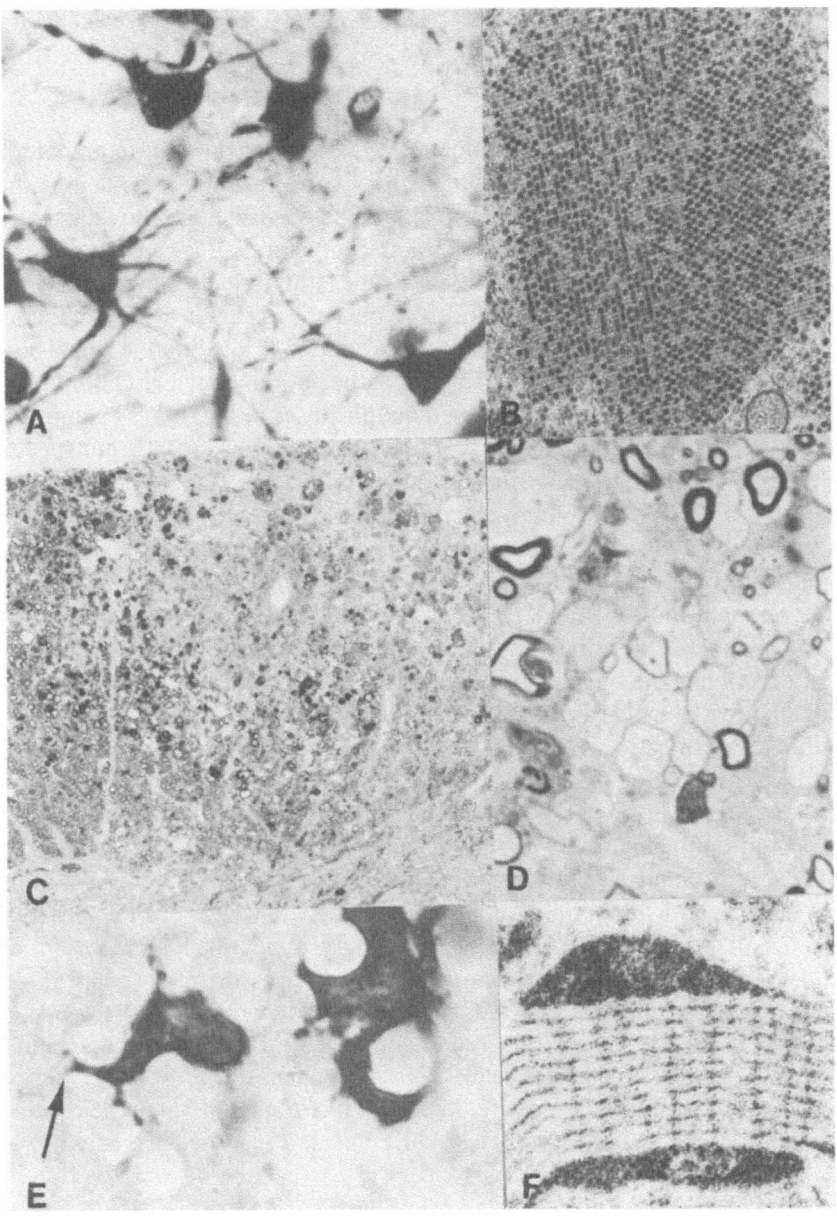

Fig. 3. Theiler's mouse virus encephalomyelitis characterized by a biphasic course that involves first neurons and later glial cells in the white matter resulting in demyelination. A. Viral antigen is localized in nerve cells in the spinal cord of a SJL mouse 7 days after infection. Immunoperoxidase ×1000. B. Paracrystalline array of picorna virus particles in a nerve cell 4 days after neonatal infection ×30,000. C. Perivascular demyelination associated with mononuclear cell infiltrates in the spinal cord 110 days after infection ×250. D. Demyelinated axons in the spinal cord 45 days after infection ×2000. E. Viral antigen localized in inner and outer glial cells in the white matter of the spinal cord 45 days after infection. Arrow points to connections with myelin sheaths. Immunoperoxidase ×1000. F. Viral antigen localized in oligodendroglial loops of a myelin sheath in the spinal cord 360 days after infection. Immunoperoxidase ×160,000.

"Bystander" Demyelination

It has been proposed that inflammatory cells participating in immunologic reactions may release substances capable of causing nonspecific demyelination. This idea runs counter to neuropathologic observations showing that in most inflammatory diseases of the brain, demyelination is not a prominent feature. However, experimental studies suggest that demyelination may occur within the inflammatory infiltrate of delayed hypersensitivity reactions [15]. Local injections of purified protein derivative (PPD) into the spinal cord of guinea pigs sensitized to this antigen induce a mononuclear cell infiltrate associated with demyelination. Based on this observation, it has been postulated that demyelination in virus infections may represent a nonspecific "bystander" effect triggered by an immunologic reaction directed against the virus. *In vitro* studies of myelinated tissue cultures showed that demyelination can be induced by neutral proteinases, such as plasminogen activator, which are secreted by activated macrophages [16]. Other studies indicate, however, that the "bystander" effect may account only for minor demyelination in nonspecific inflammation, whereas more widespread demyelination is induced when the inflammatory reaction is associated with the recruitment of lymphocytes sensitized against myelin components or when antimyelin antibodies permeate the lesion (HC Powell, 1984, personal communication). The local production of a delayed hypersensitivity reaction in the peripheral nerve in rabbits by means of purified protein derivative is associated with significant demyelination when antigalactocerebroside antibody is elevated in the serum. Similarly, rats sensitized to PPD and galactocerebroside show demyelination of the peripheral nerve after local injection of PPD, whereas rats sensitized only to PPD show no or only minor demyelination in advanced stages. "Bystander" demyelination in virus infection could therefore be quite significant in a host with latent hypersensitivity to myelin, because stimulation of immunocompetent cells by virus may simultaneously turn on cells sensitized to myelin. This mechanism of nonspecific recruitment of cells or antibody directed against myelin might explain demyelination in infections with completely different viruses such as paramyxo-, rhabdo-, arbo-, retro-, and herpes virus [11].

Virus-Induced Autoimmune Demyelination

Human post-infectious encephalomyelitis generally develops a few weeks after a virus infection and shows morphologic features that resemble acute experimental autoimmune encephalomyelitis [2]. These findings prompted speculations that the initial invasion of the brain by virus may alter myelin or oligodendrocytes. Sensitization against the altered membranes may evoke immune responses against myelin, leading to autoimmune demyelination. Experimentally it is indeed possible to demonstrate both cellular and humoral antimyelin sensitivity after virus infections, e.g., after infections with distemper or Marek's disease virus [2,11,17]. Another theory suggests that sensitization to myelin occurs because the peptide sequences of some viruses mimic those of myelin proteins. There is evidence indicating that cross-reactivity of antibody exists against measles virus and myelin basic protein [18]. Further, pertinent in this regard is the fact that following vaccination with killed virus, the host may develop a demyelinating disease that resembles post-infectious or autoimmune encephalomyelitis. As mentioned above, virus could also nonspecifically alter the immune system and thus permit the emergence of latent hypersensitivity to myelin. This idea is supported by the finding that recurrence of experimental autoimmune encephalomyelitis is precipitated by herpes virus infection [19].

Acknowledgment. The work described in this paper was supported by US PHS Grants NS-09053, 12428, and 07078.

References
1. Lampert PW (1983) Multiple sclerosis: The patient, treatment and pathology. *In* Hallpike J, Adams CWM, Tourtellotte WW (eds) Chapman Hall, London
2. Lampert PW (1978) Autoimmune and virus-induced demyelinating disease. A review. Am J Pathol 91:176–208
3. Cheaver FS, Daniels JB, Pappenheimer AM, Bailey OT (1949) A murine virus (JHM) causing disseminated encephalomyelitis with extensive destruction of myelin. 1. Isolation and biological properties of the virus. J Exp Med 90:181–194
4. Weiner LP (1973) Pathogenesis of demyelination induced by mouse hepatitis virus (JHM virus). Arch Neurol 28:298–303
5. Knobler RL, Tunison LA, Lampert PW, Oldstone MBA (1982) Selected mutants of mouse hepatitis virus type 4 (JHM stain) induce different CNS diseases. Am J Pathol 109:157–168
6. Lampert PW, Sims JK, Kniazeff AJ (1973) Mechanism of demyelination in JHM virus encephalomyelitis Acta Neuropathol 24:76–85
7. Herndon RM, Price DL, Weiner LP (1977) Regeneration of oligodendroglia during recovery from demyelinating disease. Science 195:693–694
8. Zu Rhein GM (1969) Association of papova-virus with a human demyelinating disease (progressive multifocal leukoencephalopathy). Prog Med Virol 11:185–247
9. Lampert PW, Joseph BS, Oldstone MBA (1976) Morphological changes of cells infected with measles or related viruses. *In* Zimmerman HM (ed) Progress in Neuropathology. Grune and Stratton, New York, 3:51–68
10. Lipton HL (1975) Theiler's virus infection in mice: An unusual biphasic disease process leading to demyelination. Infect Immunol 11:1147–1145
11. Dal Canto MC, Rabinowitz SG (1982) Experimental models of virus induced demyelination of the central nervous system Ann Neurol 11:109–127
12. Lipton HL, Dal Canto MC (1976) Theiler's virus induced demyelination: Prevention by immunosuppression Science 192:62–64
13. Brahic M, Stroop WG, Baringer JR (1981) Theiler's virus persists in glial cells during demyelinating disease Cell 26:123–128

14. Rodriguez M, Leibowitz JL, Lampert PW (1983) Persistent infection of oligodendrocytes in Theiler's virus induced encephalomyelitis. Ann Neurol 13:426–433
15. Wisniewski HM, Bloom BR (1975) Primary demyelination as a non-specific consequence of a cell mediated immune reaction. J Exp Med 141:346–359
16. Cammer W, Bloom BR, Norton WT, Gordon S (1978) Degradation of basic protein in myelin by neutral proteases secreted by stimulated macrophages: A possible mechanism of inflammatory demyelination. Proc Nat Acad Sci USA 75:1554–1558
17. Pepose JS, Stevens JG, Cook ML, Lampert PW (1981) Marek's disease as a model for the Landry–Guillain–Barre Syndrome. Am J Pathol 103:309–320
18. Panitch HS, Swoveland P, Johnson KP (1979) Antibodies to measles virus react with myelin basic protein. Neurology 29:548–549
19. Hochberg FH, Lehrich JR, Arnason BGW (1977) Herpes simplex infection and experimental model system for reactivation of EAE. Neurology (Minneap) 27:584

CHAPTER 38

Virus Can Alter Cell Function Without Causing Cell Pathology: Disordered Function Leads to Imbalance of Homeostasis and Disease

MICHAEL B. A. OLDSTONE*

The ability of a virus to alter the "luxury" or differentiated function of a cell, without disturbance of its vital function, i.e., ability to survive, was first noted in murine neuroblastoma cells persistly infected with lymphocytic choriomeningitis virus (LCMV) [1]. The term "luxury function" was initially coined to describe the impediment of synthesis or degradation of acetylcholine, a major neurotransmitter, while the cell continued to survive, grow, and synthesize protein, RNA, DNA, and respiratory enzymes, etc. during infection. Later it was found that viruses also perturbed the secretion of specific lymphocyte products like immunoglobulins [2,3] and hampered such functions of lymphocytes as killing of virus-infected or transformed cells [3,4]. Most recently it was noted that hormone-producing cells like somatotrophic cells, which make growth hormone (GH) [5,6], or beta cells of the islets of Langerhans, which make insulin, can falter in these activities when so infected. The evidence for this newly found mechanism whereby viruses alter the products of differentiated cells and cause disease without structurally altering or killing the infected cell is the subject of this chapter. Although the term "luxury function" was appropriate for analysis of cultured cells first described in this context, it is inappropriate when applied to the whole organism. Such "luxury" or differentiated products play an essential role in the balance of normal host physiology and thus the avoidance of disease. Without question, these differentiated products are important for the host's ability to control its internal milieu, while protecting it against potentially hostile surroundings.

Virus-induced cell injury and resultant disease occur both through direct destruction of cells by viruses and/or secondarily through lysis of infected

---
* Scripps Clinic and Research Foundation, La Jolla, California.

cells by immunologic assault on viral antigens [7–10]. In either case, the tissue damage ranges from breakdowns in cell structure to cell lysis, frequently with inflammatory infiltrates. Conventionally, these patterns have been considered characteristic of viral infections [8–10]. Now it is clear that viruses cause injury and disease by an additional mechanism—namely, by altering the cell's normal or expected functions or changing physiologic control factors without cell lysis. For this to occur, three events are necessary. First, the infecting virus or its variant generated within the host must be essentially noncytocidal and thus able to coexist with the cell it infects. Second, the infected cell's product must be essential for maintaining the host's homeostasis. Third, disrupted synthesis of this product cannot affect the cell's survival even though jeopardizing the development or well-being of its host.

Homeostatic imbalance associated with virus infections, although not understood, was first described as a clinical observation of the mid- to late 1800s. Osler, in his textbook of medicine [11], recognized primary measles virus infection as a common cause of activating bronchial or systemic tuberculosis. In experiments over 80 years ago Clemens von Pirquet [12] noted that sensitivity to tuberculin, an immune measurement of prior exposure, was lost during acute measles virus infection and reappeared when the infection cleared. These observations were amply confirmed and extended to numerous other virus infections, although the molecular basis remained largely obscure.

Other examples of virus-induced imbalance of homeostasis or altered cell physiology resulted from serendipitous experimental observations. Zabriskie et al. [13], while attempting to obtain a murine model of paramyxovirus-induced demyelination, noted that mice inoculated with canine distemper virus developed a disorder in fat metabolism. Hence, 3 months after exposure to the virus, these mice became obese and manifested a variety of hormonal abnormalities. Despite careful investigation of CNS and endocrine tissues, these authors saw no evidence of cell destruction. Similarly, while studying the regulation of gene expression, Jaenisch noted that insertion of retrovirus genes into developing mouse embryo was associated with changes in several host factors of newborn mice, including coat color [14]. Recently, Rowe [15] found deformed whiskers in mice infected with certain newly arisen MCF-type recombinant viruses, but with no evidence of accompanying cell injury.

Definitive evidence for the concept of virus-altered homeostasis recently came from studies of mice with C3H or CBA backgrounds and persistent LCMV infections. Depletion and reconstitution experiments showed that these viruses affected the synthesis of GH as they replicated in somatotrophic cells [5]. This hormone deficiency altered both growth and glucose metabolism, events normally regulated by GH. Reconstituting the infected mice with GH restored homeostasis and resulted in growth and glucose metabolism equivalent to that of uninfected age- and sex-matched controls

[6]. Ongoing analysis of this model provides the following conclusions. First, host genetic factors are important since LCMV, while infecting all strains of mice and causing persistent infection, alters the synthesis of GH only in certain strains. Those strains whose somatotrophic cells replicate the most virus have the severest deficiency in GH synthesis. Second, the strain of LCMV used to infect susceptible mice is critical. Thus, LCMV Armstrong-1371 (Arm) or LCMV E-350 causes severe disorders of growth and glucose metabolism in susceptible C3H mice but, in contrast, LCMV-Traub and WE do not. The fact that LCMV-Arm and E-350 are closely related according to T1 oligonucleotide maps, sharing over 95% of observed fingerprints [16], whereas LCMV-WE and Traub differ from Arm and E-350 by over 50% of the observed T1 oligonucleotides may be important in this regard.

Studies with LCMV-Arm showed that the virus was trophic for cells making GH and that the virus did not replicate in cells making prolactin, thyroid stimulating hormone, or adrenal cortical stimulating hormone [5,17]. Despite virus replication in somatotrophic cells in the anterior lobe of the pituitary, there was no evidence of lysis among these infected cells or of inflammatory infiltrates. Further, at high resolution, electron microscopy showed normal morphology of the GH-producing cells that replicated virus [5,17]. Yet, biochemical analysis demonstrated GH levels diminished by approximately 50% in pituitary glands of infected mice compared with those of uninfected, matched controls. Prolactin levels remained normal. Additionally, amounts of an insulin-like factor, whose concentration in serum is GH dependent and correlates with GH levels in the anterior lobe of the pituitary, were reduced in LCMV-infected mice [6].

Mice persistently infected with LCMV undergo significant decreases in body length and weight and become defective in glucose metabolism. The abnormality in glucose metabolism leads to severe hypoglycemia and death of C3H and CBA mice 20 to 30 days after the initiation of persistent infection. Their GH deficiency directly caused this metabolic disorder, based on reconstitution experiments in which LCMV-infected recipients of GH infusions grew normally and survived at rates expected for uninfected age- and sex-matched controls [6]. Further, such persistently infected mice reconstituted with GH had normal blood glucose levels and normal glucose tolerance tests, indicating restoration of normal glucose metabolism.

Thus, LCMV, an essentially noncytocidal replicating virus, is trophic for cells in the anterior lobe of the pituitary gland that make GH and coexists with the cells it infects. Even though these somatotrophic cells survive and show normal cytomorphology despite perturbation of GH synthesis, their deficiency in GH impedes the development and well-being of the virus-infected host. Often the outcome is severe hypoglycemia, which can lead to death.

LCMV contains two single-stranded RNA species, approximately 9 and 4 Kb, of negative polarity. The genome codes three known structural polypeptides, two of which are glycosylated (gp1-MW 43,000, and gp2-MW 36,000) and a third that is not (nucleoprotein-MW 63,000). The glycoproteins comprise approximately 30% of viral polypeptides formed in infected cells or virions, and gp1 and gp2 form in infected cells in equimolar amounts after proteolytic cleavage from a common precursor, GPC. The use of monoclonal antibodies directed against these three LCMV polypeptides revealed that, as persistent infection progressed, expression of these viral glycoproteins was restricted but that of nucleoproteins was not.

For virus to cause persistent infection in a host, two unique host–virus relationships are required. First, the virus *per se,* or variants arising *in vivo,* must be noncytopathic for the host cell that it infects. Second, the infected cell must escape recognition by the host's immune system, which eliminates virus and virus-infected cells. Several models of persistent infection *in vitro* have demonstrated selective loss of viral gene products normally represented at the cell surface as opposed to viral products found inside the cell (reviewed in [18]). Decreased glycoprotein expression on the surface of an infected cell correlates directly with the ability of that cell to resist killing by cytotoxic lymphocytes as well as cytotoxic antibody and complement. Using monoclonal antibodies to viral glycoprotein and nucleoprotein as a means of defining cellular expression of the corresponding molecules, we noted a selective decrease in the expression of viral glycoprotein in single cells from the pituitary's anterior lobe during persistent LCMV infection of mice *in vivo* [18]. Of numerous organs examined during the first 2 weeks after infection, neurons of the central nervous system, parenchymal cells of the liver, tubular epithelial cells of the kidney, and cells of the anterior lobe of the pituitary expressed viral glycoprotein and nucleoprotein in a manner similar to acutely infected cells from adult mice and acutely infected cells in culture. However, after 30 days and later, individual infected cells located in the central nervous system, liver, kidney, and anterior lobe of the pituitary lost viral glycoprotein but retained the expression of viral nucleoprotein. These results, first derived by using light microscopy to visualize sites where fluoroscein-labeled monoclonal antibodies bound, were confirmed by electron microscopy and immunochemical techniques. Such studies suggest that cellular synthesis of viral glycoproteins decreases during persistent infection *in vivo,* lowering their expression on the cells' surfaces. Interestingly, electron microscopic study of infected cells from persistently infected mice reacted with monoclonal antibodies confirmed that nucleoprotein accumulated inside the cell but collected as aggregates in the area of polyribosomes. Viral glycoprotein inside the cell was reduced and the virus became restricted from budding at the cell surface.

Decreased viral glycoprotein expression would allow persistently infected cells to resist assault by virus-specific immune reactants. By this means, a noncytopathic virus could persist throughout the life span of an infected cell, and the cell remain alive, unaffected by the host's immune surveillance mechanism. In addition to the host cell–virus relationship regulating viral glycoprotein expression, in certain viral infections antibody may initiate viral persistence by stripping or modulating viral glycoproteins on the surfaces of infected cells. This phenomenon, called antibody-induced antigenic modulation, is discussed in Chapter 27.

Findings *in vivo* in individual neuronal, hepatic, epithelial, and endocrine cells from mice persistently infected with LCMV extend the results observed with other viruses *in vitro*. Quantitative studies during LCMV infection *in vivo* and *in vitro* demonstrate that infected cells develop a 10- to 50-fold decrease in viral glycoprotein synthesis and glycoprotein expression compared with nucleoprotein synthesis and expression in acutely infected cells. These results reflect the overall pattern of other persistent viral infections. For example, neuronal cells in biopsies from patients with chronic measles virus infection (subacute panencephalitis) show a decrease in or lack of measles glycoprotein on their surfaces, but they accumulate measles virus nucleocapsids internally.

By avoiding immune responses directed against the cells they infect, replicating viral agents are free to alter the synthesis of cell products. How may this occur? First, as a result of infection, the cell might release a product like interferon that would modulate its function. Gresser *et al.* [19] have provided evidence that interferon produced by infecting newborn mice with LCMV harmed mesangial and glomerular cells, thereby disturbing their normal ability to clear immune complexes. However, by reconstitution and deletion experiments, interferon was shown not to play a role in altered GH synthesis. A second possible mode of changing cellular function is that the virus might structurally alter the differentiated cells' product. However, peptide maps of normal GH developed by using trypsin or chymotrypsin digests fail to show any differences structurally from those of GH isolated from pituitary glands of infected mice. Thus, this mechanism appears not to play a role in disturbed somatotroph function during LCMV infection. A third, and perhaps more likely, way is that LCMV affects the synthesis of GH at transcriptional or translational levels. Thus, the cell's capacity for membrane translocation and synthesis may be limited so that production of viral gene products competes with manufacture of GH, thereby overloading the system. Alternatively, the virus may induce a specific alteration in an enzyme or transport system needed for the production and release of GH. Whether these biochemical dysfunctions at the cellular level relate directly to the accumulation of viral products in a manner akin to that in viral storage disease is currently a matter of active investigation.

Considering the enormous variety of human diseases whose etiologies are still unknown, it is interesting to speculate that some of these could be caused by the failure of specialized or differentiated functions of cells as a result of a virus infection. For example, several progressive neurologic diseases affecting cognitive functions, mood or locomotion, such as Alzheimer's or Parkinson's diseases are characterized by deficiencies in neurohormones, neurotransmitters, or their enzymes [20-23]. Persistent LCMV infection of murine neuroblastoma cells, *in vitro*, leads to specific deficiencies in acetylcholine and transferase esterase [1]. Vahlne, Lycke, and colleagues (personal communication) find that persistent measles virus infection in such cells results in a similar enzyme deficiency and abnormal oxygen consumption. The previous associations of influenza virus infection with a Parkinson's-like syndrome [8] and of herpes viruses with mood disorders [23] are of interest in this regard. Acquired immunologic deficiency syndrome is also suspected of having a viral cause. Moreover, synthesis of immunoglobulins by murine or human lymphocytes is significantly impaired upon infection of these cells with any of several viruses, although the lymphocytes' growth rates and viability are undisturbed. After measuring antibody synthesis *in vitro,* Pelton et al. [2] showed that specific immunoglobulin responses to specific antigens were suppressed during herpes simplex virus infection. This immunosuppression resulted from preferential infection of helper T cells. Casali *et al.* showed that measles virus and influenza virus also suppressed the synthesis of immunoglobulin G and immunoglobulin M without affecting the viability of helper T lymphocytes or B lymphocytes [3,4], while we found similar alterations in the ability of fusomas to make monoclonal antibodies during persistent LCMV infection. Additionally, Rice and Casali [4] noted that measles virus and cytomegalovirus infection of natural killer cells inhibited their cytotoxic ability. Consequently, persistent viral infections similar to those described here may exist in human disease and underlie the failure of hormone secretion by endocrine glands, dysfunction of neurons and oligodendrocytes, and/or disorders of the immune system. During the next decade, researchers may discover that viruses cause significant disease in the absence of associated cytopathology by altering the function of a differentiated cell or modifying its product.

In a series of elegant experiments, H. Holtzer and his associates have shown that Rous sarcoma virus affects differentiated products of chick chondroblasts, retinal melanoblasts or muscle. Briefly, chick cultures of these specialized cells alone or infected with temperature sensitive mutant LA 24A of the Prague strain at the nonpermissive temperature showed the expected synthetic programs of differentiation. These differentiation programs were altered at the permissive temperature for the virus [24].

Acknowledgments. This is publication no. 3151-IMM from the Department of Immunology, Scripps Clinic and Research Foundation. This research was supported by US PHS Grants AI-09484, AI-07007, NS-12428, and AG-04342.

References
1. Oldstone MBA, Holmstoen J, Welsh RM (1977) Alterations of acetylcholine enzymes in neuroblastoma cells persistently infected with lymphocytic choriomeningitis virus. J Cell Physiol 91:459–472
2. Pelton BK, Duncan IB, Denman AM (1980) Herpes simplex virus depresses antibody production by affecting T-cell function. Nature 284:176–177
3. Casali P, Rice GPA, Oldstone MBA (1984) Viruses perturb functions of human

lymphocytes: Effects of measles virus and influenza virus on lymphocyte mediated killing and antibody production. J Exp Med, in press
4. Rice GPA, Casali P, Schrier RD, Oldstone MBA (1983) Cytomegalovirus and herpes simplex virus: Their effect *in vitro* on natural killer cells, mitogen induced proliferation and antibody synthesis. International Herpesvirus Workshop, University of Oxford, England (abstract)
5. Oldstone MBA, Sinha YN, Blount P, Tishon A, Rodriguez M, von Wedel R, Lampert PW (1982) Virus-induced alterations in homeostasis: Alterations in differentiated functions of infected cells *in vivo*. Science 218:1125-1127
6. Oldstone MBA, Rodriguez M, Daughaday WH, Lampert PW (1984) Viral perturbation of endocrine function: Disorder of cell function leading to disturbed homeostasis and disease. Nature 307:278-280
7. Bablanian R (1975) Structural and functional alterations in cultured cells infected with cytocidal viruses. Prog Med Virol 19:40-83
8. Blackwood W, Cornsellis J (eds) (1976) Neuropathology. Edward Arnold, London
9. Robbins S, Angell M, Kumar V (1981) Basic pathology. W.B. Saunders Co., Philadelphia
10. Notkins A (ed.) (1975) Viral immunology and immunopathology. Academic Press, New York
11. Osler W (1904) The principles and practice of medicine (5th ed). D. Appleton and Company, New York
12. von Pirquet CE (1908) Das Verhalten der kautanen Tuberculinreaktion wahrend der Masern. Dtsch Med Wochenschr 34:1297-1399
13. Lyons MJ, Faust IM, Hemmes RB, Buskirk DR, Hirsch J, Zabriskie JB (1982) A virally induced obesity syndrome in mice. Science 216:82-85
14. Jaenisch R, Detlev J, Nobis P, Simon I, Löhler J, Harbers K, Grotkopp D (1981) Chromosomal position and activation of retroviral genomes inserted into the germ line of mice. Cell 24:519-529
15. Rowe WP (1983) Deformed whiskers in mice infected with certain exogenous murine leukemia viruses. Science 221:562-563
16. Dutko FJ, Oldstone MBA (1983) Genomic and biologic variation among commonly used lymphocytic choriomeningitis virus strains. J Gen Virol 64:1689-1698
17. Rodriguez M, von Wedel RJ, Garrett RS, Lampert PW, Oldstone MBA (1983) Pituitary dwarfism in mice persistently infected with lymphocytic choriomeningitis virus (LCMV). Lab Invest 49:48-53
18. Oldstone MBA, Buchmeier MJ (1982) Restricted expression of viral glycoprotein in cells of persistently infected mice. Nature 300:360-362
19. Gresser I, Morel-Maroger L, Verroust P, Riviere Y, Guillon, JC (1978) Antiinterferon globulin inhibits the development of glomerulonephritis in mice infected at birth with lymphocytic choriomeningitis virus. Proc Natl Acad Sci USA 75:3413-3416
20. Marsden C (1982) Neurotransmitters and CNS disease. Lancet ii:1141-1147
21. Bowen DM, Smith C, White P, Davidson N (1976) Neurotransmitter-related enzymes and indices of hypoxia in senile dementia and other abiotrophies. Brain 99:459-496
22. Perry EK, Perry RH, Blessed G, Tomlinson BE (1977) Necropsy evidence of central cholinergic deficits in senile dementia. Lancet i:189

23. Albrecht P, Torrey EF, Boone E, Hicks JT, Daniel N (1980) Raised cytomegalovirus-antibody level in cerebrospinal fluid of schizophrenic patients. Lancet i:769–772
24. Holtzer H, Pacifici M, Tapscott S, Bennett G, Payette R, Dlugosz A (1982) Lineages in cell differentiation and in cell transformation. *In* Revoltella RF (ed) Expression of Differentiated Functions in Cancer Cells. Raven Press, New York, p 169

Evolving Concepts in Viral Pathogenesis Illustrated by Selected Diseases in Humans

39. Chronic Leukemia
    P. S. Sarin and R. C. Gallo 279

40. Hepatitis B Virus Diseases
    William S. Robinson 288

41. Herpesviruses and Cancer
    Fred Rapp and Mary K. Howett 300

42. Epithelial Cell Interactions of the Epstein–Barr Virus
    Joseph S. Pagano 307

43. Viral Gastroenteritis
    Albert Z. Kapikian 315

44. Hemorrhagic Fever Viruses
    C. J. Peters and Karl M. Johnson 325

45. Marburg and Ebola Viruses: New Agents on the Frontiers of Virology
    Michael J. Buchmeier 338

46. Rabies
    Hilary Koprowski 344

47. Unconventional Viruses
    D. Carlton Gajdusek 350

CHAPTER 39
Chronic Leukemia

P. S. SARIN AND R. C. GALLO*

The involvement of retroviruses in the pathogenesis of leukemia and lymphoma in a number of animal species has been well established [1]. Attempts to define a role for retroviruses in human malignancies has been a subject of active investigation in various laboratories and the efforts have been mostly unsuccessful, until the recent and repeated isolation of a human T-cell leukemia-lymphoma virus (HTLV) from adults with T-cell malignancies [2–4]. HTLV has also been isolated from tissues from several patients with acquired immunodeficiency syndrome (AIDS).

Retroviruses can be classified into endogenous viruses and exogenous viruses. The endogenous retroviruses are contained in the germline of the species, are therefore transmitted genetically and, in general, do not appear to be pathogenic. Many exogenous retroviruses, on the other hand, have been identified as etiologic agents in animal tumors, are transmitted by infection, and can be subdivided into (1) chronic leukemia viruses and (2) acute leukemia viruses.

The chronic leukemia viruses induce tumors after long latency periods and are nondefective, i.e., they can replicate successfully by integrating the provirus in the host-cell genome. Therefore, they do not need the action of a helper virus. In general, these viruses do not transform cells *in vitro* because they do not contain genes (c-*onc*) that code for proteins specific for cellular transformation. Bovine leukemia virus (BLV), avian leukosis virus (ALV), most murine leukemia viruses, and feline leukemia virus (FeLV) belong to this class. The acute leukemia retroviruses, on the other hand, have been isolated from both laboratory-induced tumors and more rarely from naturally occurring tumors. These viruses are generally replication defective

---
* National Cancer Institute, Bethesda, Maryland.

and, therefore, they require a helper virus. The genomes of these viruses contain *onc* genes, which are responsible for rapid tumor induction *in vivo* and cell transformation *in vitro*. Examples of this group of viruses are the murine sarcoma virus and the Abelson murine leukemia virus. All viral *onc* (v-*onc*) genes have been shown to be derived from normal cellular genes (c-*onc*) of their hosts of origin. C-*onc* genes are highly conserved; their homologues can be identified in all vertebrate species and some of these genes are activated in tumors induced by chronic leukemia viruses. B-cell lymphomas and erythroblastomas induced by avian leukosis virus express high levels of the *onc* genes, c-*myc* and c-*erb,* respectively, as a result of provirus integration near these genes [5]. Some chronic leukemia viruses, notably the avian leukosis virus, appear to induce leukemias in part by activation of the cellular genes, which may or may not be homologues of the viral *onc* genes, but in several systems this mechanism has not been identified.

Until recently, consideration of an etiological role of retroviruses in human cancers was frequently met with skepticism. This was partly due to the extreme difficulty of isolating and detecting low levels of retroviruses in human cells, in contrast to the animal models which were associated with abundant virus production and viremia. In addition, all mammalian retroviruses shared some common antigenic determinants so that immunologic reagents of one virus could be used to detect the presence of other viruses. One of the animal model systems, viz., bovine leukosis [6], offers several similarities to the human leukemia-lymphoma. The role of BLV in causation of the enzootic form of bovine leukosis is well known [6] despite the fact that the fresh bovine tumor tissues are usually virus negative. Virus expression, however, does occur after prolonged culture of the tumor cells *in vitro*. Since BLV is not related to other mammalian or avian retroviruses, evidence for BLV involvement in bovine leukosis could not be obtained using probes derived from known retroviruses. This experience with BLV has some similarities with the subsequent isolation and detection of a human retrovirus from patients with T-cell malignancies.

T-Cell Growth Factor

The discovery of T-cell growth factor (TCGF) [7] in 1976 made it possible for the first time to grow both normal and neoplastic human T cells routinely in long-term suspension culture. Normal T cells grown in the presence of TCGF retain their normal karyotype and function. Using highly purified TCGF [8], T cells from normal individuals can be grown after lectin or antigen stimulation. In contrast, neoplastic T cells from some adults with T-cell malignancies respond to TCGF without prior lectin or antigen activation *in vitro*. This appears to be due to constitutive expression of receptors for TCGF on the surfaces of malignant T cells from these patients. These receptors are not present in normal individuals without lectin or antigen stimulation. After a variable period of time, exogenous TCGF is no longer required for *in vitro* maintenance and growth of many cell lines, which in some cases become autonomous producers of TCGF. It was from some of these cultured T-cell lines established from sporadic patients in the United States with a subtype of adult T-cell leukemia-lymphoma that the first human retrovirus, HTLV, was isolated [2–4,9].

HTLV Isolation and Transmission

The first isolate of HTLV was obtained from a lymph node-derived T-cell line from a patient (CR) with an aggressive variant of cutaneous T-cell lymphoma [2–4,9]. Electron microscopy showed the presence of typical retroviral particles budding from cell membranes. A second isolate, $HTLV_{MB}$, was obtained from a patient (MB) with Sezary leukemia. $HTLV_{CR}$ and $HTLV_{MB}$ were extensively characterized and found to be highly related to each other.

Like other retroviruses, HTLV contains a reverse transcriptase, a high-molecular-weight RNA genome (9 Kb or 70S), and three viral core proteins (*gag* proteins) p24, p19, and p15. These proteins correspond to the p30, p15, and p10 *gag* proteins of the murine retroviruses. Molecular hybridization [10] and immunological [11] cross-reactivity studies show that HTLV is not related to any known animal retroviruses. The reverse transcriptase of HTLV is a protein of 95,000 daltons and prefers Mg^{2+} over Mn^{2+} as the divalent cation. The characteristics of HTLV reverse transcriptase and the core protein most closely resemble BLV and the avian retroviruses [2–4]. Amino acid sequence analysis of the HTLV p24 shows a distant but significant sequence homology with BLV p24, suggesting a common ancestral origin of BLV and HTLV [12].

HTLV has been clearly shown to be an exogenous human virus. HTLV-related sequences are not present in the DNA of normal *uninfected* human cells, but they are readily detected in DNA from HTLV-positive tumor cells [10]. In addition, HTLV sequences were not found in the normal Epstein–Barr virus-infected B cells of the HTLV-positive patient CR. Only his neoplastic T cells contained HTLV-specific sequences [13]. However, a small number of B cells may sometimes harbor HTLV sequences.

Since the first isolates of $HTLV_{CR}$ and $HTLV_{MB}$, many additional isolates have been obtained from cell lines established from patients with mature T-cell malignancies from different parts of the world [14]. These include patients from the United States, South America, the Caribbean, Israel, and Japan [2–4,14–18]. A retrovirus has also been detected [19] in a T-cell line established from a white male (MO) from the United States with hairy cell leukemia.

Investigators in several laboratories in Japan have recently shown the presence of retrovirus particles by electron microscopy in cultured T lymphocytes of ATL patients and have demonstrated serum antibodies directed against determinants on cells producing these particles in almost all Japanese ATL patients as well as in 26% of normal individuals from the endemic areas of Japan [20,21]. The virus produced by MT-1, a cell line established by Japanese workers, has antigenic determinants identical to those of HTLV p24 and p19 and its mRNA and provirus hybridize virtually completely to HTLV cDNA [22]. Comparison of the different HTLV proviruses present in cells of leukemic patients from the United States, Japan, the Caribbean, and elsewhere shows that all virus isolates, except $HTLV_{MO}$, are closely related and have highly conserved genomes [23]. On the other hand, poor competition of $HTLV_{MO}$ in the p24 assay and nucleic acid sequence homology with $HTLV_{CR}$ was detected only under very nonstringent hybridization conditions [19]. Therefore, this virus belongs to a distinct subgroup in the HTLV family. We have grouped these virus isolates as $HTLV-I_{CR}$, $HTLV-I_{MT1}$, $HTLV-I_{MJ}$, and $HTLV-II_{MO}$. HTLV has been transmitted into cord blood T cells, bone marrow, and adult peripheral blood T cells from a number of cell lines by cocultivation [3,4,14–17] and in a few instances by infection with cell-free virus. The infected cells resemble neoplastic T cells in many respects. They develop lobulated nuclei, some form multinucleated giant cells and the cells have potential for indefinite growth. They have a decreased or no requirement for TCGF for growth; and they express high levels of TCGF receptors [2–4,14–16].

HTLV Epidemiology

Seroepidemeological studies have been carried out to determine the distribution of HTLV-specific antibodies in patients from various parts of the world with T-cell malignancies. HTLV-specific antibodies were first detected in sera from two US adults with aggressive T-cell malignancies (CR and MJ) and some of their healthy family members [2–4,24]. Sera of numerous normal healthy donors and patients with a wide variety of malignancies in the United States were negative for antibody except for a small percentage of T-cell leukemia and lymphoma patients [2–4,25]. HTLV-positive lymphomas and leukemias consist of adult onset, with a rapid disease course, often associated with lymphadenopathy and hepatosplenomegaly, circulating large and pleomorphic lymphocytes with lobulated nuclei, mature T-cell surface markers (usually OKT4+), and frequent hypercalcemia and skin manifestations.

The clinical syndrome of T-cell malignancy in adults termed ATL (adult T-cell leukemia-lymphoma) has also been recognized in Japan [26] and is the same as that described above in sporadic US cases of HTLV-positive malignancies. Examination of sera from ATL patients and healthy donors from endemic and nonendemic areas of Japan showed that almost all ATL patients had HTLV-specific antibodies and a number of normal individuals from this region were also HTLV antibody positive [2–4,25,27]. Recently, a second area of high prevalence of HTLV has been identified in the Caribbean and regions of South America and Africa. T-cell leukemia-lymphoma appears to have a relatively high incidence in the Caribbean and antibody to HTLV is found in patients with the disease as well as some normal individuals [2–4,28]. Other cases of HTLV-associated T-cell leukemia-lymphoma have also been found in Boston, Seattle, Alaska, the southeastern United States, Central and South America, Africa, and Israel. An interesting feature of HTLV seroepidemiology is the increased incidence of HTLV infection in families of HTLV-positive patients, compared with the levels observed in surrounding population [2–4]. This may indicate a need for intimate contact for transmission.

Cloned DNA sequences derived from approximately 1 Kb of the 5' and 3' termini of the HTLV genome as well as a 4–5 Kb defective HTLV provirus flanked by cellular sequences [29], have been used as probes to look for HTLV-related DNA sequences in fresh leukemic cells and tumor tissues from patients with different lymphoid and myeloid malignancies [23]. These studies show that: (1) cells from some patients with mature T-cell malignancies, including all patients with ATL, contain one or few copies of HTLV provirus, whereas cells from other types of malignancies involving immature T cells, B cells, or myeloid cells are by and large negative; (2) the correlation of the surveys by molecular hybridization and by serology is not 100%, e.g., a small percentage of leukemic patients with HTLV proviral sequences in the DNA of these infected cells do not have detectable antibodies to HTLV proteins in their sera; and (3) the tumor cells are clonal expansions of single infected cells with respect to the provirus integration sites. In animal systems, monoclonality has been shown to be a common feature of leukemias induced by chronic leukemia viruses but not those induced by *onc* gene-carrying retroviruses (acute leukemia viruses). In addition, several of the neoplastic T cells producing HTLV contain a single copy provirus of 8.5–9.0 Kb with no suggestion of a second defective component. Therefore, in spite of its capacity to transform cells *in vitro*, HTLV probably does not carry an *onc* gene.

Mechanism of Transformation

Chronic leukemia viruses, which represent most retroviruses associated with leukemia and lymphoma in nature, sometimes induce malignancies apparently by activating specific cellular genes in the vicinity of the provirus integration sites. However, other mechanisms also appear to be involved. In the case of B-cell lymphomas induced by chicken viruses, the provirus integrates into a common domain of several kilobases, containing the cellular gene c-*myc*, which is activated as a result of integration [5]. Since HTLV specifically transforms mature T cells, integration of its provirus most likely affects directly or indirectly the expression of genes important to T-cell proliferation. Recently, a gene has been isolated at high levels in all HTLV-positive neoplastic T cells and in normal cord blood T cells after infection with HTLV, but not the uninfected counterparts [30]. This gene, termed HT-3 is a single copy cellular gene transcribed into a 2.3 Kb mRNA at high levels in all HTLV-infected cells, including cord blood T lymphocytes infected *in vitro* with HTLV. This gene is also activated in normal peripheral blood lymphocytes stimulated with lectins [30]. These cells are known to express high levels of TCGF and TCGF receptors upon lectin induction but HT-3 is unrelated to a cloned TCGF gene. HT-3 appears to be important in T-cell proliferation, and specific activation of this gene by HTLV may be linked to a primary transformation event. The TCGF gene is not expressed in some of the HTLV-transformed cell lines, suggesting that at least in these cases the virus bypasses the usual growth regulatory mechanisms. As noted earlier, HTLV-transformed cells, however, do have receptors for TCGF, which can be easily detected by an anti-TAC antibody that specifically recognizes the TCGF receptors [2–4].

HTLV In AIDS

Several HTLV isolates have recently been obtained from patients with AIDS [31] or patients with lymphadenopathy [32]. HTLV-related antibodies have also been observed in a number of AIDS patients and not in the vast majority control subjects [33]. Although it is premature to state whether HTLV infection actually causes these immunosuppressive diseases, there are several parallels between the two. Since HTLV is T-cell tropic, and the primary defect in AIDS is also in T cells, it is possible that HTLV or related virus can cause T-cell proliferation (leukemia-lymphoma) in some cases and T-cell suppression (AIDS) in others. Precedents for this phenomenon exist in nature and a prime example is the feline leukemia virus. Feline leukemia is T-cell tropic, produces T-cell leukemia, and also produces immunosuppression in infected cats [34]. HTLV is prevalent in the Caribbean (especially Haiti) and in Africa, and these two areas have been suggested as possible reservoirs for the AIDS agent. AIDS and HTLV-induced T-cell malignancies appear to spread by intimate contact and may require cell-to-cell transmission. Recent studies show that HTLV sequences in DNA of two patients with AIDS detected initially were not detected in subsequent blood samples from the same patients [35]. This observation points to the difficulties that one may encounter in virus detection and isolation in the advanced stages of the disease in patients with AIDS. Recently, a variant of HTLV, HTLV-III, has been isolated from a number of patients with AIDS and pre AIDS [36–39] and appears to be involved in the causation of the disease. With the availability of a cell line that can produce large quantities of HTLV-III, it should be possible to develop a blood test for detection of HTLV-III infection in blood donors and homosexual population at risk and to develop a vaccine to control the disease.

Future studies will focus on the mechanism of cell transformation and the role of HTLV and other related retroviruses in human neoplasms. It will be important to determine whether specific-cell gene activation is an important mechanism for immortalization of T cells and development of malignancies including Kaposi sarcoma and AIDS. Such studies may also define a role of chronic leukemia viruses in human neoplasia.

References
1. Gallo RC, Wong-Staal F (1982) Retroviruses as etiologic agents of some animal and human leukemias and lymphomas and as tools for elucidating the molecular mechanisms of leukemogenesis. Blood 60:545–557
2. Gallo RC, Wong-Staal F, Sarin PS (1984) Cellular onc genes, T cell leukemia-lymphoma virus and leukemias and lymphomas of man. *In* Dacie JV, Goldman JM, Jarrett JO (eds) Leukemia Today—Mechanisms of Viral Leukemogenesis. Churchill-Livingstone, London, p 11
3. Sarin PS, Gallo RC (1983) Human T cell leukemia-lymphoma virus (HTLV). *In* Brown EB (ed) Progress in Hematology. Grune and Stratton, New York, p 149
4. Sarin PS, Gallo RC (1983) T cell proliferation and human T cell leukemia virus (HTLV). *In* 13th International Cancer Congress, Part C. Alan R. Liss Inc., New York, p 147
5. Hayward WS, Neel BG, Astrin SM (1981) Induction of lymphoid leukosis by avian leukosis virus; activation of a cellular *onc* gene by promotor insertion. Nature 290:475–480
6. Burny AS, Bruck C, Chantrenne H, *et al.* (1983) Bovine leukemia virus: Molecular biology and epidemiology. *In* Klein G (ed) Viral Oncology. Raven Press, New York, p 231
7. Morgan DA, Ruscetti FW, Gallo RC (1976) Selected *in vitro* growth of T lymphocytes from normal human bone marrow. Science 193:1007–1008
8. Sarin PS, Gallo RC (1984) Human T cell growth factor (TCGF). CRC Critical Reviews in Immunology 4:279–305
9. Poiesz BJ, Ruscetti FW, Gazdar AF, Bunn PA, Minna JD, Gallo RC (1980) Detection and isolation of type-C retrovirus particles from fresh and cultured lymphocytes of a patient with cutaneous T-cell lymphoma. Proc Natl Acad Sci USA 77:7415–7419
10. Reitz MS, Poiesz BJ, Ruscetti FW, Gallo RC (1981) Characterization and distribution of nucleic acid sequences of a novel type-C retrovirus isolated from neoplastic human T-lymphocytes. Proc Natl Acad Sci USA 78:1887–1891
11. Kalyanaraman VS, Sarngadharan MG, Poiesz BJ, Ruscetti FW, Gallo RC (1981) Immunological properties of a type-C retrovirus isolated from cultured human T-lymphoma cells and comparison to other mammalian retroviruses. J Virol 38:906–915
12. Oroszlan S, Sarngadharan MG, Copeland TD, Kalyanaraman VS, Gilden RV, Gallo RC (1982) Primary structure analysis of the major internal protein p24 of human type C T-cell leukemia virus. Proc Natl Acad Sci USA 79:1291–1294

13. Gallo RC, Mann D, Broder S, et al. (1982) Human T-cell leukemia-lymphoma virus (HTLV) is in T but not B-lymphocytes from a patient with cutaneous T-cell lymphoma. Proc Natl Acad Sci USA 79:4680–4683
14. Popovic M, Sarin PS, Robert-Guroff M, et al. (1983) Isolation and transmission of human retrovirus (HTLV). Science 219:856–859
15. Sarin PS, Aoki T, Shibata A, et al. (1983) High incidence of human type-C retrovirus (HTLV) in family members of an HTLV-positive Japanese T-cell leukemia patient. Proc Natl Acad Sci USA 80:2370–2374
16. Popovic M, Lange-Wantzin G, Sarin PS, Mann D, Gallo RC (1983) Transformation of human umbilical cord blood T cells by T cell leukemia/lymphoma virus. Proc Natl Acad Sci USA 80:5402–5406
17. Markham PD, Salahuddin Z, Kalyanaraman VS, Popovic M, Sarin PS, Gallo RC (1983) Infection and transformation of fresh human cord blood cells by multiple sources of human T cell leukemia/lymphoma virus (HTLV). Int J Cancer 31:413–420
18. Haynes BF, Miller SE, Moore JO, Dunn PH, Bolognesi DP, Metzgar RS (1983) Identification of human T cell leukemia virus in a Japanese patient with adult T cell leukemia and cutaneous lymphomatous vasculitis. Proc Natl Acad Sci USA 80:2054–2058
19. Kalyanaraman VS, Sarngadharan MG, Robert-Guroff M, Blayney D, Golde D, Gallo RC (1982) A new subtype of human T-cell leukemia virus (HTLV-II) associated with a T-cell variant of hairy cell leukemia. Science 218:571–573
20. Miyoshi I, Kubonishi I, Sumida M, et al. (1980) A novel T cell line derived from adult T cell leukemia. Gann 71:155–156
21. Yoshida M, Miyoshi I, Hinuma Y (1982) Isolation and characterization of retrovirus from cell lines of human adult T cell leukemia and its implication in the disease. Proc Natl Acad Sci USA 79:2031–2035
22. Popovic M, Reitz MS Jr, Sarngadharan MG, et al. (1982) The virus of Japanese adult T-cell leukaemia is a member of the human T-cell leukaemia virus group. Nature 300:63–66
23. Wong-Staal F, Hahn B, Manzari V, et al. (1983) A survey of human leukemias for sequences of a human retrovirus HTLV. Nature 302:626–628
24. Kalyanaraman VS, Sarngadharan MG, Bunn PA, Minna JD, Gallo RC (1981) Antibodies in human sera reactive against an internal structural protein of human T-cell lymphoma virus. Nature 294:271–273
25. Gallo RC, Kalyanaraman VS, Sarngadharan MG, et al. (1983) The human type-C retrovirus: Association with a subset of adult T-cell malignancies. Cancer Res 43:3892–3899
26. Uchiyama, T, Yodoi J, Sagawa K, Takatsuki K, Uchino H (1977) Adult T-cell leukemia: Clinical and hematologic features of 16 cases. Blood 50:481–492
27. Robert-Guroff M, Nakao Y, Notake K, Ito Y, Sliski A, Gallo RC (1983) Natural antibodies to human retrovirus HTLV in a cluster of Japanese patients with adult T-cell leukemia. Science 215:975–978
28. Blattner WA, Kalyanaraman VS, Robert-Guroff M, et al. (1982) The human type-C retrovirus, HTLV, in Blacks from the Caribbean region and relationship to adult T-cell leukemia/lymphoma. Int J Cancer 39:257–264
29. Manzari V, Wong-Staal F, Franchini G, et al. (1983) Human T-cell leukemia-lymphoma virus, HTLV: Cloning of an integrated defective provirus and flanking cellular sequences. Proc Natl Acad Sci USA 80:1574–1578

30. Manzari V, Gallo RC, Franchini G, Westin, et al. (1983) Abundant transcription of a cellular gene in T-cells infected with human T-cell leukemia-lymphoma virus (HTLV). Proc Natl Acad Sci USA 80:11–15
31. Gallo RC, Sarin PS, Gelmann EP, et al. (1983) Isolation of human T-cell leukemia virus in acquired immune deficiency syndrome (AIDS). Science 220:865–867
32. Chermann JC, Rey F, Barre-Sinoussi F, et al. (1983) Isolation of a T-lymphotrophic retrovirus from a patient at risk for acquired immune deficiency syndrome (AIDS). Science 220:868–871
33. Essex M, McLane MF, Lee TH, et al. (1983) Antibodies to cell membrane antigens associated with human T-cell leukemia virus in patients with AIDS. Science 220:859–862
34. Trainin Z, Wernicke D, Ungar-Waron H, Essex M (1983) Suppression of the humoral antibody response in natural retrovirus infections. Science 220:858–859
35. Gelmann EP, Popovic M, Blayney D, et al. (1983) Proviral DNA of a retrovirus, human T-cell leukemia virus, in two patients with AIDS. Science 220:862–865
36. Gallo RC, Salahudin Z, Popovic M, et al. (1984) Frequent detection and isolation of a human T lymphotropic retrovirus, HTL-III from patients with AIDS and at risk for AIDS. Science, in press.
37. Popovic M, Sarngadharan M, Read E, et al. (1984) Detection, isolation and continuous production of cytopathic human T-lymphotropic retroviruses (HTLV-III) from patients with AIDS and pre AIDS. Science, in press.
38. Sarngadharan M, Popovic M, Bruch L, et al. (1984) Antibodies reactive with human T lymphotropic retrovirus (HTLV-III) in the sera of patients with acquired immune deficiency syndrome. Science, in press.
39. Schupbach J, Popovic M, Gilden RV, et al. (1984) Serologic analysis of a new subgroup of human T-lymphotropic retrovirus (HTLV-III) associated with AIDS. Science, in press.

CHAPTER 40
Hepatitis B Virus Diseases

WILLIAM S. ROBINSON*

Hepatitis B virus (HBV) has a worldwide distribution and is of great medical importance because it may be the most common cause of chronic liver disease including hepatocellular carcinoma (HCC) in humans [1]. In addition, this virus and three close relatives recently found in lower animal species (woodchuck hepatitis virus or WHV in *Marmota monax,* ground squirrel hepatitis virus or GSHV in *Spermophilus beechei,* and duck hepatitis B virus or DHBV in Pekin ducks) share very interesting ultrastructural, antigenic, molecular and biological features that distinguish them from members of all of the previously recognized virus groups (reviewed in [2]). These viruses represent a new virus group that has been called the hepadna viruses [2]. They have very narrow host ranges and have not been transmitted to convenient laboratory animals. Since they have not been propagated in tissue culture, many details of their mechanism of replication are still not known. However, several unique features are apparent and one of these is virion ultrastructure (reviewed in [2–4]). The hepatitis B virion or Dane particle is a spherical particle approximately 42 nm in diameter with an electron dense spherical inner core or nucleocapsid and an outer shell or envelope. The lipid-containing envelope bears the hepatitis B surface antigen (HBsAg), which is contained in a viral genome-specified polypeptide to which virus neutralizing antibody (anti-HBs) is directed. The nucleocapsid, which can be released from virions by detergent treatment, bears the hepatitis B core antigen (HBcAg) and contains the viral DNA, a DNA polymerase activity, and a protein kinase activity that phosphorylates the viral genome-specified major polypeptide of the core. HBcAg exists in serum only as an internal antigen of virions. In addition to virions, there are numerous small

---
* Stanford University School of Medicine, Stanford, California.

spherical and long filamentous particles that bear HBsAg in serum of HBV-infected patients. These particles are composed of lipid, protein, and carbohydrate, and lack HBcAg, nucleic acid, or other virion core components. These are considered to be incomplete viral forms.

The third HBV antigen, the hepatitis Be antigen (HBeAg), was first detected as a soluble antigen in serum and is antigenically and physically distinct from the HBsAg particles described above (reviewed in [3]). HBeAg also appears to be present in a cryptic form in the virion core and can be detected only after disruption of core particles (for example, by detergent treatment). The major polypeptide of virion cores appears to manifest HBeAg specificity when isolated from cores.

The structure of the DNA of these viruses is also unique (reviewed in [2–4]). The viral DNA consists of small circular molecules that are partly single stranded and the DNA polymerase activity in the virion can repair the DNA to make it fully double stranded. Recent studies indicate that the mechanism of viral DNA replication is unique and appears to involve an RNA intermediate (reviewed in [4]). Unique biological features include striking tropism of the viruses for hepatocytes and the common occurrence of persistent infection characterized by high concentrations of viral antigen and infectious virus in the blood and lower concentrations in certain other body fluids continuously for years. This pattern of infection accounts for the common transmission of HBV by percutaneous transfer of serum and serum-containing material. In geographic areas of the world where HBV infections are most common, such as in eastern Asia and sub-Saharan Africa, rates of persistent infection exceed 10% in some populations [1] and the associated liver disease is among the most common and important of all health problems in these populations.

One of the intriguing features of these viruses is the spectrum of disease syndromes that may occur during acute and chronic infection and the evidence that several different pathogenetic mechanisms are involved. Acute and chronic hepatitis B represent syndromes of hepatocellular necrosis and inflammatory responses associated with HBV infection of hepatocytes. Primary HBV infection may be associated with little or no liver disease or with acute hepatitis of varying severity from mild to fulminant. Persistent HBV infection is sometimes associated with a histologically normal or near normal liver and normal liver function and sometimes with syndromes designated chronic persistent hepatitis (CPH) or chronic active hepatitis (CAH). CPH is considered not to be progressive and CAH may be more severe and progress to cirrhosis. The mechanism of liver cell injury in hepatitis B has not been established, but there has been much conjecture about the role of the immune response (reviewed in [5,6]). Several factors have been identified that appear to correlate with severity of acute or chronic hepatitis B and may provide clues about pathogenesis. Among these is the infecting dose of virus, with high doses of HBV usually resulting in shorter incubation periods and more severe acute hepatitis than low infecting doses [7]. A second factor appears to be age. Hepatitis B virus infections at very young ages are usually associated with very mild initial hepatitis [8]. Ancedotal cases suggest that HBV infections are associated with milder initial disease in immunologically impaired hosts than in immunologically normal individuals [9]. These observations raise the possibility that the immune response may influence the severity of acute hepatitis B.

During acute and chronic HBV infection there is almost always a humoral immune response to HBcAg [10] and frequently to HBsAg [11] and during chronic infection to HBeAg [12]. There is also evidence that humoral immunity to hepatic (host) antigens occurs in chronic hepatitis B (reviewed in [5]). However, a role for the humoral immune response in hepatic injury would appear to be excluded by the observation that severe acute and chronic viral hepatitis can occur in the absence of an intact humoral immune response, such as in patients with agammaglobulinemia [13].

Cellular immune responses directed against HBsAg and/or liver cell antigens in acute and chronic HBV infection have also been detected by numerous investigators (reviewed in [5,6]). Specific cellular immune reactivity to HBsAg has been detected by leukocyte migration inhibition and lymphocyte transformation assays during acute hepatitis B, variably during chronic hepatitis B, and most strongly and regularly during recovery and early convalescence. Cellular sensitization to HBsAg by these assays is frequently undetectable or weak in HBsAg carriers without liver disease. The localization of HBsAg on the cell surface in patients with CAH (in contrast to its cytoplasmic location in those with little or no liver disease) by immunofluorescent staining [14] is considered to be an appropriate state for immune attack directed at HBsAg. On the other hand, circulating cytotoxic effector T cells directed against HBsAg-bearing target cells have not been clearly demonstrated. Although claims for the cytotoxicity of lymphocytes in patients with hepatitis B against artificial or nonphysiological target cells such as avian erythrocytes bearing chemically coupled HBsAg have been made, antigen-specific cytotoxic effector cell activity against human hepatocellular carcinoma cell lines producing HBsAg has not been detected (reviewed in [5,6]). In these studies, target cells were not HLA matched with the lymphocyte donors, a condition that is probably necessary for optimum detection of antigen-specific cytotoxicity. Thus, no convincing evidence of lymphocyte cytotoxicity directed against cells bearing HBsAg has been reported, although optimum testing for such cytotoxic effector cell activity has apparently not been carried out.

Cellular sensitization to host antigens in viral hepatitis has also been demonstrated by several investigators (reviewed in [6]). Sensitization to a liver-specific hepatocyte surface membrane lipoprotein (liver-specific protein, or LSP) has been detected in approximately one-half of patients with acute hepatitis B, but is transient and has not been demonstrated following recovery. Reactivity to the same antigen is also detectable in most patients with chronic active hepatitis (whether HBsAg positive or not). Suppressor T-cell activity is also depressed in the blood of patients with acute and chronic hepatitis B; this activity returns to normal with recovery [15].

The findings described here have led some to conclude that cellular reactivity to HBsAg and/or hepatocyte antigens (e.g., LSP) play a centrol role in liver injury in acute and chronic hepatitis B, but this mechanism does not appear to be proven at this time. So far, the failure to clearly demonstrate cytotoxic effector T cells directed against HBsAg-bearing target cells in patients with hepatitis B, leaves HBsAg in doubt as an important target antigen. Certain viral markers in the serum and liver of persistently infected patients appear to be present more often in the presence than in the absence of significant chronic hepatitis. HBeAg [12,16] and virion DNA polymerase activity [16] in serum, and HBcAg as well as HBsAg in the liver detected by immunofluorescent staining [14] are often found in patients with CAH or CPH. There is little difference between the prevalence of these markers, however, in CAH and CPA. In contrast, "healthy" carriers, including those with little or no abnormality documented by liver biopsy [16], appear most often to be HBeAg negative, virion DNA polymerase negative, and frequently anti-HBe positive [12,16] and liver biopsies reveal only HBsAg and no HBcAg by immunofluorescent staining [14]. These findings suggest that complete virus replication is more often proceeding in liver cells of patients with chronic hepatitis B compared with the livers of carriers with little or no hepatitis in which oftentimes only the expression of the HBsAg viral gene product can be detected.

These findings raise the possibility that it is the cellular immune response directed at HBcAg or HBeAg (or some as yet undescribed viral antigen) rather than at HBsAg that is responsible for hepatic injury during HBV infection. Indeed, a recent study has provided evidence for cytotoxicity of peripheral blood cells for autologous hepatocytes in chronic hepatitis B that is blocked by anti-HBc, suggesting that HBcAg may be the target antigen for the cytotoxic cell [17].

A prominent reason supporting the hypothesis that liver cell injury in hepatitis B is caused by immune attack is the belief that HBV is not a cytopathic virus. Evidence for this is the observation that hepatoma cell lines in culture that contain the entire HBV genome, but express only HBsAg and are not replicating complete virus, have no apparent impairment in growth or function [18]. Also, healthy carriers producing only HBsAg may have little or no liver injury [16]. These observations do not exclude the possibility, however, that when HBV-infected cells express HBcAg, HBeAg, and other viral gene products in addition to HBsAg and are replicating viral DNA, the infection is cytopathic. Again, the association of HBeAg in serum and HBcAg in liver with active hepatitis is consistent with such a mechanism. Thus, a direct cytotoxic effect of the virus must still be considered a possible mechanism for hepatic injury by HBV.

There are striking differences in liver disease associated with other hepadna virus infections. Although woodchucks with chronic WHV infection often have moderately severe active hepatitis, cirrhosis as seen in humans with HBV infection has not been observed in woodchucks (reviewed in [2]). GSHV-infected ground squirrels consistently have little or no hepatitis [2]. These differences may be important for future studies to determine which viral, host, or environmental factors may be involved in pathogenesis of the liver disease associated with infections by these viruses. With regard to the mechanism of liver cell injury, although HBeAg and DNA and DNA polymerase-containing virions in the blood, and HBcAg in liver are associated with chronic hepatitis in humans noted above, this is not the case in ground squirrels. These animals have little or no hepatitis, although the concentrations of DNA and DNA polymerase-containing virions in the blood are far higher than in patients with active hepatitis [2]. Thus at least in the ground squirrel, complete virus replication does not appear to lead to liver cell damage.

A final possible mechanism to be considered for liver cell injury during HBV infection is one that appears to be unique to HBV among known viral diseases of man. That is coinfection with a second cytopathic virus, the delta agent, which appears to be dependent in some way on HBV. Delta antigen was discovered by immunofluorescent staining as a nuclear antigen distinct from HBsAg, HBcAg, and HBeAg in hepatocytes of some HBsAg carriers in Italy, the geographic area with the highest prevalence of delta antigen (reviewed in [3]). Most patients with delta antigen in liver have antibody to delta antigen (anti-delta) in their sera. Surveys show a high prevalence of anti-delta in Italians residing in Italy and elsewhere, in intravenous drug users, and in polytransfused HBsAg carriers. Although antigen has not been found in HBsAg-negative patients, anti-delta was found in low prevalence in polytransfused HBsAg-negative patients, but only in those with anti-HBs. Delta antigen purified from hepatic tissue was found to be an approximately 68,000-molecular-weight protein and in serum it appears to be an internal component of a discrete subpopulation of HBsAg particles containing low molecular weight RNA. Inoculation of delta antigen-containing sera into HBsAg carrier chimpanzees without hepatic delta antigen resulted in the appearance of delta antigen in the hepatocyte nuclei of these animals, the disappearance of HBcAg detected by immunofluorescent staining in liver, and a rise in serum alanine aminotransferase (ALT). These findings have led to the interesting postulate that delta is an antigen of an infectious agent (delta agent) that is defective; its replication requires coinfection with HBV; phenotypic mixing results in delta antigen-containing particles with HBsAg-containing envelopes; and infection with the agent results in hepatic injury. More work will be required to determine how correct this model is.

Early studies of the role of delta agent in human hepatitis have found a higher incidence of delta antigen in HBsAg-positive patients with acute and chronic hepatitis than in asymptomatic carriers. Although researchers have speculated that simultaneous infection with HBV and delta may lead to severe or fulminant hepatitis more often than infections with HBV alone, and cases of fulminant hepatitis B positive for delta antigen have been observed, in the United States and Ireland, at least, delta antigen appears to be present in only a small fraction of fulminant hepatitis B cases. Thus, delta agent probably plays no role in most cases [19]. Delta may play a more prominent role in geographic areas where it is more common, such as in Italy where 32% of HBsAg carriers with CAH and 52% with cirrhosis were found to be delta antigen positive compared with HBsAg carriers with no liver disease in which none had detectable delta [20]. It seems likely that exacerbations of hepatitis may occur in HBsAg carriers when they subsequently acquire delta infection, but how often this mechanism accounts for such exacerbations and the precise impact of delta on the severity of acute and chronic hepatitis B in this country remain to be determined. The mechanism of liver cell injury associated with delta infection has not been studied.

A second form of disease associated with chronic HBV infection is hepatocellular carcinoma (HCC), in which there is increasing evidence supporting an etiologic role for HBV. HCC has a worldwide distribution and, numerically, is one of the major cancers in the world today (reviewed in [1]). Although rare in most parts of the world, HCC occurs commonly in sub-Saharan Africa, southeast Asia, Japan, Oceania, Greece, and Italy. In certain areas of Asia and Africa, it is the most common cancer. Geographic areas with the highest incidence of HCC are also areas where hepatitis B virus infection is common and persistent HBV infections occur at the highest known frequencies. Within the limits of the data available, there appears to be a good correlation between the worldwide geographic distribution of HCC and active HBV infection, with the highest frequency of both occurring in sub-Saharan Africa and southeast Asia. In addition, active HBV infection occurs significantly more frequently in patients with HCC than in controls in both high- and low-HCC-incidence geographic areas. A recent prospective study of HBsAg-positive and HBsAg-negative middle-aged males in Taiwan revealed an incidence of HCC over 300 times higher in the HBsAg-positive group than in the HBsAg-negative group [21].

The high incidence of persistent HBV infection in mothers of HCC patients, in contrast to fathers (reviewed in [1]), suggests that transmission from mothers to newborn or infant children may be a common mode of HBV infection in HCC patients. The finding of low serum HBsAg titers and absence of HBeAg, together with the rare occurrence of hepatic HBcAg in most HCC patients, also suggest that the persistent infections in HCC patients are long lasting. If HBV infection does occur frequently at very early ages in HCC patients in high-HCC-incidence areas, the age distribution of patients with clinically recognized tumors would suggest that these tumors appear after a mean duration of approximately 35 years of HBV infection. Very few cases of HCC occur in children. Between 60% and 90% of HCC patients have coexisting cirrhosis, suggesting that this lesion in association with persistent HBV infection may predispose a person to HCC, although clearly cirrhosis need not be present.

Hepatitis B virus infection in humans is not the only hepadna virus infection associated with HCC. In fact, a much higher incidence of HCC formation occurs in woodchuck-hepatitis-virus-infected woodchucks (reviewed in [2]) than observed in humans with HBV. Approximately one-third of infected animals in captivity develop HCC per year and no tumors have been observed in uninfected animals. There are interesting differences between humans and woodchucks in underlying liver disease and the pattern of viral markers in serum. Hepatocellular carcinoma develops in woodchucks with histological changes of acute and chronic hepatitis and never with cirrhosis, and with high levels of complete virus in the blood. In contrast, cirrhosis without active inflammatory disease and with low titers of HBsAg and no detectable HBeAg or virion DNA polymerase in serum is the rule in humans with HCC.

Ducks in some regions of China have been noted to develop HCC and DHBV has been found in the same duck populations (reviewed in [2]), although no direct correlation of virus and tumor has been reported. A recent study failed to detect DHBV DNA with detection limit down to 0.2 viral genome copies per cell in three HCC in ducks from China. A similar virus has been found in the blood of commercial duck flocks in this country, but HCC has not been observed in these flocks. No HCC has been observed in 24 GSHV-infected captive ground squirrels followed for 3 years [2]. It will be important to determine whether viral, host, or environmental factors contribute to the differences in HCC incidence in the different host-hepadna virus systems.

Several studies have evaluated the state of virus in HCC tissue from humans and woodchucks, and in tissue culture lines from human HCC. Immunofluorescent and immunoperoxidase staining of tumor tissue have demonstrated that in patients with HBsAg in the blood and in whom nontumorous liver cells are positive for HBsAg and/or HBcAg, tumor cells appear most often to be negative, although some studies have reported small numbers of HBsAg-positive cells in tumors (reviewed in [22]). HBcAg has been detected even more rarely. Thus, few tumor cells appear to express either viral gene product in amounts that can be detected by immunofluorescent staining.

The earliest studies of viral DNA in tumors utilized virion DNA radiolabeled by the endogenous virion DNA polymerase reaction as a probe for hybridization in solution with DNA extracted from HCC tissue. Viral DNA sequences were found in some but not all tumors with a lower limit of detection less than one viral genome copy per cell (reviewed in [3]). More recently, restriction endonuclease digestion of tumor cell DNA and Southern blot analysis has suggested that in addition to free episomal genome length (3200 bp) viral DNA forms, integrated viral DNA can be detected in many although not all HCC (reviewed in [4]). The evidence suggests that integration occurs in a few (usually 1 to 4) specific cellular DNA sites in human and woodchuck tumors but always at different sites in different tumors. Evidence for this is the finding of one or more DNA fragments containing viral DNA sequences that are larger than unit length viral DNA (3200 bp) after, but not before, digestion of cell DNA with a restriction enzyme (e.g., HindIII) for which no recognition sites exist in the viral DNA.

Cloning and sequence analysis of integrated viral DNA with flanking cellular DNA sequences has proven that viral sequences are integrated in host DNA of HCC (reviewed in [4]). In all cases of woodchuck and human tumors and tumor cell lines studied to date, the integrated viral DNA contains extensive deletions and rearrangements that are different for each integrated viral sequence, and the site in the viral DNA that joins cellular DNA is also different for each integrated viral sequence. HBV DNA sequences appear to be integrated in at least eight specific cellular DNA sites in one hepatoma cell line. These have been shown to be extensively methylated, unlike HBV DNA forms in virions and nontumorous liver in which there was no detectable methylation. The more extensive methylation of the coding sequences for the core polypeptide than sequences coding for the HBsAg polypeptide correlated with the expression of HBsAg but not HBcAg or other viral gene products by this cell line. This finding raises the possibility that methylation of viral DNA may be involved in regulation of viral gene expression in hepatomas.

Evidence by Southern blot analysis for integration of HBV DNA is not unique to HCC, because similar high molecular weight HindIII fragments with HBV DNA sequences have been detected in DNA of acutely and chronically infected nontumorous liver (reviewed in [4]). As in HCC, the specific high-molecular-weight HindIII DNA fragments containing viral sequences were different in livers of different chronically infected patients. The ability to detect such DNA fragment by Southern blotting has been interpreted to mean that viral DNA is integrated in the same site in many different cells of the liver of each chronically infected patient, but the site differs in different patients. The complete meaning of these findings is not clear; however, integration of other viral DNAs (e.g., retroviruses) appears to be random. Specific integration sites are not detected in tissue DNA by the experimental strategy just described unless the cells are of clonal origin (e.g., as are cells in most viral-induced tumors).

Other investigators have obtained evidence for random integration by detecting subgenomic DNA fragments with HBV DNA sequences after digestion of infected liver DNA with a restriction enzyme that cleaves HBV DNA at more than one site, and no fragments greater than genome length after HindIII digestion (reviewed in [4]). The integrated viral DNA sequences in infected liver have not been characterized by cloning and sequence analysis as have those in HCC.

Thus, although there is a strong association between prior long-standing hepadna virus infection and HCC formation in humans and woodchucks, and viral DNA is found integrated apparently at different sites in the cellular DNA of different (but not all) tumors. Exactly how these viruses are involved in HCC formation is not clear.

Superficially, at least, these viruses appear to share some features with RNA tumor viruses or retroviruses (reviewed in [4]), although many features of HBV and the related viruses are unique and define this virus group. Among the common features are similarities in genome structure, although the nucleic acid type in virions is different for the two virus groups (that is, DNA in the case of hepadna viruses and RNA in retroviruses). Separation and repair of the cohesive ends of hepadna virus DNAs result in linear molecules with direct, repeat terminal sequences of approximately 300 bp as described above and similar to those of retroviruses. All of the viral messenger RNAs appear to be transcribed from the same DNA strand and thus in the same direction for viruses of both groups. In addition, the hepadna viral DNA appears to replicate through an RNA intermediate utilizing a reverse transcriptase as described above, a mechanism with some analogy to retrovirus replication. Viruses of both groups cause persistent infection with virus and viral antigen in the blood for many years. A fourth similarity is that viruses of both groups appear to integrate readily in cellular DNA. However, it has yet to be shown that hepadna virus DNA integration is a regular and integral event in virus replication, that the viral genome in the integrated state retains its organization, that integration occurs at a specific site in the viral DNA, or that integration is essential to the mechanism of cell transformation—all of which appear to be features of retroviruses.

A fifth similarity is that when exclusively integrated in the DNA of infected cells, both hepadna viruses and retroviruses may express only the gene for their envelope protein. A sixth similarity is tumor formation during infection by at least some members of each group. The clear association between HBV and WHV infections and hepatocellular carcinoma is among the more intriguing features of these viruses, and it will be of great interest to investigate in more detail their role in formation of these tumors. It will be important to determine whether they may integrate in sites adjacent to oncogenes and function as some retroviruses are thought to function in cell transformation and tumor induction or cause tumors by some other mechanism. Although researchers have speculated that some environmental factor such as a chemical carcinogen in addition to virus may be necessary for HCC formation, there is no evidence for such a mechanism.

Several additional syndromes with extrahepatic manifestations have been associated with HBV infection and for some there is evidence that HBsAg-anti-HBs complexes play a role in pathogenesis (reviewed in [3,23,24]). The serum sickness-like syndrome consisting of fever, rash, urticaria, arthralgias, and sometimes acute arthritis, which occurs in 10%–20% of patients during the incubation period of acute hepatitis B, is accompanied by HBsAg-antibody complexes and low levels of complement components in serum, synovial fluid, and synovial membranes from involved joints.

In several series, one-third to one-half of patients with biopsy-proven polyarteritis nodosa have had persistent HBV infection (reviewed in [3,23,24,25]). Among all HBsAg carriers, however, this syndrome occurs infrequently. Such patients have low serum complement levels and circulating HBsAg-anti-HBs complexes. Immune complexes and complement components have also been regularly detected in diseased vessels by immunofluorescent staining.

A significant number of cases of membranous glomerulonephritis are associated with chronic active hepatitis and persistent HBV infection (reviewed in [3,23,24,25]). Immune complex deposits can be found along the sub-epithelia surfaces of glomerular basement membranes by electron microscopy. Nodular deposits of HBsAg, immune globulin, and C3 in glomeruli have been observed by immunofluorescent staining in these cases.

Several additional syndromes with unknown pathogenesis have also been associated with HBV infection (reviewed in [3]). Infantile papular acrodermatitis appears to be frequently associated with persistent HBV infection in Mediterranean countries and Japan. In a series of 19 patients with essential mixed cryoglobulinemia, cryoprecipitates were shown to contain HBsAg in 6 cases and anti-HBs in 11, suggesting that some cases of cryoglobulinemia may be related to HBV infection. This association, however, could not be confirmed in a subsequent study. A significant number of cases of aplastic anemia have been observed following acute viral hepatitis. Although in most cases the specific hepatitis virus has not been identified, in a few cases acute hepatitis B has been documented. The basis of this apparent association is not clear. Further work is needed to clarify the pathogenesis of these syndromes and to determine whether HBsAg-antibody complexes or HBV infection in some other way plays an important role.

Clearly, infections with HBV and other hepadna viruses have protean manifestations; different distinct pathogenetic mechanisms appear to be responsible for a number of different disease syndromes.

References
1. Szmuness W (1978) Hepatocellular carcinoma and the hepatitis B virus: Evidence for a causal association. Prog Med Virol 24:40–69
2. Marion PL, Robinson WS (1984) Hepadna viruses: Hepatitis B and related viruses. In Curr Top Microbiol Immunology, in press
3. Robinson WS (1983) Hepatitis B virus. In Deinhardt F, Deinhardt J (eds) Viral Hepatitis—Laboratory and Clinical Science. Marcel Dekker Inc., New York, p 57
4. Robinson WS, Miller R (1984) Structure and replication of hepatitis B virus. Ann Int Med, in press
5. Edginton TS, Chisari FV (1975) Immunological aspects of hepatitis B infection. Am J Med Sci 270:213–227
6. Dienstag JL, Kahn AK, Klingenstein RJ, Savarese AM (1982) Immunopathogenesis of liver disease associated with hepatitis B. In Szmuness W, Alter HJ, Maynard JE (eds) Viral Hepatitis. Franklin Institute Press, Philadelphia, p 221

7. Barker JF, Murray R (1972) Relationship of virus dose of incubation time of clinical hepatitis and time of appearance of hepatitis-associated antigen. Am J Med Sci 263:27
8. Schweitzer IL, Dunn AEF, Peters RL, Spears RL (1973) Viral hepatitis in neonates and infants. Am J Med 55:762
9. London WT, DiFiglia M, Sutnick AI, Blumberg BS (1969) An epidemic of hepatitis in a chronic hemodialysis unit. N Engl J Med 281:571–578
10. Hoofnagle JH, Seef LB, Bales ZB, Gerety RJ, Tabor E (1978) Serologic responses in hepatitis B. *In* Vyas GN, Cohen SN, Schmid R (eds) Viral Hepatitis: A Contemporary Assessment of Etiology, Epidemiology, Pathogenesis, and Prevention. Franklin Institute Press, Philadelphia, p 219
11. Nydegger UE, Miescher PA (1980) Detection and clinical implication of circulating immune complexes in hepatitis B virus infection. *In* Bianchi W, Gorok W, Sickinger K, Stalder GS (eds) Virus and the Liver. MTP Press Ltd., London, p 183
12. Aldershville J, Skinhoj P, Frosner GG, Black G, Deinhardt F, Nielsen JO (1980) The expression pattern of hepatitis B and antibody in different ethnic and clinical groups of hepatitis B surface antigen carriers. J Infect Dis 142:18
13. Good RA, Page AR (1960) Fatal complications of virus hepatitis in two patients with agammaglobulinemia. Am J Med 29:804
14. Gudat F, Bianchi L, Sonnabend W, Theil G, Aenishaenslin W, Stadler GA (1975) Pattern of core and surface expression in liver tissue reflects state of specific immune response in hepatitis B. Lab Invest 32:1–18
15. Chisari FW, Castle KL, Xavier C, Anderson DS (1981) Functional properties of lymphocyte subpopulations in hepatitis B virus infection. I. Suppressor cell control of T-lymphocyte responsiveness. J Immunol 126:38
16. Hess G, Arnold W, Shih JWK, Kaplan PM, Purcell RH, Gerin JL, Myer zum Bueshenfelde KH (1977) Expression of hepatitis B virus-specific markers in asymptomatic hepatitis B surface antigen carriers. Infect Immun 17:550
17. Eddleston ALWF, Mondelle M, Mieli-Vergani G, Williams R (1982) Lymphocyte cytotoxicity to autologous hepatocytes in chronic hepatitis B virus infection. Hepatology 2:122
18. Marion PL, Salazar FH, Alexander JJ, Robinson WS (1979) Polypeptides of hepatitis B virus specific antigen produced by a human cell line. J Virol 32:796
19. Tabor E, Ponzetto A, Gerin JL, Gerety RJ (1983) Does delta agent contribute to fulminant hepatitis? Lancet 1:765
20. Shattock A, Morgan B, Peutherer J, Inglis J, Fielding T, Kelly M (1983) High incidence of delta antigen in serum. Lancet 2:104–105
21. Columbo M, Cambieri R, Rumi M, Ronchi G, Del Ninno E, DeFranchis R (1983) Long-term delta superinfection in hepatitis B surface antigen carriers and its relationship to the course of chronic hepatitis. Gastroenterology 85:235–239
22. Beasley RP, Lin CC, Hwang LY, Chien CS (1981) Hepatocellular carcinoma and hepatitis B virus: A prospective study of 22,707 men in Taiwan. Lancet 2:1129–1133
23. Kew MD (1978) Hepatoma and HBV. *In* Vyas GM, Cohen SN, Schmid R (eds) Viral Hepatitis: A Contemporary Assessment of Etiology, Epidemiology, Pathogenesis, and Prevention. Franklin Institute Press, Philadelphia, p 439

24. Gocke D (1975) Extrahepatic manifestations of viral hepatitis. Am J Med Sci 270:49
25. Gocke DJ (1978) Immune complex phenomena associated with hepatitis. *In* Vyas GN, Cohen SN, Schmid R (eds) Viral Hepatitis: A Contemporary Assessment of Etiology, Epidemiology, Pathogenesis and Prevention. Franklin Institute Press, Philadelphia, p 277

CHAPTER 41
Herpesviruses and Cancer

FRED RAPP AND MARY K. HOWETT*

Herpesviruses represent a ubiquitous class of DNA viruses, of which many individual members have been implicated in the development of specific neoplasms. In fact, of the five known human herpesviruses, four are associated to varying degrees with human cancers. The incidence of one of these, herpes simplex virus type 2 (HSV-2) is rapidly increasing in today's sexually permissive society. This review discusses the evidence that HSV-2 plays a causal role in human cervical cancer. An additional review of herpesviruses and cancer can be found in the two-volume treatise edited by Rapp [1].

Cervical cancer, an intraepithelial neoplasm, is the most common cancer of the female reproductive system, responsible for approximately 7200 deaths in the United States in 1982. However, in lower socioeconomic groups and Third World countries, regular Papanicolaou testing is not carried out and mortality from this tumor is significantly higher. This cancer, therefore, represents a major oncologic disease and the recent increase in infection with HSV-2 may eventually result in an even higher incidence of cervical cancer.

The attractive notion that cervical cancer could be caused by a transmissible agent first arose in the mid-19th century when Rigoni-Stern postulated that occurrence of this neoplasm was significantly increased in sexually active females. In the 150 years following this observation, serologic and epidemiologic studies have strongly suggested that genital herpetic infection may play a causal role in this tumor. This article highlights the molecular evidence that HSV-2 can transform cells and specifically discusses current knowledge of virus involvement in the transformation of cervical cells.

---

* The Pennsylvania State University College of Medicine, Hershey, Pennsylvania.

For details of the seroepidemiologic data implicating HSV-2 in this cancer, we refer readers to an excellent review by Nahmias and Norrild [2]. Evidence for involvement of a second virus group, the papilloma viruses, in cervical cancer has been reviewed by Howley [3]. In order to limit the bibliography, this review includes only very recent and/or key references.

Role of HSV-2 in Transformation of Cells

Notwithstanding the accumulation of seroepidemiologic evidence associating HSV-2 with cervical anaplasia, dysplasia, and neoplasia, the first direct evidence that this virus could transform cells from a normal to an oncogenic phenotype was accomplished when Duff and Rapp [4] transformed hamster cells in culture by infecting them with partially inactivated (ultraviolet-irradiated) virus. Since that time, transformation of other cells in culture has been achieved by a number of investigators using other methods of virus inactivation for type 1 (HSV-1) as well as for HSV-2. Human cells have not yet been successfully transformed to a malignant phenotype, although there have been reports that altered cell morphology has been observed after infection of such cells with transforming doses of virus. In addition, a number of investigators have reported that transformation of rodent cells in culture can be achieved by transfecting cells with either sheared virus DNA, virus DNA cleaved by certain restriction endonucleases, or in a few cases with isolated fragments of virus DNA. The details of such experiments have been recently reviewed [5].

The ability to carry out such transformations opened avenues of investigation into the mechanism by which HSV-2 transforms cells. Investigators began to examine transformants for persistence of HSV-2 nucleic acids and/or proteins. Since a variety of HSV proteins have been reported in the transformants, but no single one has been assigned a role in the transformation process, these data are not discussed.

The question of whether HSV-2 genetic information is necessary for maintenance of the transformed state remains unanswered. Early studies with transformed cell lines indicate that animals bearing tumors induced by these cells produce HSV-2-neutralizing antibodies even though these cells do not synthesize infectious virus. The original transformed lines generated by Duff and Rapp contain HSV-2-specific messenger RNA and HSV-2 DNA, usually less than one virus genome equivalent per cell. Because (1) the HSV-2 genome has a large protein coding capacity, (2) HSV-2-transformed lines do not contain rescuable virus, and (3) clearly less than one virus genome per cell genome is present in the cells, the hypothesis that a subset of the virus genome is sufficient for initiation and/or maintenance of transformation seems likely to be correct. Galloway and McDougall [6] reported that a fragment of HSV-2 DNA (BglII-N) of 4.8×10^6 daltons and mapping between 0.58 and 0.63 on the virus genome is capable of morphologically transforming rodent cells. These investigators used focus formation in low serum and anchorage independence as criteria of transformation, and derived cell lines capable of producing tumors. This work extended the observation by Reyes and coworkers [7] that this region of the genome may have morphologic transforming potential. The latter group, however, has not reported tumor formation by their transformants. Southern blot analysis of the cell DNA of such transformants demonstrated that less than one genome copy of this region per cell was present; therefore, the exact role of this region in maintenance or initiation of transformation is unclear. Jariwalla [8] reported oncogenic transformation of hamster embryo cells with a DNA fragment of 16.5×10^6 daltons and mapping on the genome between 0.41 and 0.52. The reason for the disparity between this report and others is also unclear, although differences in laboratory protocols for selecting transformants may have played a role.

Recently Galloway and McDougall [9] reported that a 2.1-Kb fragment from the left half of the HSV-2 *Bgl*II-N fragment was sufficient to initiate transformation. Sequences from this region could not be detected in the established transformed lines, or even in lines transformed after transfection with the entire *Bgl*II-N fragment. In the latter cell lines, only sequences from the right half of the *Bgl*II-N have been found. These investigators suggest that sequences of HSV-2 DNA able to transform cells *in vitro* need not persist in the transformants; but, they have not defined a mechanism by which these sequences initiate transformation. A recent report [10] has suggested that infection with neutral red-inactivated HSV-1 is mutagenic; however, prolonged virus inactivation negated mutational enhancement, suggesting that at least some virus gene expression is required. It was reported that virus DNA does not persist in the mutated cells, but the data were not presented for scrutiny. One can cite several examples in the herpesvirus literature where initial reports of the absence of virus nucleic acids were subsequently found to be incorrect because of the insensitivity of detection methods. Caution should be exercised before adoption of the "hit and run" hypothesis for HSV transformation.

In line with the *in vitro* transformation experiments described above, limited attempts have been made to induce tumors *in vivo* with HSV-2. The lack of experimental animal models is mainly due to the ability of the virus to cause disseminated disease unless very low doses are administered. However, Nahmias and his coworkers [11] observed low-level tumor development (less than 4%) in baby hamsters inoculated with less than 10^3 TCD$_{50}$ of live or inactivated HSV-2. The resulting tumors failed to demonstrate HSV-2 markers. More interestingly, investigators in several laboratories have now reported the effects of intravaginal administration of HSV-2. Muñoz [12] reported induction of a few cervical cancers in mice preimmunized with inactivated HSV-2 and subsequently exposed vaginally to HSV-2 and hormones. This study is difficult to interpret because the number of animals used was small and because a few controls also developed cancer. Another study examined intravaginal infection of mice with formalin-inactivated HSV-2. After insertion for extended time periods (5 times per week for 88 weeks) of tampons soaked with inactivated virus, neoplastic disease of the cervix was noted. Many of the animals (77%) developed preinvasive lesions as early as 15 weeks after the first administration of virus, and a large number developed carcinoma of the cervix and adenocarcinoma of the uterine horn.

The same investigators [13] published studies using HSV-1 and HSV-2 inactivated with formalin or ultraviolet light. After insertion of virus-soaked tampons 5 times per week for 20–90 weeks, premalignant and malignant cervical lesions were detected in 78%–91% of the animals and dysplasia and invasive carcinoma were detected in 18%–66% and 24%–60% of the animals, respectively. The controls remained normal. Furthermore, the frequency of invasive cancer was twice as high after exposure to ultraviolet-inactivated virus. In addition to rodent studies, one group of investigators detected an increased frequency of dysplasia in Cebus monkeys exposed to HSV-2 [14]. The *in vivo* data at this time, though suggestive, require independent confirmation.

Direct Evidence of HSV-2 Genetic Information in Cervical Cancer Biopsies

The first report of HSV-2 DNA in a cervical cancer biopsy was published by Frenkel and her colleagues [15] using DNA–DNA hybridization to measure reassociation of cervical biopsy DNA with a radioactively labeled HSV-2 DNA probe. Attempts to repeat this result have been unsuccessful, probably because the complexity of the whole virus DNA probe is so great that persistence of a small portion of the genome would not be detected in such experiments. Recent studies have turned to the detection of transcribed and, therefore, presumably amplified HSV-specific messenger RNA in biopsy material. Using *in situ* hybridization, several laboratories have reported that cervical dysplasia and carcinoma samples are positive for HSV-2 RNA [16–19] (R. Hyman, personal communication). The percentage of positive biopsies varies in different reports with as many as 70% positive. The quantitative intensity of the hybridized probe also varies. McDougall and his coworkers [19] mapped the HSV-2-specific RNAs from dysplasias and carcinomas *in situ* to three regions of the HSV-2 genome (0.07–0.40, 0.58–0.63, and 0.82–0.85); none of these was found to be present in every positive sample. Recently, these investigators reported [9] that three out of nine invasive carcinomas were positive by this technique. One sample contained RNA hybridizing to probe DNA derived from the 0.58–0.63 region of virus DNA, one contained RNA hybridizing to probe DNA at 0.32–0.40, and the third hybridized to both probes. Meaningful conclusions cannot be drawn until more biopsies are examined. It is also necessary to note that at least one group [17] has reported that data obtained by *in situ* hybridization may be artifactual because a large percentage of biopsies that hybridized to HSV-2 probe also hybridized to adenovirus DNA probe. Identification and characterization of the HSV-2-specific sequences are clearly required.

In addition to searching for nucleic acids, researchers have looked for virus protein markers in cervical tumors. Aurelian and her coworkers [20] detected an HSV-2 antigen, AG-4 (ICP 10; 160k), in 90% of cervical tumor biopsies compared with 10% of controls. This protein may also be expressed in cells transformed by the *Bgl*II-C (0.41–0.52 fragment). In addition, 91% of patients with carcinoma *in situ* had positive antibody titers against AG-4 compared with 5%–9% of control patients [21]. Kawana and his coworkers [22] also examined the frequency of AG-4 antibodies in patients with uterine cervical cancer and found a 47% incidence of antibody compared with 14% in controls. Aurelian suggests that antibody to AG-4 is prognostic in cervical cancer patients since in one study AG-4 antibody became undetectable in women successfully treated for cervical carcinoma.

Levels of AG-4 antibody have been shown to correlate positively with antibody against another HSV tumor-associated antigen (HSV-TTA) [23]. Melnick and his coworkers [24] reported an increased incidence of reactivity against an early nonstructural protein (VP134) found in HSV-2-infected cells; 93% of cervical cancer patients demonstrated antibody to VP134 compared with 30% of breast cancer patients and 40% of controls. The major DNA-binding proteins, ICSP11/12 and ICSP34/35 have also been demonstrated in cervical tumor cells [19,25,26].

More recently, Gilman and coworkers [27] demonstrated antibodies to two HSV-2-associated proteins with molecular weights of 38,000 and 118,000 in cervical cancer patients. The 38,000-MW protein may also be coded by a region of HSV-2 DNA capable of *in vitro* transformation. Cabral and coworkers [28], by use of immunoperoxidase staining, found an HSV-2-specific polypeptide (VP143) in a significant number of dysplasias (31%), carcinomas *in situ* (29%), and invasive squamous cell carcinomas (41%), but not in normal biopsies.

We are left then with a strong indication that HSV-2 is at least one of the causative agents of cervical neoplasia. The questions that remain to be answered include determination of whether the agent acts singularly or in concert with another factor(s) and determination of whether virus genetic influence persists in the tumor. As in many other virus/tumor associations, absolute proof that HSV-2 is a cause of cervical cancer will only be obtained long after widespread administration of an antiviral vaccine or effective antiviral therapy. The social concern about widespread venereal infection with this virus will force researchers to develop preventive measures.

Acknowledgments. The authors' research is supported by Grants CA-18450, CA-27503, and CA-25305 awarded by the National Cancer Institute.

References

1. Rapp F (ed) (1980) Oncogenic herpesviruses (2 Vols). CRC Press, Inc., Boca Raton, Florida
2. Nahmias AJ, Norrild, B (1980) Oncogenic potential of herpes simplex viruses and their association with cervical neoplasia. In Rapp F (ed) Oncogenic Herpesviruses. CRC Press, Inc, Boca Raton, Florida, p 25
3. Howley PM (1982) The human papillomaviruses. Arch Pathol Lab Med 106:429–432
4. Duff R, Rapp F (1971) Properties of hamster embryo fibroblasts transformed in vitro after exposure to ultraviolet-irradiated herpes simplex virus type 2. J Virol 8:469–477
5. Rapp F, Howett MK (1984) Involvement of herpes simplex virus in cervical carcinoma. In Chandra P (ed) Antiviral Mechanisms in the Control of Neoplasia, Vol 21. NATO Advanced Study Institute Series, Plenum Press, New York, in press
6. Galloway DA, McDougall JK (1981) Transformation of rodent cells by a cloned DNA fragment of herpes simplex virus type 2. J Virol 38:749–760
7. Reyes GR, La Femina R, Hayward SD, Hayward GS (1979) Morphological transformation by DNA fragments of human herpes viruses: Evidence for two distinct transforming regions in HSV-1 and HSV-2 and lack of correlation with biochemical transfer of the thymidine kinase gene. Cold Spring Harbor Symp Quant Biol 44:629–641
8. Jariwalla RJ (1981) Neoplastic transforming activity of defined HSV-2 DNA fragments representing 6.5–13% of the viral genome. In Proceedings of the International Workshop on Herpesviruses, July 29–31, Bologna, Italy
9. Galloway DA, McDougall JK (1983) The oncogenic potential of herpes simplex viruses: Evidence for a "hit-and-run" mechanism. Nature 304:21–24
10. Schlehofer JR, zur Hausen H (1982) Induction of mutations within the host cell genome by partially inactivated herpes simplex virus type 1. Virology 122:471–475
11. Nahmias AJ, Del Buono I, Ibrahim I (1975) Antigenic relationship between herpes simplex viruses, human cervical cancer and HSV-associated hamster tumours. IARC Sci Publ 11:309–313
12. Muñoz M (1973) Effect of herpes simplex virus type 2 and hormonal imbalance on the uterine cervix of the mouse. Cancer Res 33:1504–1508
13. Wentz WB, Reagan JW, Heggie AD, Fu Y, Anthony DD (1981) Induction of uterine cancer with inactivated herpes simplex virus, types 1 and 2. Cancer 48:1783–1790
14. Palmer AE, London WT, Nahmias AJ, Naib ZM, Tunca J, Fuccillo DA, Ellenberg JH, Sever JL (1976) A preliminary report on investigation of oncogenic potential of herpes simplex virus type 2 in Cebus monkeys. Cancer Res 36:807–809
15. Frenkel M, Roizman B, Cassai E, and Nahmias A (1972) A DNA fragment of herpes simplex 2 and its transcription in human cervical cancer tissue. Proc Natl Acad Sci USA 69:3784–3789
16. Eglin RP, Sharp F, MacLean AB, MacNab JCM, Clements JB, Wilkie NM (1981) Detection of RNA complementary to herpes simplex virus DNA in human cervical squamous cell neoplasms. Cancer Res 41:3597–3603

17. Maitland NJ, Kinross JH, Busuittel A, Ludgate SM, Smart GE, Jones KW (1981) The detection of DNA tumour virus-specific RNA sequences in abnormal human cervical biopsies by *in situ* hybridization. J Gen Virol 55:123–137
18. McDougall JK, Galloway DA, Fenoglio CM (1980) Cervical carcinoma: Detection of herpes simplex virus RNA in cells undergoing neoplastic change. Int J Cancer 25:1–9
19. McDougall JK, Crum CP, Fenoglio CM, Goldstein LC, Galloway DA (1982) Herpesvirus-specific RNA and protein in carcinoma of the uterine cervix. Proc Natl Acad Sci USA 79:3853–3857
20. Aurelian L, Davis HJ, Julian CG (1973) Herpesvirus type 2 induced tumor specific antigen in cervical carcinoma. Am J Epidemiol 98:1–9
21. Aurelian L, Cornish JD, Smith MF (1975) Herpesvirus type 2-induced tumor specific antigen (AG-4) and specific antibodies in patients with cervical cancer and controls. IARC Sci Publ 11:79–87
22. Kawana T, Sakamoto S, Kasamatsu T, Aurelian L (1978) Frequency of anti-AG-4 antibodies in patients with uterine cervical cancer and controls. Gann 69:389–391
23. Notter MFD, Docherty JJ (1976) Comparative diagnostic aspects of herpes simplex virus tumor-associated antigens. J Natl Cancer Inst 57:483–488
24. Melnick JL, Courtney RJ, Powell KL, Schaffer PA, Benyesh-Melnick M, Dreesman GR, Anzai T, Adam E (1976) Studies on herpes simplex virus and cancer. Cancer Res 36:845–856
25. Dreesman GR, Burek J, Adam E, Kaufman RH, Melnick JL (1980) Expression of herpesvirus induced antigens in human cervical cancer. Nature 283:591–593
26. McDougall JK, Galloway DA, Purifoy DJM, Powell KL, Richart RM, Fenoglio CM (1980) Herpes simplex virus expression in latently infected ganglion cells and in cervical neoplasia. *In* Essex M, Todaro G, zur Hausen H (eds) Viruses and Naturally Occurring Cancer. Cold Spring Harbor Laboratory, New York, p 101
27. Gilman SC, Docherty JJ, Clark A, Rawls WE (1980) The reaction patterns of HSV type 1 and type proteins with the sera of patients with uterine cervical carcinoma and matched controls. Cancer Res 44:4640–4647
28. Cabral GA, Courtney RJ, Schaffer PA, Marciano-Cabral F (1980) Ultrastructural characterization of an early, nonstructural polypeptide of herpes simplex virus type 1. J Virol 33:1192–1198

CHAPTER 42
Epithelial Cell Interactions of the Epstein–Barr Virus

JOSEPH S. PAGANO*

The identity and unity of the herpesviruses derive from a commonality of structural features and biology. Remarkably diverse, inhabiting the entire animal kingdom, herpesviruses have a morphology which, while as complex as any of the viruses, is similar both superficially and at the level of the structural organization of their genomes. Linked to structure is a conserved biologic expression, the ability to produce latent as well as active infection. These biologic hallmarks of the herpesviruses are tied to cell type. The virus has a tendency to replicate in one cell type and to cause latent infection in another cell type, carrying the implication that there are both primary and secondary cell targets of herpesvirus infections and that the cell type determines or contributes to the type of virus–cell interaction: dormant or vegetative.

Among several of the human and nonhuman herpesviruses, the primary cell infected, in which there is active replication of the virus, is an epithelial cell. The secondary target is a nerve cell, which serves as a repository for latent virus infection. This functional segregation is age-related; it breaks down *in utero* with CMV, in the immediate neonatal period with HSV, but also in immunosuppressed hosts at any age in whom HSV, VZV, and CMV disseminate widely and replicate in tissues and organs ordinarily sheltered from the virus. Of the five human herpesviruses, HSV type 1, HSV type 2, and varicella-zoster virus provide the best examples of this scheme of virus–cell interaction. Human cytomegalovirus and Marek's disease virus of chickens seem to follow a similar scheme, but the secondary cell type is a white blood cell rather than a nerve cell. The Epstein–Barr virus, while retaining its structural identity with the human herpesviruses, has seemed to

---
* University of North Carolina, Chapel Hill, North Carolina.

depart from this general biologic scheme having, until recently, been supposed to infect only lymphoid cells.

There is, however, much to suggest that the Epstein-Barr virus may in fact adhere to the general pathogenetic course adopted by the other herpesviruses. The evidence is both positive and negative. We have known for years that the Epstein-Barr virus can gain access to epithelial cells. The transformed epithelial cells of carcinoma of the posterior nasopharynx, a malignancy intimately associated with EBV infection throughout the world, contain EBV genomes in their episomal form [1–3]. On the other hand, active replication of EBV has never been demonstrated *in vivo* in the cell type with which the virus is familiarly associated, namely, the B lymphocyte. Neither virus nor viral antigens connected with replication of virus have been found in lymphocytes, either circulating in blood or at the site of viral replication in the oropharynx, whether normal lymphocytes or transformed lymphocytes as in Burkitt's lymphoma have been examined [4]. Such lymphocytes do in fact contain EBV genomes as indicated by the presence of the EBV nuclear antigen (EBNA) [4], so that the B lymphocyte seems to represent the cell type, presumably secondary, in which the virus remains dormant under normal conditions.

EBV and Infection of Epithelial Cells *in Vivo*

In 1977 the first evidence that EBV might be like the other human herpesviruses and replicate in epithelial cells was reported by Lemon et al. [5]. In squamous epithelial cells recovered from the oropharynx of persons with acute infectious mononucleosis, large amounts of EBV DNA were demonstrated intracellularly by *in-situ* cytohybridization with tritiated EBV-specific cRNA and autoradiography. The intensity of hybridization suggested that there was active replication of EBV DNA in the cells rather than simply latent infection. The hybridization results correlated well with presence of infectious EBV in oropharyngeal washings as detected by the conventional assay, transformation of cord-blood lymphocytes. An interesting feature was the apparent asynchronous nature of the infection. Clumps of epithelial cells contained different amounts of hybridizable DNA, some heavily laden and others showing hybridization only over nuclei. Shedding of EBV from the oropharynx persists for a long time [6], and the observations suggested that infection might course through the oropharynx from cell to cell rather than infecting all cells simultaneously. Inasmuch as there is constant exfoliation and replenishment of cells on mucosal and epithelial surfaces, second and subsequent waves of infection might be expected. Thus, the characteristic pharyngitis of IM as well as the persistence of virus shedding were provided with a direct pathogenetic explanation.

Lemon et al. [5] also found hybridization in clumps of epithelial cells in secretions from Stensen's duct, which suggested that the source of virus excreted from the parotid gland [7] might also be epithelial. Whether the infection takes place only in ductal cells or also in epithelial elements in the body of the gland remains unknown. Cytomegalovirus also replicates in salivary glands where ductal cells may be the type supporting CMV replication. Wolf *et al.* have contributed additional evidence obtained by sectioning parotid gland and carrying out *in-situ* cytohybridization for EBV. Although the cytoarchitecture was indistinct, the distribution of the grains in the autoradiograms seemed to follow a ductal configuration [8]. These observations, incidentally, raise the possibility that EBV may be associated with malignancies of the parotid gland.

The original observations have been confirmed recently by Sixbey *et al.* [9]. Again, with *in-situ* cytohybridization large amounts of EBV DNA were detected in epithelial cells recovered from the oropharynx. In order to substantiate the specificity of hybridization, a variety of new EBV probes including tritiated cRNAs transcribed from either purified virion DNA or from cloned fragments of EBV DNA, as well as biotinylated EBV DNA (BAM HI V) probes with hybridization detected by fluorescinated avidin were applied. Over a two-year period, specimens from a substantial number of patients with acute infectious mononucleosis gave quite consistent results with hybridizable DNA detected in almost all cases. Hybridization was generally more sensitive than the lymphocyte transformation assay [4] for detection of virus. In only one case was there no hybridization in epithelial cells when virus was detected in the throat-washings. Uninfected persons without EBV antibodies, including a case with active pharyngitis resulting from another cause, never disclosed hybridization. SV40 cRNA probe used as a control with some of the EBV-positive samples gave negative results. Patients enrolled in a trial of Acyclovir for treatment of infectious mononucleosis [10] who were shedding EBV from the oropharynx stopped shedding the virus while the drug was being given and resumed excretion of virus after the drug was stopped. Cells in throat-washings showed a parallel pattern of cytohybridization with hybridizable EBV DNA detected before and after drug administration, but not when Acyclovir was being given.

Finally and perhaps conclusively, EBV RNA was demonstrated in epithelial cells from patients with acute IM, but not in cells from a normal uninfected control subject [9]. The EBV RNA was detected by biotinylated, nick-translated, cloned BamHI H EBV DNA probe under conditions favoring DNA-RNA hybrid formation; in control studies in superinfected Raji cells in which abundant EBV RNA is synthesized, hybridization was abolished by an RNAse treatment step in the *in-situ* procedure. The quality and distribution of fluorescence differed depending upon whether DNA or RNA was detected with intense, punctate fluorescence indicating DNA, or suffused, less intense nuclear and perinuclear fluorescence indicating RNA. Thus the evidence pointing to oropharyngeal epithelial cells as an early site of EBV replication has now become compelling and needs only visualization of intracellular virus by electron microscopy for a final line of proof.

EBV Infects Epithelial Cells *in vitro*

The experimental *in vitro* systems for EBV have revolved around infection of B lymphocytes. However, several years ago we discovered that it might be possible to infect, with some difficulty, human epithelial cells *in vitro*. The results, considered preliminary, were briefly described [11,12]. In the meantime Glaser *et al.* [13] succeeded in inducing EBV early antigen by exposing explanted NPC tumor cells to the P3HR1 strain of EBV. Since NPC cells already contain EBV genomes and express EBNA, the results suggested that superinfection and initiation of an abortive viral cycle had taken place.

It followed from these observations that some epithelial cells must bear receptors for EBV, at least after the cells had become transformed into NPC. EBV receptors are linked to, if not identical to the complement receptor, C3D [14].

Nevertheless, attempts in other laboratories to infect epithelial cells with laboratory strains of EBV were quite unsuccessful [15]. However, nonlymphoid cell types can support viral expression and perhaps even replication once the EBV genome gains access to such cells as shown by transfection, microinjection, and receptor-tranplantation experiments [16–19].

Extensive new data are now available. Sixbey *et al.* [20] have shown clearly that human epithelial cells explanted from ectocervical tissue and grown as monolayers [21] can be infected with EBV reproducibly with production of viral antigens including EBNA and EA/VCA and probable viral DNA replication as indicated by *in-situ* cytohybridization. The appearance of the antigens takes at least 8 days and is abolished if the infecting virus is pretreated with serum containing EBV antibodies. Finally, important is the fact that only wild type EBV freshly obtained from throat-washings seems able to infect the epithelial cells *in vitro*, whereas laboratory strains of EBV are totally without effect. The P3HR1, B95-8, and MCUV5 strains of EBV, commonly used laboratory strains, have been long adapted to growth in lymphoid cells, which may help to explain their inability to infect epithelial cells. The molecular basis for such a change, for example, whether it is due to a host-range deletion mutation, is completely unknown. Thus far, infectious virus has not been recovered from the infected epithelial cell cultures; however, the insensitivity of the assay for EBV is such that small amounts of virus could easily be missed.

One feature of the *in vitro* infection system, while vexing at first, provided a fascinating lead to possible cellular mechanisms. Only a fraction of the cells in the monolayer appeared to be infected, usually less than 1%. Cells cultivated *in vitro* may lose receptors, and such a loss might explain the results. Also, epithelial cells in different body sites have differentiated features; possibly oropharyngeal cells or epithelial cells from Waldeyer's ring where NPC arises may be richer in receptors for the virus than epithelial cells from other sites. However, epithelial cells undergo maturation *in vitro*, which suggests another explanation. Depending upon the site of origin, such cells may begin to produce a variety of proteins including fibronectin and keratin [22]. The cervical epithelial cells, once they have formed monolayers, are largely static. They then begin to synthesize DNA. The final maturational step exhibited by the cervical cells in culture is senescence marked by exfoliation. Differentiation in primary explant cultures is also reflected by the somewhat stratified pattern of cell growth [22,23] that indicates rapid cell turnover characteristic of the tissue of origin. The cell sheets themselves contain epithelial populations with morphologic features of basal, parabasal, and desquamating cell layers [23].

Suspecting that EBV expression and replication may be linked to cellular maturational phase, we have analyzed loose cells shed into the culture media and shown that they are greatly enriched for EBV antigens [20]. Furthermore, the delay in appearance of antigens corresponds to the interval between the onset of cellular DNA synthesis and desquamation of those cells. The hypothesis that flows from these observations is that epithelial cells may be widely infected with EBV, but expression of viral antigens and replication of the virus is triggered by some step in the cell maturational or differentiation process. The virus would be expected to lie dormant in most of the cells, awaiting a cellular signal for replication. The expected consequence would be asynchronous infection with most of the virus present in sloughing cells, which seems to be what actually happens *in vivo* [5,9]. Such a sequence would provide a perfect mechanism for persistent infection and the continued virus excretion so characteristic of EBV infection.

The replication of other viruses is linked to cell differentiation both *in vivo* and in experimental systems. Perhaps the most striking examples are provided by the papillomaviruses: Human wart virus apparently infects basilar epithelium, but virus is produced only in keratinized cells [24]. Mouse CMV, polyomavirus, and SV40 although they are each capable of adsorbing to and penetrating undifferentiated cells, do not begin to replicate in certain systems until there is differentiation of the infected cells [25–27]. Finally, recently Epstein and colleagues succeeded in infecting explanted nasopharyngeal epithelial cells *in vitro* with a strain of EBV derived from NPC. By electron microscopy they also observed that exfoliated cells in this system seemed to be the preferred cells containing virus (A. Epstein, personal communication).

On the molecular level there is virtually no information relating to the influence of epithelial cell type on EBV gene expression. There is a suggestion in the *in vitro* infection system that in the majority of cells the EBV replicative cycle is abortive, but this is also the case in lymphoid cells. Further, in at least some of the infected epithelial cells EBNA is produced [18]; this antigen ordinarily signals either latent EBV infection or the transforming relation. In a slowly coursing infection, generating both defective and complete virus, when EBV gains access to epithelial cells it may either become latent—perhaps a necessary prelude to transformation—or virus replication may ensue. We have not seen morphologic changes of growth transformation after exposure of epithelial cells to intact virus. However we have transfected epithelial cells with EBV DNA with production of at least transient stimulation of growth and induction of EBNA under the proper conditions, thus giving the impression that EBV may, in fact, be able to enter the first stages of a transforming relation in such cells (Sixbey and Pagano, unpublished results).

Finally there is now some evidence from the work of Raab-Traub, *et al.* that transcription of EBV genes in transformed epithelial cells may differ from EBV transcription in lymphoid cells, with an "activated" state of transcription detected in tissues of some nasopharyngeal carcinomas [28]. Whether this distinctive transcriptional pattern is peculiar to epithelial cells or to nasopharyngeal carcinoma is at this point quite unclear.

Conclusion: Pathogenesis of EBV-Related Diseases

The normal course of EBV infection may, therefore, have unexpected outcomes. If, as seems likely, EBV first infects epithelial cells in the oro- and nasopharynx, and virus is produced in these cells (primary target cells for viral replication) a few cells might not be destroyed by infection, and viral genomes might persist and create progenitor cells for nasopharyngeal carcinoma. Persistent infection might depend upon the failure of an occasional basilar or epithelial cell to differentiate and thus permit virus replication. Whether or not a monoclonal expansion of such latently infected epithelial cells ever takes place may depend upon a variety of cofactors including genetic predisposition and insertion of EBV genes into the cellular chromosome.

The peculiar localization of EBV epithelial transformation to the posterior nasopharynx may result from a population of receptor-rich cells in that site. Since, clearly, T lymphocytes in some way mediate and indeed seem to be necessary for herpesvirus latency, modulation of the infection by the abundant lymphocytes in the Waldeyer's ring may also be a factor. Finally, rarer instances of epithelial transformation may occur in other tissues, near if not actually the preferred or special sites for EBV replication, such as the supraglottal area [29], the parotid [8], and epithelial rests in the thymus (W. Henle, personal communication).

In the usual course of events, EBV replicated in epithelial cells infects B lymphocytes bearing receptors for EBV; whether these cells represent a subset of B lymphocytes is unknown. Infection of B lymphocytes settles into a latent state in persons with normal immune mechanisms, but with the side-effect of polyclonal growth transformation of the lymphocytes. This proliferation of cells is ordinarily held in check by normal immune surveillance. If these mechanisms are impaired, as sometimes occurs in allograft recipients, or break down entirely, as in acquired immunodeficiency syndrome, polyclonal proliferation of EBV-DNA-containing cells resurges, sometimes with invasive and lethal effects [30,31]. Occasionally, B lymphocytes of monoclonal lineage emerge and become dominant, producing both diffuse and localized lymphocytic malignancies. African Burkitt's lymphoma, because of its endemic pattern of incidence, represents a special instance, never adequately explained, of such an outcome.

References
1. Wolf H, zur Hausen H, Becker Y (1973) EB viral genomes in epithelial nasopharyngeal carcinoma cells. Nature New Biol 244:245–257
2. Nonoyama M, Pagano JS (1972) Separation of Epstein–Barr virus DNA from large chromosomal DNA in nonvirus-producing cells. Nature New Biol 238:169–171
3. Raschka-Dierich C, Adams A, Lindahl T, Bornkamm GW, Bjursell G, Klein G, Giovanella BC, Singh S (1976) Intracellular viruses of EBV DNA in human tumor cells in vivo. Nature 260:302–306
4. Klein G, Svedmyr E, Jondal M, Perrson PO (1976) EBV-determined nuclear antigen (EBNA) positive cells in the peripheral blood of infectious mononucleosis patients. Int J Cancer 17:21–26
5. Lemon SM, Hutt LM, Shaw JE, Li JLH, Pagano JS (1977) Replication of EBV in epithelial cells during infectious mononucleosis. Nature 268:268–270
6. Miller G, Niederman JC, Andrews LL (1973) Prolonged oropharyngeal excretion of Epstein–Barr virus after infectious mononucleosis. N Engl J Med 288:229–232
7. Niederman JC, Miller G, Pearson HA, Pagano JS, Dowaliby JM (1976) Infectious mononucleosis: Epstein–Barr virus shedding in saliva and the oropharynx. N Engl J Med 294:1355–1359
8. Wilmes E, Wolf H (1981) Der Nachweis von Epstein–Barr-genomen in der ohrspeicheldrüse. Laryn Rhinol Otol 1:7–10
9. Sixbey JW, Nedrud JG, Raab-Traub N, Hanes RA, Pagano JS (1984) Epstein–Barr virus replication in oropharyngeal epithelial cells. N Engl J Med 310: May 1984
10. Pagano JS, Sixbey JW, Lin JC (1983) Acyclovir and Epstein–Barr virus infection. J. Antimicrob Chemother 12 (Suppl B): 113–121
11. Pagano JS (1978) Epstein–Barr virus infection of epithelial cells and lymphocytes. In Stevens JG, Todaro GJ (eds) Persistent Viruses. ICN–UCLA Symposia on Molecular and Cellular Biology. Academic Press, New York, 1:787–804
12. Pagano JS, Nedrud JG (1980) Latency of the Epstein–Barr virus and cytomegalovirus. In Nahmias AJ, Dowdle WR, Schinazi RF (eds) The Human Herpesviruses, An Interdisciplinary Perspective. Elsevier-North Holland Publishing Company, Part IV:206–218.
13. Glaser R, de The' G, Lenoir G, Ho JHC (1976) Superinfection of nasopharyngeal carcinoma epithelial tumor cells with Epstein–Barr virus. Proc Natl Acad Sci USA 73:960–963
14. Hutt-Fletcher LM, *et al.* (1983) Studies of Epstein–Barr virus receptor found on Raji cells. II. A comparison of lymphocyte binding sites for Epstein–Barr virus and C3d. J Immunol 130:1309–1312
15. Shapiro IM, Volsky DJ (1983) Infection of normal human epithelial cells by Epstein–Barr virus. Science 219:1225–1228

16. Graessmann A, Wolf H, Bornkamm GW (1980) Expression of Epstein-Barr virus genes in different cell types after microinjection of viral DNA. Proc Natl Acad Sci USA 77:433-434
17. Grogan E, Miller G, Henle W, Niederman JC (1981) Expression of Epstein-Barr viral early antigen in monolayer tissue cultures after transfection with viral DNA and DNA fragments. J Virol 40:861-869
18. Volsky DJ, Shapiro IM, Klein G (1980) Transfer of EBV receptors to receptor-negative cells permits virus penetration and antigen expression. Proc Natl Acad Sci USA 77:5433-5457
19. Stoerker J, Parris D, Yajima Y, Glaser R (1980) Pleiotropic expression of Epstein-Barr virus DNA in human epithelial cells. Proc Natl Acad Sci USA 78:5852-5855
20. Sixbey JW, Vesterinen EH, Nedrud JG, Raab-Traub N, Walton LA, Pagano JS (1983) Epstein-Barr virus replication in human epithelial cells infected in vitro. Nature 306:480-483
21. Vesterinen EH, Nedrud JG, Collier AM, Walton LA, Pagano JS (1980) Explantation and subculture of epithelial cells from human uterine ectocervix. Cancer Res 40:512-518
22. Alitalo K, Halila H, Vesterinen E, Vaheri A (1982) Endo- and ectocervical human uterine epithelial cells distinguished by fibronectin production and keratinization in culture. Cancer Res 42:1142-1146
23. Vesterinen EH, Carson J, Walton LA, Collier AM, Keski-Oja J, Nedrud JG, Pagano JS (1980) Human ectocervical and endocervical epithelial cells in culture: A comparative ultrastructural study. Am J Obstet Gynecol 137:681-686
24. Grussendorf EI, zur Hausen H (1979) Localization of viral DNA-replication in sections of human warts by nucleic acid hybridization with complementary RNA of human papilloma virus type 1. Arch Dermatol Res 264:55-63
25. Dutko FJ, Oldstone MBA (1981) Cytomegalovirus causes a latent infection in undifferentiated cells and is activated by induction of cell differentiation. J Exp Med 154:1636-1651
26. Boccara M, Kelly F (1978) Expression of polyoma virus in heterokaryons between embryonal carcinoma cells and differentiated cells. Virology 90:147-150
27. Swartzendruber DE, Friedrick TD, Lehman JM (1977) Resistance of teratocarcinoma stem cells to infection with simian virus 40: Early events. J Cell Physiol 93:25-30
28. Raab-Traub, N, Hood, R, Yang, C-S, Henry, B, Pagano, JS (1983) Epstein-Barr virus transcription in nasopharyngeal carcinoma. J Virol 48:165-175
29. Brichacek, B, Hirsch, I, Sibl, O, Vilikusova, E, Vonka, V (1983) Association of some supraglottic laryngeal carcinomas with EB virus. Int J Cancer 32:193-197
30. Hanto, DW (1982) Epstein-Barr virus-induced B-cell lymphoma after renal transplantation. Acyclovir therapy and transition from polyclonal to monoclonal B-cell proliferation. N Engl J Med 306:913
31. Pagano, JS, Sixbey, JW (1984) New perspectives on the Epstein-Barr virus in the pathogenesis of lymphoproliferative disorders. *In* Remington J, Schwarz, M (eds) Current Clinical Topics in Infectious Diseases McGraw-Hill, New York, in press.

CHAPTER 43
Viral Gastroenteritis

ALBERT Z. KAPIKIAN*

The history of medicine, and medical virology in particular, is replete with impressive examples of the relatively rapid discovery of specific viruses as etiologic agents of frequently occurring illnesses such as smallpox, chicken pox, measles, poliomyelitis, mumps, rubella, the common cold, and influenza. Such discoveries led to the development of vaccines that prevented or modified certain of these diseases. Conspicuously absent from this list of medical conquests had been the etiologic agents of viral gastroenteritis, not only one of the most frequently occurring illnesses worldwide—second only to acute respiratory illnesses—but also one of the most important since it appears to be a major cause of mortality in infants and young children in developing countries [1]. However, with the discovery in 1972 of the 27-nanometer (nm) Norwalk virus, which is now known to be a major etiologic agent of a self-limited form of epidemic nonbacterial gastroenteritis of adults and children, and the discovery in 1973 of the 70-nm rotavirus, which is the single most important etiologic agent of severe diarrhea of infants and young children, a major segment of an important disease category was finally on the verge of exposure etiologically [1–5]. It is remarkable that in this era of tissue culture virology both of these agents were identified initially without the benefit of any *in vitro* tissue culture system since neither could be propagated directly in cell cultures from clinical specimens. Each of these viruses was detected initially by examination of clinical material by electron microscopy. This method has been termed "direct virology" since it entails the examination of material directly from clinical specimens without the conventional intermediate *in vitro* or animal *in vivo* systems characteristically required for virologic studies [5].

---

* National Institutes of Health, Bethesda, Maryland.

Viral gastroenteritis consists of two distinct clinical entities with quite different epidemiological characteristics. One, which is associated with the 27-nm Norwalk group of agents, is designated epidemic viral gastroenteritis and occurs in family or community-wide outbreaks affecting adults, school age children, family contacts, and some young children as well. The clinical characteristics, which usually last only 24 to 48 hours, are self limited and mild and are comprised of a variety of signs and symptoms such as nausea, vomiting, diarrhea, abdominal pain, and low-grade fever or a combination of these [5–7]. Parenteral fluid replacement therapy is not usually required.

The 27-nm Norwalk virus was visualized by immune electron microscopy (IEM) in an infectious stool filtrate derived from a gastroenteritis outbreak in an elementary school in Norwalk, Ohio, which affected 50% of the students and teachers and had a secondary attack rate of 32% [2]. Attempts to recover an agent from this outbreak were unsuccessful. However, a rectal swab filtrate derived from a secondary case induced the disease in volunteers; a pathological lesion characterized by broadening and blunting of the villi of the jejunum was observed by light microscopy [6,6a]. Serial passages also induced illness [7]. In spite of the experimental transmission of the disease, attempts to cultivate an agent were once again unsuccessful [6,7]. However, with the IEM technique, antibody-coated 27-nm particles were visualized in an infectious stool filtrate following reaction with a convalescent serum [2,5]. Serologic evidence of infection with this particle was also demonstrated by IEM with paired sera from certain volunteers and individuals from the original outbreak [2]; in addition, a close temporal relationship between virus shedding and illness was observed by IEM [8]. From these and other studies, the 27-nm Norwalk virus was established as the etiologic agent of the Norwalk outbreak without the benefit of any *in vitro* system.

It soon became apparent that the role of the Norwalk agent in epidemic viral gastroenteritis was not limited to a localized area in the United States. The development of a radioimmunoassay blocking technique enabled the testing of paired serum specimens obtained from numerous outbreaks of nonbacterial gastroenteritis: 24 (34%) of 70 such outbreaks were associated serologically with the Norwalk agent [4,9]. Family and community outbreaks predominated, but several common source outbreaks were also observed including contaminated shell food, drinking water, and swimming water. Recently, a major outbreak of Norwalk gastroenteritis was associated with the ingestion of contaminated cake frosting [10]. Affected individuals in these outbreaks ranged from 4-year-old children to elderly adults [4,9].

Limited studies in infants and young children in both developed and developing countries show that the Norwalk virus is clearly not the long-sought agent of severe infantile gastroenteritis [5,11,12]. Further evidence of this was revealed from antibody prevalence studies in the United States, which demonstrated that Norwalk antibody was acquired gradually, beginning slowly in childhood and reaching 50% by the fifth decade. In contrast, rotavirus antibody was acquired early in life as over 90% of the same individuals developed such antibody by the age of 36 months [5]. However, Norwalk antibody was acquired quite rapidly by infants and young children in several developing countries [9]; and in a study in Bangladesh, Norwalk seroresponses were observed in this age group and 1%–2% of the episodes of diarrhea were associated with Norwalk virus infection [13].

The nature of immunity to Norwalk virus remains an unusual and unresolved feature of the natural history of such infection. Adults are highly susceptible to both natural and experimental challenge. The attack rate reached over 80% in certain outbreaks and in addition, about 50% of unselected volunteers developed illness following oral administration [4,7,13a,b]. Volunteer studies reveal both short- and long-term immunity [14]. The long-term immunity is rather perplexing as volunteers challenged with Norwalk virus on two separate occasions 27 to 42 months apart either developed illness following both challenges or were resistant to both challenges [14]. The difference in susceptibility could not be explained on the basis of pre-challenge Norwalk virus antibody levels. In further studies, volunteers who developed illness tended to have higher serum antibody titers and significantly higher jejunal fluid antibody titers to the Norwalk virus than did volunteers who remained healthy following challenge [9].

Classification of the Norwalk agent has not been possible since it does not grow in cell cultures. However, study of the proteins of Norwalk virus by polyacrylamide gel electrophoresis following purification of antigens from stools revealed that the virus had a single structural protein with a molecular weight of 59,000 daltons and an additional soluble protein of 30,000-dalton molecular weight [15]. Such a pattern is consistent with that of the calicivirus. Although the Norwalk virus does not possess the distinctive cup-like surface morphologic features of the caliciviruses, it does have surface indentations in certain preparations and thus could be considered, at least, to be calicivirus-like. However, a final classification of this virus awaits further study including such properties as its nucleic acid content. It should be noted that other fastidious agents that are morphologically similar but antigenically distinct from the Norwalk virus have been detected by electron microscopy and associated with gastroenteritis outbreaks. These include the Hawaii, Ditchling or W, Cockle, Parramatta, Snow Mountain (Colorado), and Marin County agents [5].

Rotaviruses are the major known etiologic agents of severe diarrhea of infants and young children worldwide and are associated with the other form of viral gastroenteritis—sporadic infantile gastroenteritis, which differs from epidemic viral gastroenteritis in three major ways: (1) it is associated with a severe diarrheal illness in infants and young children, which may require hospitalization and fluid replacement therapy; (2) it does not usually occur in sharp outbreaks; and (3) the illness rate among family contacts of patients with this type of gastroenteritis is low (although the infection rate may be high) [5,16].

The human rotaviruses were discovered in 1973 in Australia by electron microscopic examination of thin sections of duodenal biopsies obtained from infants and young children hospitalized with gastroenteritis [3]. Characteristically, pathological changes in the mucosal epithelial cells included shortening and blunting of the villi and the presence of virus-like particles within distended cisternae of the endoplasmic reticulum. Soon thereafter, rotaviruses were detected readily in stool suspensions by negative stain electron microscopy [17]. Initially, human rotaviruses could not be propagated in cell culture from clinical specimens and thus once again electron microscopy became the mainstay for the diagnosis of an enteric infection. However, second and third generation assays have been developed that also exploit the direct virologic approach. These include radioimmunoassay, ELISA, CIEOP, gel electrophoresis of RNA, dot hybridization, and many others [5,18]. The most widely used method worldwide is the ELISA.

The word, rotavirus, is derived from the Latin word *rota* meaning wheel, because the sharply defined circular outline of the outer capsid layer gives the appearance of the rim of a wheel placed on short spokes radiating from a wide hub [16]. Rotaviruses belong to the reoviridae family and possess a genome consisting of 11 segments of double-stranded RNA. Complete double-shelled particles have a density of 1.36 g/cm^3 in cesium chloride [16]. In infected cells, rotaviruses code for 10 to 12 polypeptides of which 5 to 9 are considered structural; at least 3 are on the inner, and 2 on the outer capsid [16]. Rotaviruses also contain an RNA-dependent RNA polymerase, which is activated upon removal of the outer shell polypeptides [16].

In studies in Australia, United States, and Japan, rotaviruses have been associated with about 35% to 45% of acute diarrheal disease requiring hospitalization of children [11,12,19]. For example, in the United States study, from January 1974 to July 1982 rotaviruses were detected in the stool of 34.5% of 1537 pediatric in-patients with gastroenteritis [12]. In the temperate climates, rotaviruses display a unique temporal pattern of infection characterized by peaks, which occur almost exclusively in the cooler months of the year [11,12,19]. For example, in the United States study, between January 1974 and December 1981, rotavirus infections were not observed in the months of June, July, August, and September, whereas the major peaks of infection occurred in January and February [12]. The reason for the peaks during the cooler months is not known, although the association with cold weather periods has been shown; the influence of relative humidity has not been consistent [11,12]. In tropical climates, rotavirus infections are observed throughout most of the year, although less pronounced peaks can occur [1,11]. It should be noted that rotavirus can also cause mild gastroenteric illness not requiring hospitalization [1,5].

Rotaviruses have also been shown to be important etiologic agents of severe diarrheal illness in less-developed countries [1,20]. For example, in a study of the etiology of moderately severe or severe gastroenteritis in 6352 patients treated or hospitalized in a treatment center in Dhaka, Bangladesh, during a one-year period, rotaviruses were the most important pathogen in patients two years of age or younger; 46% of such patients shed this agent [20]. It was of interest that toxigenic *E. coli* strains were next in importance in this age group, with 28% of a sample of patients shedding such strains. Bacterial agents were detected more frequently than rotaviruses among children 2 years of age and older.

Rotaviruses have also been implicated as important etiologic agents of diarrhea in numerous animals including calves, pigs, mice, foals, lambs, and many others [5]. Historically, it is of interest that the mouse, calf, and simian rotaviruses and the "0" agent (from intestinal washings of sheep and cattle) were described prior to the discovery of human rotaviruses [5,16]. The animal strains characteristically share common CF and immunofluorescent antigen(s) with each other and with human rotaviruses, as well, although recently a few human and animal strains have been described that do not share this common antigen(s) [5,16]. These have been designated tentatively as pararotaviruses. Rotaviruses also have a subgroup antigen that separates most strains into 2 subgroups [21]. The group and subgroup specificities are coded for by the sixth gene, the product of which is the major inner capsid polypeptide with a molecular weight of 42,000 daltons [16,21].

A major advance in the understanding of the natural history of rotavirus infection was made when Japanese investigators discovered that human rotaviruses could be grown efficiently in tissue culture following trypsin treatment of the virus prior to and during cultivation in roller tube cultures at 37°C [22,23]. The cell culture propagation of rotaviruses has allowed the conventional classification of rotaviruses according to neutralization specificity [24,25]. At present there is agreement on at least four serotypes, which are designated serotype 1 (Wa), serotype 2 (DS-1), serotype 3 (P or M), and serotype 4 (Saint Thomas No. 4) [24]. (A strain of each serotype is shown in parentheses.) It is striking that certain animal rotaviruses share neutralization specificity with human rotaviruses [24,25]. For example, rhesus (SA-11, MMU18006), canine (CU-1), feline (Taka), and equine (H-2) rotaviruses are indistinguishable from human rotavirus type 3 by neutralization [25]; however, except for the canine and feline strains, which have identical RNA gel electrophoretic patterns, each of these animal strains differs in its RNA pattern from the others and from the human strain, and in addition, differs in its subgroup specificity from the human strain [25]. The major protein responsible for neutralization specificity is coded for by the eighth or ninth gene, the product of which is an outer capsid protein that is glycosylated post-translationally and has a molecular weight of 34,000 daltons [16]. It is noteworthy that serotypes 1, 3, and 4 have subgroup 2 specificity whereas serotype 2 has subgroup 1 specificity [24,25].

Information on the mechanism of immunity to rotaviruses in humans is not clear. Volunteer studies reveal a correlation of serum antibodies with resistance to illness following challenge; in this same study the role of intestinal fluid neutralizing activity was not clearcut [26]. However, animal studies indicate that antibody in the lumen of the small intestine is the prime determinant of protection against illness [1,5,16]. A paradoxical aspect of immunity to rotaviruses is the high rate of subclinical infection in neonates undergoing initial infection in newborn nurseries [16,27]. It is striking that such neonatal infections conferred significant protection against illness (but not infection) following documented second natural rotaviral infections for a period of about 3 years following the initial neonatal infection [27].

Since it is clear that a rotavirus vaccine is needed in the developing as well as the developed countries, various approaches are being pursued in producing such a vaccine. (1) *A live attenuated human rotavirus vaccine.* The Wa strain, which was successfully cultivated in primary African green monkey kidney cell cultures following prior serial passage in piglets, was studied as a vaccine candidate initially in animals, then in volunteers [26,28]. This vaccine appeared promising as none of the volunteers developed illness and 6 of 10 with low levels of prechallenge antibody developed a seroresponse [26]. However, studies with this vaccine lot were suspended when unexpected serum transaminase elevations were observed in three volunteers, only one of whom had developed a rotavirus seroresponse [26]. The etiology of these elevations is not yet known, although various possibilities are described [26]. (2) *A live attenuated vaccine derived from a heterologous host.* A bovine rotavirus (Lincoln strain of NCDV), which had been attenuated by multiple passages in calf kidney cells at low temperature (this strain is used as a vaccine in newborn calves and in cows) was administered orally to adult volunteers initially and later to children [29]. It did not cause illness and in children induced frequent homotypic and infrequent heterotypic rotavirus serum antibodies. This approach was based on previous animal studies that demonstrated that prior administration of bovine rotavirus to calves while *in utero* protected against challenge with human rotavirus (serotype 1) at birth [30]; in addition, prior administration of this candidate vaccine to piglets significantly decreased the shedding of human rotaviruses following such challenge [29]. Recent studies indicates that the calves developed not only homotypic neutralizing antibody, but also heterotypic antibody to human rotaviruses following the *in utero* administration of bovine rotavirus [30]. A rhesus rotavirus is also a possible vaccine candidate since it not only represents a strain from a heterologous host but is serotypically identical to human rotavirus type 3 [24,25]. (3) *An attenuated human-animal rotavirus reassortant vaccine.* This approach was used successfully for the propagation of noncultivatable human rotavirus strains [31]. It took advantage of the well-known characteristic of the reoviridae to undergo genetic reassortment with efficiency during mixed infection in cell culture [31]. Noncultivatable human rotaviruses were cocultivated with a *ts* bovine rotavirus mutant at permissive temperature and harvests were subjected to selection by growth at a restrictive temperature in the presence of bovine rotavirus antiserum. The resulting progeny virus not only grew efficiently in cell culture but also had the neutralization specificity of the human rotavirus parent [31]. Thus, noncultivable human rotavirus strains were "rescued" for study of the serotypic diversity of naturally occurring strains. This basic approach is now being used for development of reassortant rotavirus vaccines that possess ten animal rotavirus genes and the major neutralizing gene from a human rotavirus parent except that wild type bovine rotavirus or rhesus rotavirus (i.e., *ts* +) is used as the donor of the heterologous genes [32]. (4) *Cloning the rotavirus genome by DNA technology.* Major advances have been made in sequencing various genes of several rotaviruses. Of special note has been the sequencing of the genes that code for the major SA-11 and UK rotavirus, neutralization protein; in both instances the gene was 1062 bases long [33,34]. Insertion of a complete DNA copy of such a gene, or that portion responsible for its major antigenic sites in an acceptable vector and its expression in a suitable vehicle might provide an efficient method for immunization [33,34a]. (5) *Production of synthetic vaccines.* Since the major neutralizing gene of at least two rotaviruses has been sequenced, it should be possible to define the region that codes for type-specific antibody. It may then be feasible to produce a synthetic vaccine containing the peptide(s) responsible for inducing protective antibodies.

It should be noted that administration of rotavirus antigens to expectant mothers as a means of boosting breast milk antibody is being considered as a means of imparting protective antibodies passively to the infant. In addition, significant passive protection of neonates against rotavirus-induced gastroenteric symptoms has been demonstrated following oral administration of human gamma globulin containing rotavirus antibody [35]. Such approaches have been shown to be effective in animal studies [1,5]. Also, rotavirus antibody-containing milk has been administered as a therapeutic measure to immunodeficient patients with chronic rotavirus infection [36].

Remarkable progress has been made in elucidating the major etiologic agents of viral gastroenteritis. Not considered in this discussion has been the role of other enteric viral agents associated with gastroenteric illnesses, such as enteric adenoviruses, calciviruses, astroviruses, small round viruses other than Norwalk, putative coronaviruses, and the Otofuke agent [5]. With the exception of the enteric adenoviruses, which are associated with approximately 5% of diarrheal illnesses of infants and young children requiring hospitalization, the contribution of these other agents in the etiology of severe infantile diarrhea appears minimal. Although numerous agents—viral, bacterial, and parasitic—are detectable in stools, it should be emphasized that even with the most advanced techniques, an etiologic agent cannot be found in approximately 40% of all pediatric patients hospitalized with diarrhea. Thus, elucidation of the etiology of these undiagnosed diarrheal illnesses remains an important area of research that must be pursued vigorously if the morbidity and mortality from diarrheal diseases is to be controlled effectively and comprehensively. Although much progress has been made in the field of viral gastroenteritis, it is clear that major efforts are still needed to conquer this very common and important affliction.

References

1. Kapikian AZ, Wyatt RG, Greenberg HB, Kalica AR, Kim HW, Brandt CD, Rodriguez WJ, Parrott RH, Chanock, RM (1980) Approaches to immunization of infants and young children against gastroenteritis due to rotaviruses. Rev Inf Dis 2:459–469
2. Kapikian AZ, Wyatt RG, Dolin R, Thornhill TS, Kalica AR, Chanock RM (1972) Visualization by immune electron microscopy of a 27 nm particle associated with acute infectious nonbacterial gastroenteritis. J Virol 10:1075–1081
3. Bishop RF, Davidson GP, Holmes IH, Ruck BJ (1973) Virus particles in epithelial cells of duodenal mucosa from children with acute gastroenteritis. Lancet 2:1281–1283
4. Greenberg HB, Valdesuso J, Yolken RH, Gangarosa E, Gary W, Wyatt RG, Konno T, Suzuki H, Chanock RM, Kapikian AZ (1979) Role of Norwalk virus in outbreaks of nonbacterial gastroenteritis. J Infect Dis 139:564–568
5. Kapikian AZ, Greenberg HB, Kalica AR, Wyatt RG, Kim HW, Brandt CD, Rodriguez WJ, Flores J, Singh N, Parrott RH, Chanock RM (1981) New developments in viral gastroenteritis. *In* Holme T, Holmgren J, Merson MH, Molby R (eds) Acute Enteric Infections in Children. New Prospects for Treatment and Prevention. Elsevier/North Holland, Biomedical Press, p 9–57
6. Dolin R, Blacklow NR, DuPont H, Formal S, Buscho RF, Kasel JA, Chames RP, Hornick R, Chanock RM (1971) Transmission of acute infectious nonbacterial gastroenteritis to volunteers by oral administration of stool filtrates. J Infect Dis 123:307–312
6a. Agus SG, Dolin R, Wyatt RG, Tousimis AJ, Northrup RS (1973) Acute infectious nonbacterial gastroenteritis: intestinal histopathology. Histologic and enzymatic alterations during illness produced by the Norwalk agent in man. Ann Int Med. 79:18–25
7. Dolin R, Blacklow NR, DuPont H, Buscho RF, Wyatt RG, Kasel JA, Hornick R, Chanock RM (1972) Biological properties of Norwalk agent of acute infections nonbacterial gastroenteritis. Proc Soc Exp Biol Med 140:578–583
8. Thornhill TS, Kalica AR, Wyatt RG, Kapikian AZ, Chanock RM (1975) Pattern of shedding of the Norwalk particle in stools during experimentally induced gastroenteritis in volunteers as determined by immune electron microscopy. J Infect Dis 132:28–34
9. Greenberg HB, Wyatt RG, Kalica AR, Yolken RH, Black R, Kapikian AZ, Chanock RM (1981) New insights in viral gastroenteritis. Perspect Virol 11:163–187
10. Kuritsky JN, Osterholm MT, Greenberg HB, Korlath JA, Godes JR, Hedberg CW, Forfang JC, Kapikian AZ, McCullough JC, White KE (1984) Norwalk gastroenteritis: A community-wide outbreak associated with bakery product consumption Ann Int Med 100:519–521
11. Konno T, Suzuki H, Katsushima N, Imai A, Tazawa F, Kutsuzawa T, Kitaoka S, Sakamoto M, Yazakin N, Ishida N (1983) Influence of temperature and relative humidity on human rotavirus infection in Japan J Infect Dis 147:125–128
12. Brandt CD, Kim HW, Rodriguez WJ, Arrobia JO, Jeffries BC, Stallings EP, Lewis C, Miles AJ, Chanock RM, Kapikian AZ, Parrott RH (1983) Pediatric viral gastroenteritis during eight years of study. J Clin Microbiol 18:71–78
13. Black RE, Greenberg HB, Kapikian AZ, Brown KH, Becker S (1982) Acquisition of serum antibody to Norwalk virus and rotavirus in relation to diarrhea in a

longitudinal study of young children in rural Bangladesh. J Infect Dis 145:483–489

13a. Kaplan JE, Gary GW, Baron RC, Singh N, Schoenberger LB, Feldman R, Greenberg HB (1982) Epidemiology of Norwalk gastroenteritis and the role of Norwalk virus in outbreaks of acute nonbacterial gastroenteritis. Ann Int Med 96:756–761.

13b. Wyatt RG, Dolin R, Blacklow NR, DuPont HL, Buscho RF, Thornhill TS, Kapikian AZ, Chanock RM (1974). Comparison of three agents of acute infectious nonbacterial gastroenteritis by cross-challenge in volunteers. J Inf Dis 129:709–714

14. Parrino TA, Schreiber DS, Trier JS, Kapikian AZ, Blacklow NR (1977) Clinical immunity in acute gastroenteritis caused by the Norwalk agent. N Engl J Med 297:86–89

15. Greenberg HB, Valdesuso JR, Kalica AR, Wyatt RG, McAuliffe VJ, Kapikian AZ, Chanock RM (1981) Proteins of Norwalk virus. J Virol 37:994–999

16. Holmes IH (1983) Rotaviruses. In Joklik WK (ed) The Reoviridae. Plenum Press, New York, p 359–423

17. Flewett TH, Bryden AS, Davies H (1973) Virus particles in gastroenteritis. Lancet 2:1497

18. Flores J, Boeggeman E, Purcell R, Sereno M, Perez I, White L, Wyatt RG, Chanock RM, Kapikian AZ (1983) A dot hybridisation assay for detection of rotavirus. Lancet 1:555–557

19. Davidson GP, Bishop RF, Townlee RRW, Holmes IH, Ruck BJ (1975) Importance of a new virus in acute sporadic enteritis in children. Lancet 1:242–246

20. Black RE, Merson MH, Mizanur Rahman ASM, Yanus MD, Alim ARMA, Huq I, Yolken RH, Curlin GT (1980) A 2 year study of bacterial, viral and parasitic agents associated with diarrhea in rural Bangladesh. J Infect Dis 142:660–664

21. Kalica AR, Greenberg HB, Wyatt RG, Flores J, Sereno MM, Kapikian AZ, Chanock RM (1981) Genes of human (strain Wa) and bovine (strain UK) rotaviruses that code for neutralization and subgroup antigens. Virology 112:385–390

22. Sato K, Inaba Y, Shinozaki T, Fujii T, Matumoto M (1981) Isolation of human rotavirus in cell cultures. Brief report. Arch Virol 69:155–160

23. Urasawa T, Urasawa S, Taniguchi K (1981) Sequential passages of human rotaviruses in MA-104 cells. Microbiol Immunol 25(10):1025–1035

24. Wyatt RG, James HD Jr, Pittman AL, Hoshino Y, Greenberg HB, Kalica AR, Flores J, Kapikian AZ (1983) Direct isolation in cell culture of human rotaviruses and their characterization into four serotypes. J Clin Microbiol 18:310–317

25. Hoshino Y, Wyatt RG, Greenberg HB, Flores J, Kapikian AZ (1984) Serotypic similarity and diversity of human and animal rotaviruses as studied by plaque reduction neutralization. J Inf Dis 149:694–702

26. Kapikian AZ, Wyatt RG, Levine MM, Black RE, Greenberg HB, Flores J, Kalica AR, Hoshino Y, Chanock RM (1983) Studies in volunteers with human rotavirus. Devel Biol Standard 53:209–218

27. Bishop RF, Barnes GL, Cipriani E, Lund JS (1983) Clinical immunity after neonatal rotavirus infection. A prospective longitudinal study in young children. N Engl J Med 309:72–76

28. Wyatt RG, James WD, Bohl EH, Theil KW, Saif LJ, Kalica AR, Greenberg HB,

Kapikian AZ, Chanock RM (1980) Human rotavirus type 2: Cultivation *in vitro*. Science 207:189–191
29. Vesikari T, Isolauri E, Delem A, D'Hondt E, Andre FE, Zissis G (1983) Immunogenicity and safety of live oral attenuated bovine rotavirus vaccine strain RIT 4237 in adults and young children. Lancet 2:807–811
30. Wyatt RG, Kapikian AZ, Mebus CA (1983) Induction of cross reactive serum neutralizing antibody to human rotavirus in calves after *in vitro* administration of bovine rotaviruses. J Clin Microbiol 18:505–508
31. Greenberg HB, Kalica AR, Wyatt RG, Jones RW, Kapikian AZ (1981) Rescue of noncultivatable human rotaviruses by gene reassortment during mixed infection with ts mutants of a cultivatable bovine rotavirus. Proc Natl Acad Sci USA 78:420–424
32. Greenberg H, Midthun K, Wyatt R, Flores J, Hoshino Y, Chanock R, Kapikian A (1983) Use of reassortant rotaviruses and monoclonal antibodies to make gene-coding assignments and construct rotavirus vaccine candidates. *In* Chanock RM and Lerner RA (eds) Modern approaches to vaccines. Molecular and chemical basis of virus virulence and immunogenicity. Cold Spring Harbor Laboratory, New York p 319–327
33. Both GW, Mattick JS, Bellamy AR (1983) Serotype specific glycoprotein of simian 11 rotavirus: Coding assignment and gene sequence. Proc Natl Acad Sci USA 80:3094–3095
34. Elleman TC, Hoyne PA, Dyall-Smith ML, Holmes IH, Azad AA (1983) Nucleotide sequence of the gene encoding the serotype-specific glycoprotein of UK bovine rotavirus. Nucl Acids Res 11:4689–4701
34a. Flores J, Sereno M, Kalica A, Keith J, Kapikian A, Chanock RM (1983) *In* Chanock RM and Lerner RA (eds) Modern approaches to vaccines. Molecular and chemical basis of virus virulence and immunogenicity. Cold Spring Harbor Laboratory, New York p 159–164
35. Barnes G, Doyle LW, Hewson PH, Knoches AML, McLellan JA, Kitchen WH, Bishop RF (1982) A randomised trial of oral gamma globulin in low-birth-weight infants infected with rotavirus. Lancet 1:1371–1373
36. Saulsbury FT, Winkelstein JA, Yolken RH (1980) Chronic rotavirus infection in immunodeficiency. J Pediat 97:61–65

CHAPTER 44
Hemorrhagic Fever Viruses

C. J. Peters and K. M. Johnson*

The viral hemorrhagic fever (VHF) syndrome has received increasing attention in recent years and several "new" etiologic agents have been isolated. Unfortunately, because of socioeconomic conditions in areas where these diseases are most prevalent and the hazard of manipulating their causative agents in the laboratory, our knowledge of their pathogenesis has lagged behind that of many other viral infections. Nevertheless, a number of intriguing biological properties of these novel agents have been recognized. While hemorrhagic phenomena and perhaps even the fully developed syndrome have been reported as rare manifestations of a number of viral infections (e.g., varicella, measles, smallpox), only a few viruses are known to cause VHF regularly when they infect humans (Table 1).

New Clues to Pathogenesis of Certain VHF

Human response to infection with different HF viruses varies widely. In the case of Argentine hemorrhagic fever (AHF) and Bolivian hemorrhagic fever (BHF) (or collectively the South American HF), human infections usually result in expression of hemorrhagic disease of varying severity. In contrast, when yellow fever, dengue, Rift Valley Fever (RVF), Congo Crimean hemorrhagic fever (CCHF), or Lassa viruses infect humans milder clinical manifestations occur in the majority of patients, although severe and fatal VHF develops in others [1]. The reasons for the variation in disease manifestations are not well understood. In infants experiencing hemorrhagic fever, as a result of dengue-2 virus infection there is increasing epidemiological and

---

*US Army Medical Research Institute of Infectious Diseases, Frederick, Maryland.

Table 1. Viral hemorrhagic fevers (VHF) of humans

Virus family	Disease	Source of human infection	Current known distribution
Togaviridae (*Flavivirus* genus)	Yellow fever	Mosquito	Africa, South America
	Dengue HF	Mosquito	Asia, Caribbean
	Omsk HF	Tick	Russia
	Kyasanur forest disease	Tick	India
Bunyaviridae	Rift Valley fever	Mosquito	Africa
	Congo Crimean HF	Tick	Africa, Asia, eastern Europe
Arenaviridae	Argentine HF (Junin virus)	Rodent	Argentina
	Bolivian HF (Machupo virus)	Rodent	Bolivia
	Lassa fever (Lassa virus)	Rodent	Africa
Unassigned (? Bunyaviridae)	Korean hemorrhagic fever, hemorrhagic fever with renal syndrome, (Hantaan virus and relatives)	Rodent	Asia, Europe, Americas, Africa
Filoviridae (proposed)	Marburg disease	Unknown	Africa
	Ebola Hemorrhagic fever	Unknown	Africa

laboratory evidence that enhancement of virus replication by passive antibody is a major determinant of the development of hemorrhage and shock. Dengue virus is thought to replicate *in vivo* exclusively in mononuclear phagocytes. *In vitro,* limiting quantities of antibody act through an Fc-mediated mechanism to greatly enhance infection and virus yield from human monocyte cultures. Furthermore, similar passive antibody concentrations result in higher viremias in rhesus monkeys. Thus, infants born in dengue-endemic regions of southeast Asia are initially protected from infection and disease by high levels of maternal antibody. Ig catabolism proceeds, virus-neutralizing capacity of the child's serum wanes and infection may then occur in the presence of trace quantities of virus group-reactive IgG, resulting in more widespread involvement of the "target organs," which by some mechanism (see below) converts simple dengue fever into a serious disease characterized by hemorrhage and shock [2]. Dengue viruses do have the potential to cause primary HF, but the available evidence supports the importance of this secondary antibody-mediated enhancement either by maternal antibodies or by antibodies acquired from previous infection with other dengue serotypes.

Rift Valley fever also is a self-limited temporarily prostrating disease; however, a small proportion of patients develops one of two life-threatening complications: HF or encephalitis [3,4]. It is not known why each of these mutually exclusive manifestations, which fall completely outside the clinical spectrum of uncomplicated RVF, may occasionally develop in humans. In the laboratory rat, manipulating the host genotype can result in either fatal encephalitis, fatal hepatitis, or no overt disease after virus inoculation [5]. The fulminant hepatitis is characterized not only by early liver destruction, but also by viral involvement of endothelium, kidney, adrenal, and other organs. Virus replication is greater than that in other genotypes as early as 16–24 hours after infection. Resistance to the fulminant disease, which in many ways resembles human HF, is controlled by a single Mendelian dominant gene that appears to function by determining the sensitivity of the host cells (particularly macrophages) to induction of the antiviral state by low concentrations of interferon (J Rosebrock, GW Anderson, TW Slone, H Shellekens, and CJ Peters, unpublished observations). There is no evidence concerning the existence of such a gene in humans but one could imagine the importance of early interferon-mediated protection of vascular endothelium as well as the liver in preventing lethal HF.

Host–virus interactions independent of the immune response also may be important in determining the outcome of Lassa fever. Patients presenting with very high viremias ($\geq 10^4$ $TCID_{50}$/ml) have a greatly enhanced probability of dying, particularly if liver involvement (SGOT ≥ 150) is also prominent. The high viremia is present in the early stages and persists throughout the course of fatal disease. There is a suggestion that race could be a marker for the host factor. Serosurveys and prospective studies of Black Africans have shown that mild or asymptomatic infections with Lassa virus are common, and indeed may be the rule. Retrospective questioning of white missionaries having Lassa fever antibodies revealed that the majority recalled a clinically compatible severe febrile disease [6].

Clinical Features of VHF

The fully developed HF syndrome has similar clinical features in all the diseases listed in Table 1. Any single patient with one of these infections could well be indistinguishable from the others with the exception of the HFRS syndrome, in which renal failure and diuresis after the febrile phase are pathognomonic markers. However, there are diverse clinical findings which, although not formally documented, seem to be characteristic of hospitalized patients with these nosologic entites (Table 2). A major obstacle to

Table 2. Characteristics emphasized in diagnosed hospitalized cases of VHF

Disease	Clinical	Pathogenetic
Yellow fever	Black vomit; liver involvement	B-cell necrosis; terminal hepatorenal syndrome in monkeys.
Dengue HF	Vascular permeability dominates	Antibody enhancement of macrophage infection; macrophage as target organ
Tick-borne flavivirus HF	Pulmonary involvement; second wave of encephalitis may occur	
Rift Valley fever	Liver necrosis extensive	DIC in monkey model
Congo Crimean HF	Hemorrhage extensive and occasionally life-threatening; spreading ecchymoses	
South American HF	Prominent neurological involvement; hemorrhage very common but not important in volume	Bone marrow necrosis
Lassa fever	Deafness and male-dominant pericarditis common in late stages; hemorrhage and neurological involvement seen only in most severe cases in contrast to South American arenaviral HF	High viremia predicts fatal outcome
Hemorrhagic fever with renal syndrome	Renal involvement dominates. Long incubation period	Random tubular lesion; onset of disease appears to correlate with immunologic control of viral dissemination
Marburg–Ebola	High mortality; abdominal pain and diarrhea usually found; maculopapular rash common.	DIC implicated as one of mechanisms.

unraveling these syndromes is the lack of histopathological correlates: fever, weight loss, and focal or more extensive hepatic necrosis are the only near-common denominators.

Patients presenting with any VHF usually exhibit fever, malaise, myalgia, and prostration. There are signs of multisystem involvement including variable damage to the gastrointestinal and respiratory systems. Vascular involvement is signaled by petechiae and changes in permeability. Neurovascular involvement manifests itself as flushing, hyperemia, and postural hypotension. Most severely ill patients have obvious neurological involvement, and indeed virtually all patients with the South American HF present tremor, dysarthria, and changes in personality. There has not been sufficient study of the various syndromes to implicate viral invasion of the nervous system and, indeed, an indirect encephalopathy such as postulated for Reyes syndrome may be operative in some cases. Similarities among the diseases and their pathophysiological manifestations suggest that they may share common pathways; prime candidates include the interrelated mechanisms of endothelial injury, mediator secretion, and neurological participation.

Effector Mechanism in VHF

Few data are available on the status of the endothelium in VHF. Direct viral invasion of endothelium may well occur in RVF and Marburg–Ebola infections, leading to local clotting and secondary disseminated intravascular coagulation (DIC). A recent report of endothelial tropism by Hantaan virus in rodents also deserves attention [7]. In the case of arenavirus infections, evidence for DIC is scant or lacking and this also suggests that direct endothelial damage may not be a major inciting event. Fluorescent antibody studies in fatal human cases have failed to demonstrate impressive viral involvement of endothelium in DHF [8], CCHF [9], and AHF [10] although significant limitations exist in the interpretation of those negative results. It is possible that viral infection with a highly cytopathic virus such as RVF leads to direct destruction of endothelial cells. However, it is also possible that infection of these cells by nonlytic viruses may interfere with production or regulation of physiologically important molecules elaborated by endothelium. This could alter the balance of the poised coagulation cascade, which is maintained by factors such as charge, thrombomodulin, and prostaglandins resulting in microvascular DIC. Alternately, there could be changes in mediators such as prostaglandins or leucotrienes, which would lead to abnormal vascular regulation and permeability.

The lack of evidence for direct vascular involvement (although certainly not definitive) leads one to consider the participation of physiologically active mediator molecules in the pathogenesis of some of these disorders. Careful studies of DHF have implicated complement (C) consumption and release of vasoactive C fragments (but not kinins) in the evolution of the shock syndrome associated with DHF [11,12]. In dengue this condition is more clearly related to vascular permeability changes and hypovolemia than in any other VHF. Careful studies of AHF cases have detected evidence for DIC only in a few severely and terminally ill patients. Although AHF patients have decreased levels of C components, there is no clear pattern of classic or alternate pathway activation [13]. The presence of circulating serum cathepsins in conjunction with elevated immunological to functional ratios of C4 and factor VIII [14] suggest that circulating leucocytic enzymes may be critical in the South American HF. In spite of the lack of direct evidence for classic or alternate pathway C activation in the pathogenesis of most VHF there are phenomena in addition to increased vascular permeability that are suggestive. The prominent occurrence of pneumonitis in several VHF is reminiscent of the hemoperfusion syndrome associated with complement activation. Lassa fever also has been associated with terminal increases in leucocyte count, which may be due to C3e. There have been no studies to probe the role of leucotrienes or prostaglandins in the evolution of VHF.

The source of these hypothesized mediators is also unknown, although candidate cells exist. An obvious source, endothelium, is still under consideration but the common denominator for the VHF remains the macrophage [2,9,10]. This is the only cell type repeatedly implicated *in vivo* and *in vitro* as a site of viral replication and it also serves as a pivotal source of secretion of complement and clotting factors, as well as enzymes and low-molecular-weight mediator molecules.

The vascular phenomena (flushing, postural hypotension) present in the prodrome, evolution, and convalescence of VHF in conjunction with neurological involvement and shock raise the question of CNS or neurotransmitter involvement. The unexpected importance of centrally controlled neurotransmitters such as endorphins and related molecules in endotoxin and hemorrhagic shock [15] suggests that their participation in VHF should be evaluated. An exclusively neurological syndrome is seen in a minority of patients with the South American HF. These patients usually die with coma, convulsions, and intractable shock. Indeed, the occurrence of CNS involvement is recognized as a poor prognostic sign in Lassa, AHF, BHF, CCHF, HFRS, Marburg, and Ebola.

These infections are usually globally immunosuppressive to some degree. For exmple, in AHF there is a 20% incidence of bacterial superinfections, T cells and B cells are markedly reduced, and delayed hypersensitivity skin responses to other antigens are temporarily abolished [16]. Secondary bacterial infections are such a regular feature in severe cases that one must recognize the role of these organisms and endotoxin in exacerbating the clinical course as well as speculate on their regular participation in pathogenesis of some of the manifesttions of the VHF. Virus-specific immunosuppression is discussed later.

Virus–Reservoir Host Biology

This discussion has centered on the pathogenesis of these diseases in humans, their unnatural hosts; however, there are fascinating questions to be asked about the relation of these viruses to their reservoir hosts. Several of the viruses listed in Table 1, or agents closely related to them, are vertically transmitted in the host, which infects humans. Dengue and yellow fever viruses have been shown to pass from infected *Aedes* mosquitoes to their eggs at a low rate, which as yet has no clearly proven significance in nature. Transovarial transmission (TOT) is a regular and important feature of flaviviruses and CCHF virus in reservoir ticks [17]. Similarly, La Crosse virus and several phleboviruses that are close relatives of RVF virus exhibit TOT as a quantitatively significant feature of their maintenance cycle in nature. Virus can persist throughout the developmental cycle from ovum to post-reproductive adult with little or no compromise of arthropod development or function. Virus–host specificity of TOT is a well documented but poorly understood fact [18]. It also has been shown that serial selection of mosquitoes can result in lines of insects in which virtually all mosquitoes and their progeny are infected [19]. What the genes and gene functions are that control this phenomenon, what their occurrence in nature is, and what the differences are in molecular events that control noncytolytic infection in invertebrates as compared with cytolytic infection in vertebrates remain to be discovered.

Hantaan and related agents are thought not to utilize arthropods in their maintenance in nature. Available evidence suggests that infected rodents transmit virus to their offspring or other adult rodents by excretion of the causative agent in saliva, respiratory secretions, or urine. Viremia and dissemination to various organs occur in the infected reservoir. High-titered neutralizing antibodies are detected in serum after virus disappears, but virus excretion continues with no sign of acute or chronic disease. The failure to eradicate infectious virus secretion from mucosal surfaces is unexplained. If Hantaan virus is the prototype member of a new genus within the family Bunyaviridae, as increasing biochemical evidence suggests, then it may mature and be exocytosed via internal smooth endoplasmic reticulum in a fashion resembling other Golgi-mediated processes. How virus escapes mucosal IgA is unclear, but some Bunyaviridae exhibit a paucity of cell surface antigen [20], providing a possible mechanism for the failure of immune elimination of the productively infected cell. Alternately, viral secretion on these epithelial surfaces may proceed by a polarization mechanism akin to that described for certain budding viruses in MDCK cells [21]. There are other as yet unexplored possibilities.

Like Hantaan and related agents, Arenaviridae do not involve arthropods in their natural cycle but rather maintain themselves through chronic infection of a specific rodent reservoir for each virus. This infection is frequently manifest by life-long viremia as well as chronic virus shedding and is usually vertically transmitted. The virus is capable of inducing specific suppression of the immune response, which allows the rodent to circulate virus without serious acute disease. The best-studied arenavirus, lymphocytic choriomeningitis virus (LCMV), produces acute disease in its natural host (*Mus musculus*) only when the immune response is triggered to eradicate virus artificially introduced into the central nervous system. Even direct intracranial inoculation of virus is innocuous if the murine T-cell response is ablated. Extensive viral infection of neurons and other cell types occurs, but only subtle behavorial effects can be detected as a result. The explanation for this neuropsychiatric dysfunction may well be in the "luxury function" hypothesis espoused elsewhere in this volume. In laboratory mouse strains chronically viremic with laboratory-manipulated viruses (in contrast to available evidence for infected wild mice), there is a persistent humoral response capable of resulting in chronic immune complex disease.

Immunologic Parameters

Human HFRS is unique among the VHF clinically in its long incubation period and severe renal involvement and imunologically in the disappearance of viremia at the time of onset of disease. Most patients come to medical attention with no detectable viremia and with serum antibody present and rising. This circumstantial evidence coupled with demonstration of serum complement activation and deposition of Ig and C in kidney tissue [22] suggests that this HF is precipitated by the immune response. Why the immune control of viral dissemination is not associated with overt disease in the rodent reservoir but is pathogenic in humans is unexplained.

In contrast to the situation in human HFRS or in the arenavirus' natural host, there is no evidence that immunopathology is important in arenaviral HF of humans or of realistic primate and guinea pig models of acute HF. Detectable humoral immune responses may precede, coincide with, or follow clinical improvement. Furthermore, cyclosporin A or cyclophosphamide immunosuppression does not ameliorate disease but enhances the pathogenic potential of partially attenuated arenaviruses. Neither circulating nor tissue-deposited immune complexes were detected in detailed studies of human AHF cases [10,13]. The potential for immune complex disease exists in Lassa fever and its experimental analogues: non-neutralizing antibodies reacting with virus-coded antigens may be detected simultaneously with viremia. There is no clinical evidence to suggest that this antibody response is helpful or harmful. In rhesus and cynomolgous monkeys, its onset has not been correlated with clinical events or fall in viremia although, in this setting, serum viral antigen levels decrease and infectious virus may be precipitated with anti-IgM reagents. The generalized arteritis seen in the squirrel monkey may, however, be due to immune complex deposition.

We have mentioned the role of the immune response in enhancing dengue virus infection and perhaps inducing tissue damage in HFRS but have not discussed its participation in protection or recovery from VHF (Table 3). Neutralizing antibody has received the most attention, probably because sensitive methods for its measurement exist and because of the suggestive application of the *in vitro* phenomenon to the sick patient. The flavivirus

Table 3. Immune response parameters for the VHF

Virus	Efficient in vitro N by convalescent sera *	Brisk appearance of N ab**	Homologous N ab thought to protect †	Suspected role of effector T cells in recovery and/or protection	Interferon sensitivity in vitro ‡
Yellow fever	+	+	+	?	Yes
Dengue	+	+	+	?	Yes
Tick-borne flavivirus	+	+	+	?	Yes
Rift Valley fever	+	+	+	?	Yes
Congo Crimean HF	0	+	?	?	Probable
South American HF	+	±	+	Important	Low
Lassa	0	0	+	Important	Low
Hantaan	+	±	?	?	Probable
Marburg-Ebola	0	0	?	?	Low

* In most virus diseases (+) serum dilution neutralization tests with convalescent sera yield titers in the hundreds or thousands. In others (0) such antibodies are undetectable or in the tens.

** Usually N antibodies appear just after cessation of viremia (+) but in some diseases they are late or do not occur (0). In the South American AF and HFRS (Hantaan virus) they appear relatively soon after termination of viremia but 3 weeks or more after infection.

† In every case, efficient N *in vitro* is thought to imply protection. Lassa virus is not efficiently neutralized *in vitro* but protection or therapy is successful in monkeys if large quantities of plasma are given.

‡ Although not discussed in the text, most of the viruses tested with alpha or beta interferon *in vitro* are inhibited. Arenaviruses are resistant to the antiviral effects of 10^3 or greater VSV-inhibitory units. Virtually no studies are available with macrophages or gamma interferon.

hemorrhagic fevers and RVF all develop efficient *in vitro* serum-neutralizing (N) antibodies at the termination of viremia and often there is evidence that passive transfer of convalescent plasma is protective or therapeutic. There seems little doubt that N antibody is sufficient for recovery and protection in these diseases, although the role of T-cell responses has not been assessed. An interesting situation exists in the rhesus monkey model of RVF. Viral antigen (thought to be nucleocapsid) continues to circulate after termination of viremia, leading to the speculation that late immune clearance of this antigen may play a role in the "saddle-back" fever seen in many uncomplicated RVF cases or the late complications of retinal vasculitis or encephalitis [23].

In contrast to the diseases mentioned in the preceding paragraph, viremia is prolonged and the appearance of N antibody clearly delayed in the arenaviral HF. After an incubation period of 7 to 10 days and an acute phase of 1–2 weeks, serum antibodies initially appear (1–2 weeks later) in patients with AHF or BHF. This time lapse, compared with the brief delay with most viral infections, may well be related to the general or virus-specific immunosuppressive properties of the infection. In Lassa fever the serum-N antibody response is considerably delayed, up to 6 months after recovery. Even at its peak, *in vitro* neutralization by convalescent sera is of low titer (in serum dilution, plaque reduction neutralization tests) and is intensely complement dependent. These cell culture phenomena are mirrored in the results of passive protection tests [24]. It is not known if this apparent anomaly is related to a peculiarity of the virus, suppression of host response to critical epitopes of viral glycoproteins, induction of low-affinity antibody, or another mechanism; but understanding it is of obvious importance to immunotherapy and vaccine production for arenaviruses.

Observations of the South American HF bear on this relationship. Both Junin (AHF) and Machupo (BHF) virus cell culture plaques are neutralized efficiently by sera from recovered patients. Administration of convalescent plasma to AHF patients during the first week of disease terminates viremia and reduces mortality from 15% to 1% [25]. Unfortunately about 10% of those receiving immunotherapy return 4–6 weeks later with a poorly understood neurological syndrome. This unexpected complication may result from an imbalance between humoral and cellular recovery mechanisms or may be a reflection of CNS involvement in severely infected patients otherwise destined to die. There are in fact emerging data showing significant Junin (AHF) virus strain differences in neurovirulence for monkeys. Additionally it has been demonstrated that homologous immunoglobulins protect monkeys from lethal infection by the related Machupo (BHF) virus, but that very high doses of antibodies led to a late fatal encephalitis.

Other properties of arenaviruses are of importance for understanding the participation of the immune response in protection and recovery with this subset of the VHF. They do not usually cause extensive damage to cultured cells, although they induce extensive cell surface antigen production and bud from the plasma membrane. Thus the immune response can be particularly effective if directed against infected cells during the protracted course of these HF. Furthermore, arenaviruses extensively involve macrophages, which are particularly effective antigen-presenting cells and are of established sensitivity to lethal effects of cytotoxic T cells or physiological alterations by the subset of T cells responsible for delayed type hypersensitivity. These considerations in conjunction with mouse studies of LCM have strongly supported the idea that T-cell responses, either through their effects on macrophages or their ability to kill virus-infected cells selectively, are critical in recovery and, at times, protection from arenavirus infection [26]. Laboratory studies in realistic guinea pig or primate HF models to investigate these notions are not available.

In the case of CCHF, the role of N antibody is not clear. Available studies suggest it is not greatly delayed, although titers, even in hyperimmune sera, are not high. Convalescent sera apparently do not neutralize Filoviridae *in vitro* nor efficiently protect *in vivo*. Hyperimmune sera clearly neutralize infectivity of Zaire strains of Ebola virus as determined by plaques in cell culture or intra-cranial infectivity for suckling mice and offer some protection to infected guinea pigs on passive transfer. Insufficient data are available for the HFRS-like agents, although *in vitro* and rodent *in vivo* neutralization of virus is efficient.

Further Generalizations

It must be clear by now in this discussion that we understand relatively little about these viruses, even the classical agents such as yellow fever. The generalizations hazarded above are scarcely more than speculations awaiting further work on the natural maintenance cycles, immunology, and virology of these agents. Nevertheless, we propose a second set of generalizations. Humans are not important to these viruses, with the probable exception of the mosquito-borne flaviviruses (yellow fever and dengue) which seem to require a primate amplifier for efficient maintenance. All VHF have a mechanism of transmission from one generation of reservoir to the next and, once again with the exception of the mosquito-borne flaviviruses, this is of major importance to their propagation in nature. The infection of humans generally represents an extension outside their basic cycle and for important but poorly understood reasons results in serious consequences for the aberrant host. Although there are compelling clinical similarities, these diseases are very different in their pathogenesis and immunology. They range from RVF virus, which is an intensely cytopathic agent that induces a brisk short-incubation disease that appears to be largely regulated by serum-neutralizing antibody, to Lassa virus, which is virtually noncytolytic *in vitro*, results in human disease evolving over a much longer time period and is thought to be controlled by nonhumoral mechanisms.

Unfortunately there are few "docile" viruses related to those of VHF that cause human-like disease in animals. Thus, further progress in study of these entities will be largely restricted to those few centers in the world possessing biological high-containment facilities.

Acknowledgments. We wish to express our appreciation to our colleagues who have shared their experimental data and ideas without which this review would be meaningless: arenaviruses, P. Jahrling, R. Kenyon, K. McKee, and E. Genovesi; Rift Valley fever, J. Rosebrock, G. Anderson, and H. Schellekens; Marburg–Ebola viruses, H. Lupton, and E. Johnson; arthropod–virus interactions, C. Bailey.

References

1. Johnson KM, Halstead SB, Cohen SN (1967) Hemorrhagic fevers of Southwest Asia and South America: A comparative appraisal. Prog Med Virol 9:105–158
2. Halstead SB (1981) The pathogenesis of dengue: Molecular epidemiology in infectious disease. Am J Epidemiol 114:632–648

3. Van Velken DJJ, Meyer JD, Oliver J, Gear J, McIntosh B (1977) Rift Valley fever affecting humans in South Africa. A clinicopathological study. S Afr Med J 51:867–871
4. Laughlin LW, Meegan JM, Strausbaugh LJ, Morens DM, Watten RH (1979) Epidemic Rift Valley fever in Egypt: Observations of the spectrum of human illness. Trans Soc Trop Med Hyg 73:630–633
5. Peters CJ, Slone TW (1982) Inbred rat strains mimic the disparate human response to Rift Valley fever virus infection. J Med Virol 10:45–54
6. Frame, JD (1975) Surveillance of Lassa fever in missionaries stationed in West Africa. Bull WHO 52:593–598
7. Kurata T, Tsai TF, Bauer SP, McCormick JB (1983) Immunofluorescence studies of disseminated Hantaan virus infection of suckling mice. Infect Immun 41:391–398
8. Boonpucknavig S, Boonpucknavig V, Bhamarapravati N, Nimmannitya S (1979) Immunofluorescence study of skin rash in patients with dengue hemorrhagic fever. Arch Pathol Lab Med 103:463–466
9. Karmysheva VYA, Leshchinskaya EV, Butenko AM, Savinov AP, Gusarev AF (1973) Results of laboratory and clinical–morphological investigations of Crimean hemorrhagic fever. Arkh Patol 2:17–22
10. Maiztegui JI, Laguens RP, Cossio PM, Casanova MB, de la Vega MT, Ritacco V, Segal A, Fernandez NJ, Arana RM (1975) Ultrastructural and immunohistochemical studies in five cases of Argentine hemorrhagic fever. J Infect Dis 132:35–43
11. Pathogenetic mechanisms in dengue haemorrhagic fever: Report of an international collaborative study. *In* Pathogenesis of Dengue Haemorrhagic Fever (1973) Bull WHO 48:117–132
12. Edelman R, Mimmannitya S, Colman RW, Talamo RC, Top Jr FH (1975) Evaluation of the plasma kinin system in dengue hemorrhagic fever. J Lab Clin Med 86:410–421
13. de Bracco MME, Rimoldi MT, Cossio PM, Rabinovich A, Maiztegui JI, Carballal G, Arana RM (1978) Argentine hemorrhagic fever. Alteration of the complement system and anti-Junin-virus humoral response. N Engl J Med 299:216–221
14. Molinas FE, Maiztegui JI (1977) Factor VIII: C and factor VIII R: Ag in Argentine hemorrhagic fever. Thrombos Haemostas 46:525–527
15. Faden AI, Holaday JW (1980) Experimental endotoxin shock: The pathophysiologic function of endorphins and treatment with opiate antagonists. J Infect Dis 142:229–238
16. Arana RM, Rittacco GV, de la Vega MT, Egozcue J, Laguens RP, Cossio PM, Maiztegui JI (1977) Immunological studies in Argentine hemorrhagic fever. Medicina (Buenos Aires) 37:186–189
17. Hoogstraal H (1981) Changing patterns of tickborne diseases in modern society. Ann Rev Entomol 26:75–99
18. Tesh RB, Chaniotis BN, Johnson KM (1972) Transovarial transmission of vesicular stomatitis virus (Indiana serotype) in phlebotomine sand flies. Science 175:1477–1479
19. Tesh RB, Shroyer DA (1980) The mechanism of arbovirus transovarial transmission in mosquitoes: San Angelo virus in *Aedes albopictus*. Am J Trop Med Hyg 29:1394–1404

20. Smith JF, Pifat DY (1982) Morphogenesis of Sandfly fever virus (Bunyaviridae Family). Virology 121:61–81
21. Compans RW, Roth MG, Alonso FV, Srinivas RV, Herrler G, Melsen LR, Meier-Ewert H (1982) In Mackenzie JS (ed) Viral Diseases in Southeast Asia and the Western Pacific. Academic Press, Sydney, p 328
22. McKee KT, Peters CJ, Craven RB, Francey DB (1984) Other viral hemorrhagic fevers and Colorado tick fever. In Belshe RB (ed) Textbook of Human Virology. PSG, Littleton, ch 22, p 649–678
23. Niklasson B, Grandien M, Peters CJ, Gargan, III, TP (1983) Detection of Rift Valley fever virus antigen by enzyme-linked immunosorbent assay. J Clin Microbiol 17:1026–1031
24. Jahrling, PB, Peters, CJ (1984) Passive antibody therapy of Lassa fever in cynomolgus monkeys: Importance of neutralizing antibody and Lassa virus strain. Infect Immun 44:528–533
25. Maiztegui JI, Fernandez NJ, de Damilano AJ (1979) Efficacy of immune plasma in treatment of Argentine hemorrhagic fever and association between treatment and a late neurological syndrome. Lancet 2:1216–1217
26. Buchmeier MJ, Welsh RM, Dutko FJ, Oldstone MBA (1930) The virology and immunobiology of lymphocytic choriomeningitis virus infection. Adv Immunol 30:275–331

CHAPTER 45
Marburg and Ebola Viruses: New Agents on the Frontiers of Virology

MICHAEL J. BUCHMEIER*

Marburg virus was first isolated in 1967 in association with a devastating outbreak of hemorrhagic fever among laboratory workers in Germany and Yugoslavia [1].Thirty-one people in two institutes were stricken and seven died. All of these deaths were among individuals having primary contact with the infected monkey tissues and no deaths occurred that could be attributed to human-to-human spread. The source of infection was traced to vervet monkeys (*Cercopithecus aethiops*) imported by air from Uganda for the purpose of preparing primary cell cultures. Two additional outbreaks of Marburg disease have been described [2,3]. In one, occurring in 1975, an Australian man was stricken while hitchhiking through South Africa and subsequently died in a Johannesburg hospital [2]. His traveling companion and an attending nurse also became ill but recovered. Although the precise source of the primary infection was not determined, exposure probably occurred while traveling through Rhodesia.

Ebola virus made an equally precipitous and more devastating entry onto the scene in 1976. Within a time span of 3 months, focal outbreaks of epidemic hemorrhagic fever occurred approximately 850 km apart in southwestern Sudan and northeastern Zaire [4,5] and claimed 150 and 211 lives, respectively. It was quickly recognized that the etiologic agents responsible for these outbreaks were morphologically similar to the previously described Marburg agent, however no evidence of antigenic relationship between the Ebola and Marburg viruses was found [6].

Clinical features of the three infections are similar [7–9]. Patients typically presented with an abrupt onset of severe frontal headache, high fever, abdominal pain and cramps, watery diarrhea, and vomiting. A macropapular

---
*Scripps Clinic and Research Foundation, La Jolla, California.

rash appeared around 5-7 days and persisted 3-4 days followed by desquamation. Severe bleeding from multiple sites including gastrointestinal tract, lungs, and mucous membranes began approximately 5-7 days after onset and death occurred between days 7 and 16, preceded by blood loss and shock. In both the Sudan and Zaire outbreaks, disease was associated predominantly with regional hospitals, and in Zaire, infections were transmitted parenterally by means of contaminated hypodermic equipment [5]. The total case fatality rate in the documented outbreaks of Marburg disease was 23%. Reported case fatality rates for Ebola virus epidemics were 53% for Sudan and 88% for Zaire, and on this basis the Sudan strain has been described as the less virulent of the Ebola viruses; a finding that has extended to animal studies in monkeys, guinea pigs, and mice [10,11].

Pathologic features of these hemorrhagic fevers include extensive hepatic and splenic necrosis without accompanying inflammatory response. Virus is readily visualized by electron microscopy in the affected tissues and in the plasma [12,13]. Both Rhesus monkeys and guinea pigs appear to provide reasonable experimental parallels to the human infection and have been useful in limited pathogenesis and protection studies [10].

The pathologic picture of rapid onset, fulminating infection with extensive liver destruction and terminal bleeding has led some investigators to suggest a role for disseminated intravascular coagulation (DIC). DIC is a feature of other hemorrhagic fevers such as Rift Valley fever and Congo Crimean hemorrhagic fever where massive destruction of liver parenchyma also occurs (see Chapter 44, this volume). Evident from these parallels is the need for careful and detailed studies of the effect of virus infection on the coagulation and complement systems if we are to understand the pathogenesis of these infections.

Marburg-Ebola Viruses: Natural Reservoir Is Unknown

Documented outbreaks of Marburg and Ebola hemorrhagic fevers have in common abrupt onset and rapid spread within medical or institutional personnel at risk. In the case of Marburg, monkeys were the source of infection, however no such association was made with the Ebola viruses. Further, it is highly unlikely that monkeys in Zaire are a significant reservoir of virus since the disease in monkeys is fatal. Efforts to define the natural reservoir of Ebola and Marburg viruses have focused upon peridomestic rodents and the possible involvement of arthropod vectors has been considered [4,9,14]. One epidemiologic study done following a 1977 outbreak of Ebola hemorrhagic fever in northwestern Zaire found that 7% of the residents of the village where the disease occurred had significant antibody titers against the Zaire agent. Thus, it appears that less severe or perhaps even inapparent infections may occur. In this outbreak, as in all previous ones, no evidence for a reservoir was found.

Structural Features of Marburg–Ebola Agents: Unique among Animal Viruses

Early descriptions of the Marburg agent detailed a virus of unprecidented morphology [12,15]. High concentrations of virus in the blood and organs of humans and experimental animals infected by Marburg virus allowed direct electron microscopic visualization of virions. The particles observed were bizarre in consisting of long filamentous forms, often exceeding 14,000 nm, torus forms resembling the number six, circles, and branched filaments. These varied morphologic types shared common features of an electron dense axial nucleocapsid core structure some 20-nm in width and bearing striations. The nominal length of these nucleocapsids observed both in the cytoplasm of infected cells *in vitro* and within virions was approximately 700 nm [16]. Branched and filamentous forms appeared to have arisen by co-envelopment of multiple nucleocapsid strands upon exit from the infected cell. More recently, the unit lengths of Marburg and Ebola virions grown in Vero cells were shown to be 790 nm and 970 nm, respectively, and the particle sedimentation coefficients estimated to be 1300 to 1400 S [17]. Such tissue-culture-propagated virus showed fewer bizarre structures than did earlier specimens obtained directly from patient or animal material. These unique morphologic features and extreme length of the filamentous virions distinguish Marburg and Ebola viruses from other known taxonomic groups, and have prompted the proposal of a new taxonomic group, the filoviridae from the Latin *filo,* filament [18].

Molecular Techniques Defined the Filoviridae

Studies of the viral RNAs of Marburg and Ebola viruses have yielded fundamental information about the nature of the genetic material, the stability of each viral strain, and the phylogenetic relationships among members of the group. Regnery *et al.* [19] demonstrated that the genome of Ebola virus (Zaire strain) consisted of a single-stranded RNA molecule with a molecular weight of 4.2×10^6 daltons, of negative (nonmessage) polarity. Both Marburg and Ebola viruses were shown to have genomes of similar size and characteristics. Cox *et al.* [20] have extended the analysis of virion RNA by RNase T_1 oligonucleotide mapping of the Ebola virus strains isolated during outbreaks in Zaire and Sudan from 1976 to 1979. Molecular evidence for two Ebola virus subtypes emerged by RNA mapping. One was associated with the disease in Zaire in 1976 and 1977, while the other was responsible for epidemics in Sudan in 1976 and 1979. Within each subtype, independently obtained isolates were remarkably stable differing by two or fewer oligonucleotides, while the Sudan and Zaire biotypes differed by approximately 60 oligonucleotides. Demonstration of shared nucleotides between the Sudan and Zaire strains led to speculation that these viruses arose from a common phylogenetic precursor. Recent studies of the polypeptides and antigens of the Ebola viruses further support this concept. Using conventional antisera, no antigenic relationship was observed between Marburg and the Ebola viruses by immunofluorescence or complement fixation assays. Radioimmune assay revealed both shared and unique antigens in comparisons between Ebola Sudan and Ebola Zaire [21]. Ebola viruses have a similar polypeptide composition. Both have a large protein, L (MW $\simeq$ 190,000), a single glycoprotein VP-1 (MW 125,000) two nucleocapsid associated proteins VP-2 and VP-3 (MW 104,000 and 40,000, respectively) and a small internal protein VP-4 (MW 26,000), which is not structurally associated with the nucleocapsid [22]. We have studied the molecular basis for the observed antigenic relationship between Ebola strains by tryptic peptide analysis of these viral polypeptides [23]. Our results indicate that Sudan and Zaire strains of Ebola virus are unique and that within each subgroup multiple isolates show a high degree of stability; findings consist with the oligonucleotide mapping studies. Major differences between these Ebola strains were evident in the VP-1 glycoprotein and VP-3 nucleocapsid protein of the virion. We have also examined Marburg viruses isolated in 1967 in the German outbreak, and in 1980 in Kenya (Buchmeier and Kiley, unpublished). These viruses were nearly identical by peptide mapping, suggesting stability similar to that observed by peptide and oligonucleotide mapping of Ebola virus subtypes. Thus, molecular and antigenic analyses have demonstrated existence of two major subdivisions of the filoviridae; the first consisting of Marburg virus alone and the second of the related but distinguishable Ebola viruses from Sudan and Zaire. Each virus is stable as evidenced by a high degree of similarity among individual isolates. Although data are incomplete, the viruses appear to be endemic in nonoverlapping geographic areas of Africa. Marburg virus appears to be endemic in Uganda [1], Rhodesia [2], and Kenya [3], while the Ebola viruses have only been isolated in Sudan [4] and Zaire [5,14].

Field and molecular studies in the past decade have yielded a wealth of new information about Marburg and Ebola viruses. Despite this, important unanswered questions remain. The most pressing of these are the identity of the natural reservoir of infection, precise definition of the endemic areas of infection, the extent of naturally occurring human infections in these areas, and finally, the pathogenic mechanisms leading to hemorrhagic fever. It is hoped that the next decade will yield answers to these questions.

Note Added in Pages

Bouree and Bergmann [Bouree P, Bergmann J-F (1983) Ebola virus infection in man: A serological and epidemiological survey in the Cameroons. Am J Trop Med Hyg 32:1465–1466] have recently demonstrated further evidence of subclinical infection among a study population of 1517 apparently healthy individuals in the Cameroons. An overall antibody positive rate of 9.7% was observed with peak incidence in young adults and rain forest farmers. In Pygmies of the Lolodorf-Bidindi region, the incidence of antibody positive individuals ranged as high as 33% among young adults.

Acknowledgments. This is publication no. 3002-IMM from the Department of Immunology, Scripps Clinic and Research Foundation. This effort was supported by US PHS Grants AI-16102, NS-12428, and US Army Research Development Command Contract No. DAMD17-83-3013. M.J. Buchmeier is an Established Investigator of the American Heart Association.

References
1. Martini G, Siegert R (eds) (1971) Marburg virus disease. Springer-Verlag, New York
2. Gear JSS, Cassel GA, Gear AJ, Trappler B, Clausen L, Meyers AM, Kew MC, Bothwell TH, Sher R, Miller GB, Schneider J, Koornhoff HJ, Gomperts ED, Isaacson M, Gear JHS (1975) Outbreak of Marburg virus disease in Johannesburg. Br Med J iv:489–493
3. Morbidity and Mortality Weekly Report, Vol 29, p 145 (US Department of Health and Human Services, Centers for Disease Control, Atlanta, 1980)
4. Ebola haemorrhagic fever in Sudan, 1976: Report of a WHO/International Study Team (1978) Bull WHO 56:247–270
5. Ebola haemorrhagic fever in Zaire, 1976: Report of an International Commission (1978) Bull WHO 56:271–293
6. Pattyn S, van der Groen G, Jacob W, Piot P, Courteille G (1977) Isolation of Marburg-like virus from a case of haemorrhagic fever in Zaire. Lancet i:573–574
7. Kissling RE, Murphy FA, Henderson BE (1970) Marburg virus. Ann NY Acad Sci 174:932–939
8. Brés P (1978) The epidemic of Ebola hemorrhagic fever in Sudan and Zaire in 1976: Introductory note. Bull WHO 56:245
9. Simpson DIH (1978) Viral hemorrhagic fevers of man. Bull WHO 56:819–832
10. Bowen ET, Platt GS, Lloyd G, Raymond RT, Simpson DI (1980) A comparative study of strains of Ebola virus isolated from southern Sudan and northern Zaire in 1976. J Med Virol 6:129–138
11. McCormick JB, Bauer SP, Elliott LH, Webb PA, Johnson KM (1983) Biologic differences between strains of Ebola virus from Zaire and Sudan. J Infect Dis 147:264–267
12. Almeida JD, Waterson AP, Simpson DIH (1971) Morphology and morphogenesis of the Marburg agent. *In* Martini GA, Siegert R (eds), Marburg Virus Disease. Springer-Verlag, New York, p 84
13. Siegert R, Slenczka W (1971) Laboratory diagnosis and pathogenesis. *In* Martini GA, Siegert R (eds) Marburg Virus Disease. Springer-Verlag, New York, p 157
14. Heymann DL, Weisfeld JS, Webb PA, Johnson KM, Cairns T, Berquist H (1980) Ebola hemorrhagic fever: Tandala, Zaire, 1977–1978. J Infect Dis 142:372–376
15. Murphy FA, Simpson DIH, Whitfield SG, Zlotnik I, Carter GB (1971) Marburg virus infection in monkeys—ultrastructural studies. Lab Invest 24:279–291
16. Ellis DS, Stamford S, Tovey DG, Lloyd G, Bowen ETW, Platt GS, Way H, Simpson DIH (1979) Ebola and Marburg viruses. II. Their development within

Vero cells and the extracellular formation of branched and torus forms. J Med Virol 4:213–225
17. Regnery RL, Johnson KM, Kiley MP (1981) Marburg and Ebola viruses: Possible members of a new group of negative-strand viruses. *In* Bishop DHL, Compans RW (eds) the Replication of Negative-Strand Viruses. Elsevier/North-Holland, New York, p 971
18. Kiley MP, Bowen ETW, Eddy GA, Isaacson M, Johnson KM, McCormick JB, Murphy FA, Pattyn SR, Peters D, Prozesky OW, Regnery RL, Simpson DIH, Slenczka W, Sureau P, van der Groen G, Webb PA, Wulff H (1982) Filoviridae: Taxonomic home for Marburg and Ebola viruses? Intervirology 18:24–32
19. Regnery RL, Johnson KM, Kiley MP (1980) Virion nucleic acid of Ebola virus. J Virol 36:465–469
20. Cox NJ, McCormick JB, Johnson KM, Kiley MP (1983) Evidence for two subtypes of Ebola virus based on oligonucleotide mapping of RNA. J Infect Dis 147:272–275
21. Richman DD, Cleveland PH, McCormick JB, Johnson KM (1983) Antigenic analysis of strains of Ebola virus: Identification of two Ebola virus serotypes. J Infect Dis 147:268–271
22. Kiley MP, Regnery RL, Johnson KM (1980) Ebola virus: Identification of the virion structural proteins. J Gen Virol 49:333–341
23. Buchmeier MJ, DeFries RU, McCormick JB, Kiley MP (1983) Comparative analysis of the structural polypeptides of Ebola viruses from Sudan and Zaire. J Infect Dis 147:276–281

CHAPTER 46
Rabies

HILARY KOPROWSKI*

> We have left numerous loose ends, but after
> a brief resume, we must plough ahead.
> J.L. Austin: *How To Do Things with Words*

When Ajax and Achilles called Hector "a rabid dog" in the Illiad (before 700 B.C.) they referred metaphorically to a condition associated with a disease first mentioned before 1800 B.C. in Mesopotamian "Laws of Eshnunna." In the course of centuries, Democritus, Aristotle, Aulus Cornelius Celsus, Girolamo Fracastoro (who first described the "incurable" course of the disease) and scores of others have studied and described rabies not only because of particular interest in the disease but also because of morbid fascination with this infection to which all warm-blooded animals, including bats, are susceptible. Rabies causes infected animals and sometimes humans to behave in such a frenzied and unnatural way that popular superstition claimed them to be "possessed" by a demon that can be transmitted to new victims. The millennia of rabies studies culminated in the late part of the 19th century with Pasteur, who ordered his disciples to search for rabies protective treatment, since the infection, which was thought then to be transmitted only by bite, presented the unique attribute that the exact time of exposure (day/hour) could be ascertained. This resulted not only in the production of the first vaccine for treatment of humans [1], but also in "dissemination" throughout the world of a score of "Pasteurians" who, in order to perpetuate work of the master, modified in various ways the original version of the vaccine and even modified the original post-exposure treatment by addition

---

* The Wistar Institute of Anatomy and Biology, Philadelphia, Pennsylvania.

of anti-rabies serum [2]. Unfortunately, they contributed little to our scanty knowledge of pathogenesis of rabies.

Although infection with rabies by animal bites (Laws of Eshnunna) was known for more than 3000 years, infection of animals and humans by inhalation has been observed only within the last decades, and the tragic transmission by corneal transplants has been described in two instances. In experimental animals it was shown that following intramuscular injection, the virus is capable of replication in striated muscle cells [3]. However, once it reaches the exposed nerve ending, its centripetal spread is strictly in the axoplasm at a rate of about 3 mm/hour. This nerve transit is apparently passive, the first replicative phase occurring in the neurons of the spinal ganglia where the virus rapidly ascends to the brain, where its presence is again strictly limited to certain groups of neurons [3]. Since the virus buds from neuronal cell membrane, it may spread in CNS through interstitial fluids and also by cell-to-cell transfer and neuronal processes. Peripheral spread of rabies again involves axoplasmic transit to peripheral nerve endings [4]. Virus has been demonstrated in chromaffin cells of the adrenals and in the pancreas. It is also present in sensory nerve endings of the skin of the head and of the eye. Since saliva is the most common source of rabies transmission from one host to another, the virus must lodge in the salivary gland. It actually buds out from the membrane of the mucous acinar cells. The cells are surrounded by the basement lamina, which "protects" them from the effect of antibody.

Concentration and time of appearance of anti-rabies neutralizing antibodies in serum of an animal or a human after exposure to rabies are of no significance as far as the outcome of the infection is concerned. High titers of antibodies were observed in those who died as well as in those who did not develop the disease. Although recovery from the disease caused by rabies infection is a rare exception, it undoubtedly occurs, perhaps more frequently in animals than in humans. Of the three cases of humans recovered from rabies, two were exposed to street virus by bite and one inhaled a large dose of fixed virus in the laboratory. In absence of any specific treatment, the cases were treated through the relief of brain edema and maintenance of respiratory functions through tracheotomy. Two out of three recovered without sequela. It is also quite certain that infection with rabies may result in an inapparent infection followed by resistance to subsequent homologous challenges. This has been shown experimentally in hamsters injected with street virus who did not show signs of disease and yet became immune to subsequent challenge with virulent street virus [5].

The histologically determined lesions of CNS do not justify the almost universal lethal outcome of rabies infection once the disease process sets in [3]. Obviously, functional impairment of vital functions of the organism is much more extensive than can be accounted for by the morphologic evidence of damages to tissue of the body. One interesting aspect of lethal infection with rabies is "pan suppression" of cell-mediated immunity in experimental animals [6]. After inoculation with street virus, the cytotoxic T-cell response is suppressed not only in CMC assay against rabies-infected cells but also against influenza virus. In addition, homotransplants of skin are not rejected by rabies-infected mice. Since rabies virus cannot infect any immunocompetent cells or their precursors, the study of "pan suppression" of cell-mediated immunity warrants further attention. More precise identification of the T-cell population involved in the immunosuppression and the study of the mechanism leading to this defect may present us with a very valuable model of centrally originated immunosuppression.

The role in the pathogenesis of rabies of the complement-mediated cytolytic effect of rabies antibody on rabies-infected cells is also unclear [7]. Does it serve to decrease spread of the virus from lysed cells by binding it to the antibody or does it harm by destroying rabies-infected cells that are essential for vital functions of the organism?

Development of monoclonal antibodies [8] not only facilitated production of variants and determination of antigenic sites defining areas of rabies glycoprotein molecule, but also permitted selection of variants [9] with altered pathogenicity. For instance, rabies variants, which were selected after interaction with one monoclonal antibody and which became resistant to neutralization by this antibody, lost their ability to kill adult mice injected intracerebrally [10]. The absence of neutralizing effect by this monoclonal antibody seemed to characterize attenuated strains, since two such strains developed decades ago by chance could also not be neutralized by this antibody. The attenuation of the strains for adult mice could be traced by analysis of tryptic peptide of the viral glycoprotein to the amino acid substitution at position 333 of arginine of the virulent strain by either isoleucine or glutamine in the attenuated strain [10]. Following few sequential passages in newborn mice, the attenuated variant reverts to virulence for adult mice, arginine replacing either isoleucine or glutamine at position 333. Thus, one draws a conclusion that presence of arginine 333 is essential for rabies virus to exercise its lethal function [10].

Antigenic variants of rabies virus cannot effectively immunize mice against challenge with either parental strains or with another variant, although they engender excellent protection when exposed to the same variant. [Interestingly enough, antibodies produced in sera of mice, neutralized the strains equally well including those which spared the vaccinated mice and those that killed them.] Not unexpectedly, antigenically distinct virus variants were found throughout the world. Strains isolated from bats were singularly distinct in the reaction with a panel of monoclonal antibodies from the so-called standard strains. Of mice vaccinated with routinely prepared variants and challenged with one of the maximally deviated field strains, only about half survive challenge. Whether those findings can be extrapolated to failures of standard vaccine preparation to protect in those geographic areas where distinct variants are present remains to be seen. It is certain that rabies vaccines do not protect against exposure to the so-called rabies-associated viruses. Recently, the death of vaccinated animals in Bulawayo, Zimbabwe was explained by isolation of rabies-associated virus and not rabies from animals exhibiting signs of rabies.

Rabies is a universal infection (Fig. 1). Not only does the virus infect all warm-blooded organisms, but its presence has also been described in most parts of the world with the exception of Australia, British Isles, Japan, and several other islands where drastically enforced protective measures against animal movements have maintained these territories rabies-free to this day. Thorough programs of vaccination of domestic animals, principally dogs, may effectively control the spread of rabies among domestic animals. Spread of rabies among wildlife, which is occurring in Europe (foxes) and North America (raccoons, skunks, bats), is much more difficult to control. Eradication of an animal population, not indicated from an ecological point of view, has been relatively ineffective as a rabies-control measure. If one considers that rabid bats have been observed in all parts of the North American continent since 1973, one realizes the futility of controlling the spread of rabies among members of this species. Exposure of humans to rabies through contact with bats is relatively infrequent, but human cases of rabies caused by exposure to other wildlife species are increasing. One must face the issue that vaccination of humans against rabies may continue forever.

How effective are rabies vaccines in post-exposure treatment? The animal brain tissue-derived original Pasteur vaccine and its derivatives were probably quite immunogenic. Their use, however, presented the vaccinated individual with two dangerous complications. If the concentration of the live

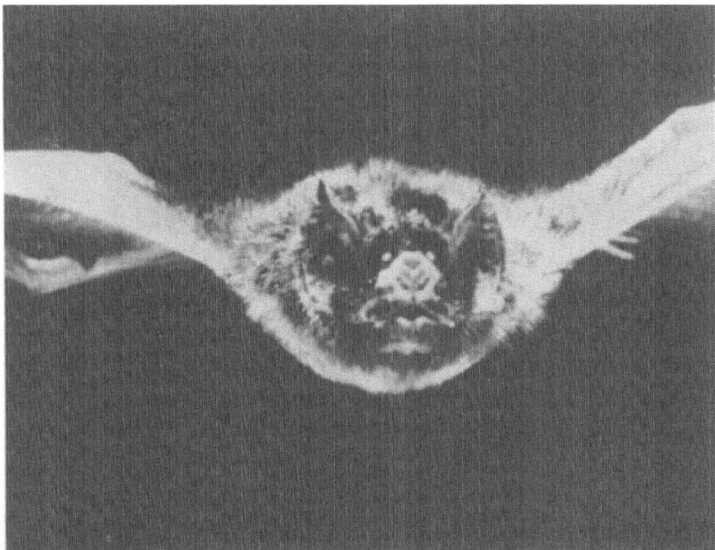

Fig. 1. A vampire bat (*Desmodus Rotundus*). A common transmitter of rabies in Central and South America.

fixed virus exceeded certain limits, it could cause rabies instead of preventing it. Furthermore, extracts of myelinated (brain) tissue used for production of the vaccine not infrequently caused a paralytic syndrome akin to the experimental allergic encephalitis induced by inoculation of animals with brain tissue extracts. Production of rabies vaccine in brain tissue of suckling mice was an attempt to circumvent the neuroparalytic consequences of vaccination; such vaccines are used extensively in South America today.

With the development of more sophisticated methods of growing rabies virus in tissue culture, it became possible to produce a vaccine from virus grown in either human diploid fibroblasts or green monkey tissue. This inactivated virus vaccine is much more antigenic than its predecessors. Hence it was possible to decrease the number of post-exposure treatments from 14–21 injections to 4–6 injections. There are no immediate or delayed reactions following vaccination. Its protective effect had to be determined in light of susceptibility of humans to rabies. This was unknown until 1953 when a failure to vaccinate a group of Iranian villagers bitten by a rabid wolf provided information. Fifty percent of those bitten on either the head or neck and 37% of those bitten on other parts of the body died of rabies [11]. In contrast, when 45 Iranians exposed to severe bites by wolves and dogs in 1975 and 1976 were treated with rabies antibody [12] and the new tissue culture vaccine, none died in the course of 6 years of observation.

The tissue culture product is not the last word in the development of rabies vaccine. Since the glycoprotein of the virus is the immunogenic component for protection against challenge [13], recent advances in the biotechnology of rabies are geared either toward possible use of an immunogenically active protein biosynthesized by *E. coli* or toward chemically synthesized immunogenic peptides.

The rapid advances in biotechnology are not paralleled by similar progress in our knowledge of pathogenesis of rabies. Although much is known about individual components of the organism reacting to rabies infection, this knowledge represents only pieces of a puzzle, which wait to be assembled in order to obtain a clear picture of the whole problem.

Writing about rabies at present, one can feel argumentative like the writers in Thurber's "the porcupines and the artichokes" who shout at each other, "you are right, you are right, you are absolutely right! The trouble is, you don't have the vaguest idea why you are."

References
1. Pasteur L (1887) Lettre sur la rage. Ann Inst Pasteur 1:1–18
2. Marie AC (1908) Recherches sur le serum antirabique. Ann Inst Pasteur 22:271
3. Murphy FA (1977) Rabies pathogenesis. Arch Virol 54:279–297
4. Schneider LG (ed) (1975) Spread of virus from the central nervous system. Natural History of Rabies. Academic Press, New York, p 199–215
5. Koprowski H, Black J (1954) Studies on chick-embryo-adapted rabies virus. V. Protection of animals with antiserum and living attenuated virus after exposure to street strain of rabies virus. J Immunol 72:85–93

6. Wiktor TJ, Doherty PD, Koprowski H (1977) Suppression of cell-mediated immunity by street rabies virus. J Exp Med 145:1617–1622
7. Wiktor TJ, Kuwert E, Koprowski H (1968) Immune lysis of rabies virus-infected cells. J Immunol 101:1271–1282
8. Koprowski H, Wiktor T (1980) Monoclonal antibodies against rabies virus. *In* Kennett RH, McKearn TJ, Bechtol KB (eds) Monoclonal Antibodies. Hybridomas: A New Dimension in Biological Analyses. Plenum Press, New York, p 335–351
9. Wiktor TJ, Koprowski H (1980) Antigenic variants of rabies virus. J Exp Med 152:99–112
10. Dietzschold B, Wunner W, Wiktor TJ, Lopes DA, Lafon M, Smith CL, Koprowski H (1983) Characterization of an antigenic determinant of the glycoprotein that correlates with pathogenicity of rabies virus. Proc Natl Acad Sci USA 80:70–74
11. Gremliza L (1953) Kasuistik zum Lyssa-Problem. Zeitschrift fur Tropenmedizin und Parasitologie 4:382–389
12. Bahmanyar M, Fayaz A, Nour-Salehi S, Mohammadi M, Koprowski H (1976) Successful protection of humans exposed to rabies infection. J Am Med Assoc 236:2751–2754
13. Wunner W, Dietzschold B, Curtis PJ, Wiktor TJ (1984) Rabies subunit vaccine. Miami Winter Symposium January 1983 meeting. Academic Press, New York, in press

CHAPTER 47
Unconventional Viruses

D. CARLETON GAJDUSEK*

Kuru and the tranmissible virus dementias have been classified in a group of virus-induced slow infections that we have described as subacute spongiform virus encephalopathies because of the strikingly similar histopathologic lesions they induce. Scrapie, mink encephalopathy, and the chronic wasting disease with spongiform encephalopathy of captive mule deer and captive elk all appear, from their histopathology, pathogenesis, and the similarities of their infectious agents, to belong to the same group [1-8]. The basic neurocytologic lesion in all these diseases is a progressive vacuolation in the dendritic and axonal processes and cell bodies of neurons and, to a lesser extent, in astrocytes and oligodendrocytes; an extensive astroglial hypertrophy and proliferation; and, finally, spongiform change or status spongiosus of gray matter [9-11]. These atypical infections differ from other diseases of the human brain, which have been subsequently demonstrated to be slow virus infectious in that they do not evoke a virus-associated inflammatory response in the brain (i.e., no perivascular cuffing or invasion of the brain parenchyma with leucocytes); they usually show no pleocytosis nor do they show marked rise in protein in the cerebrospinal fluid throughout the course of infection [12-14]. Furthermore, they show no evidence of an immune response to the causative virus and, unlike the situation in the other virus diseases, there are no recognizable virions in sections of the brain visualized by electron microscopy. Instead, they show ultrastructural alteration in the plasma membrane that lines the vacuoles [10].

The pursuit of the transmissibility and virus etiology of kuru [12,13] and the presenile dementia of the Creutzfeldt-Jakob disease (CJD) type [1,15] has led to the definition of the unconventional viruses as a new group of

---
* National Institutes of Health, Bethesda, Maryland.

microbes. Because of their very atypical physical, chemical, and biological properties, these viruses have stimulated a worldwide quest to elucidate their structures and resolve the many paradoxes they present to the basic tenants of microbiology and to solve the enormous clinical and epidemiologic problems they pose. The unanticipated ramifications of the demonstration of such slow infections by unconventional viruses and the peculiar properties of the unconventional viruses, which have even challenged the basic tenets of modern molecular biology, have led to a series of discoveries, each of which has wide implications with respect to microbiological and neurobiological research and research on neurological diseases of humans [1,4,15,16,17–20]. These may be summarized in the following sections.

The First Recognized Group of Microbes Provoking No Immune Response and Evidencing No Nonhost Antigen

These slow viruses first invade the reticuloendothelial cells and particularly low-density lymphocytes in the spleen. Yet, they provoke no antibody response that can be demonstrated using live virus preparation of infectious titers over 10^{11} ID$_{50}$/g. With the inability to demonstrate any antiviral antibody response or any immune response directed against nonhost viral components or capability of neutralizing the virus activity, these unconventional viruses become unique in their immunological behavior in microbiology. Natural and experimental infections with these viruses elicit no antibody response in the host, nor does immunosuppression with whole body radiation, cortisone, antileukocytic serum, or cytotoxic drugs alter the incubation period, progress, or pattern of disease, or duration of illness to death. Finally, *in vivo* and *in vitro* study of both B-cell and T-cell function revealed no abnormality early or late in the course of illness and in *in vitro* no sensitization of the cells taken from diseased animals to high-titer preparations of these viruses [1,15]. Since high-titer infective material, both in crude suspension and highly purified also fails to elicit an immunologic response against nonhost components, even when used with adjuvants, this becomes the first group of microbes in which such immunologic inertness has been demonstrated, and evokes the speculation that the replication of these viruses does not involve production of a virus-specified nonhost antigen [15,16].

Enormous Resistance to Physical and Chemical Inactivation

The demonstration of the resistance of the unconventional viruses to high concentrations of formaldehyde or gluteraldehyde and most other antiviral and antiseptic substances and to ultraviolet and ionizing radiation, to ultrasonication, and to heat and the further demonstration of iatrogenic transmission through implanted surgical electrodes, contaminated surgical instruments, corneal transplantation, and possibly through dentistry has led to the necessity of changing autopsy room and operating theater techniques as well as the precautions used in handling older and demented patients. Many of the gentle organic disinfectants, including detergents and the quarternary ammonium salts, often used for disinfection and even hydrogen peroxide, formaldehyde, ether, chloroform, iodine, phenol, and acetone, are inadequate for sterilizing the unconventional viruses, as is the use of the ethylene oxide sterilizer. Thus, previously acceptable procedures for decontamination and disinfection must be revised [21,22].

These unconventional viruses are also resistant, even when partially purified, to all nucleases, to beta-propiolactone, ethylenediaminetetraacetic acid (EDTA), and sodium deoxycholate. They are moderately sensitive to most membrane-disrupting agents in high concentration such as phenol (60%), chloroform, ether, urea (6 M), periodate (0.01 M), 2-chloroethanol, alcoholic iodine, acetone, chloroform-butanol, hypochlorite and alkalai, to caotropic ions such as thiocyanate, guanadinium, and trichloroacetate, and to proteinase K and trypsin when partially purified [16]; but these only inactivated 99% to 99.9% of the infectious particles, leaving behind highly resistant infectivity, [18]. Sodium hydroxide (0.1 to 1.0 N) and hypochloride (5%) however quickly inactivate over $10^5 ID_{50}$ of the virus [22]. They have a UV-inactivation action spectrum with a sixfold increased sensitivity at 237 nm over that at 254 nm or 280 nm, and 50-fold increased sensitivity at 220 nm [23-26]. Virions are not recognized on electron microscopic study of infected cells *in vivo* or *in vitro*, nor are they recognized in highly infectious preparations of virus concentrated by density-gradient banding in the zonal rotor. This has led to the speculation that the infectious agents lack a nucleic acid, and that they may be a self-replicating protein (and a derepressor of cellular DNA bearing information for their own synthesis), even a self-replicating membrane fragment that serves as a template for laying down abnormal plasma membrane, including itself [10,15,16].

Single-Mendelian Autosomal Dominant Inheritance Determines Expression in Familial CJD

CJD became the first human infectious disease in which a single gene was demonstrated to control susceptibility and occurrence of the disease. The autosomal dominant behavior of the disease in such families, including the appearance of the disease in 50% of siblings who survive to the age at which the disease usually appears, has evoked the possibility of virus etiology in other familial dementias. The presence of CJD patients in the families of well-known familial Alzheimer's disease, and the familial occurrence of the spinocerebellar ataxic form of Creutzfeldt–Jakob disease, the Gerstmann–Sträussler syndrome, which is also transmissible, have led to renewed interest in familial dementias of all types [4,5,14].

The Autoimmune Antibody to 10-nm Neurofilament Appearing in the SSVE Subjects

The demonstration by Sotelo *et al.* of a very specific autoimmune antibody directed against 10-nm neurofilaments and no other component of the CNS in over 60% of the patients with kuru and CJD as a phenomenon appearing late in the disease, was the first demonstration of an immune phenomenon in the SSVEs and an exciting new avenue of study of the transmissible dementias [27,28]. This autoimmune antibody behaves like many other autoimmune antibodies, such as the rheumatoid factor and the anti-DNA antibody in lupus and the anti-thyroglobulin antibody in Hashimoto's thyroiditis, in that it is occasionally present in normal subjects, more often present in subjects closely related to the patients. Although found in more than half of patients with transmissible virus dementia, it does not appear in 40% of patients with classical CJD. It does develop in other grey-matter diseases, including Alzheimer's and Parkinson's diseases, but at far lower incidence than in CJD. Furthermore, it is not present in patients with other immune diseases, such as disseminated lupus erythematosis and chronic rheumatoid arthritis [9]. Bahmanyar *et al.* have demonstrated that this autoimmune response is directed specifically against the 200,000-dalton protein component of the three proteins comprising the 10-nm neurofilament triad [30].

Unconventional Viruses—Subviral Pathogens, Perhaps Devoid of a Nucleic Acid or a Nonhost Protein

The scrapie virus has been partially purified by density-gradient sedimentation in the presence of specific detergents. Rohwer has succeeded in a 1000-fold purification of scrapie virus relative to other quantifiable proteins in the original brain suspension [19]. In such preparations, the virus is susceptible to proteinase K and trypsin digestion but it is not inactivated by any nuclease. Sedimented, washed, and resuspended virus has been banded into peaks of high infectivity with the use of cesium chloride, sucrose, and metrizamide density gradients in the ultracentrifuge. Sucrose–saline density-gradient banding of scrapie virus in mouse brains produced wide peaks of scrapie infectivity at densities of 1.14 to 1.23 g/cm^3 [19]. Attempts to demonstrate a nonhost nucleic acid in scrapie virus preparations using DNA homology and transfection and nuclease inactivation have been unsuccessful [31,32, Borras and Gibbs, unpublished observation].

The atypical action spectrum for inactivation of scrapie virus by UV should not be taken as proof that no genetic information exists in the scrapie virus as nucleic acid molecules, since Latarjet has demonstrated similar resistance to ultraviolet and a similar UV action spectrum for microsomes [23–26]. Ultraviolet resistance also depends greatly on small RNA size, as has been shown by the high resistance of the purified, very small, tobacco ring spot satellite virus RNA (about 80,000 daltons).

On the other hand, the unconventional viruses possess numerous properties in which they resemble classical viruses, and some of these properties suggest far more complex genetic interaction between virus and host than one might expect for genomes with a molecular weight of only 10^5 daltons. Rohwer has shown that the scrapie virus replicates in hamster brain at a constant exponential rate, with no eclipse phase, and with a doubling time of 5.2 days [20]. It grows yet slower in sheep or mouse brain. Examination of the kinetics of its inactivation and the demonstrated association or aggregation of scrapie virus particles into polymers or clusters that can be disrupted by ultrasonication have cast doubt on the calculation of its small size from ionizing radition inactivation data and from inferences about its structure from resistance to chemical inactivating agents. Thus, aggregates make necessary "multiple hits" for inactivation, while free virus is killed by a single event [18].

In plant virology we have recently been forced to modify our concepts of a virus to include subviral pathogens such as the newly described viroids causing 11 natural plant diseases—potato spindle tuber disease, chrysanthemum stunt disease, citrus exocortis disease, Cadang-Cadang disease of coconut palms, cherry chlorotic mottle, cucumber pale fruit disease, hop stunt disease, avocado sunblotch disease, tomato bunchy top disease, tomato "planta macho" disease, and burdock stunt disease—and the virusoids of four natural plant diseases (velvet tobacco mottle virus, solanum nodiflorum mottle virus, lucerne transient streak virus, subterranean clover mottle virus) to which we may turn for analogy [33,34]. All of the viroids are small circular RNAs containing no structural protein or membrane and they have all been fully sequenced and their fine structures determined. They have only partial base pairing as the circle collapses on itself. They contain only 246 to 574 ribonucleotides and replicate by a "rolling circle" copying of their RNA sequences in many sequential rotations to produce an oligomeric copy, which is then cut into monomers, sometimes dimers. No protein is synthesized from their genetic information and only the replication machinery of the cell is used. These subviral pathogens have caused us to give much thought to possible similarities to the unconventional viruses. However, we have shown that the unconventional viruses differ from the plant viroids on many counts [33,34].

Thus, the intellectually stimulating analogies of the unconventional viruses to viroids and virusoids prove to be spurious, yet these subviral pathogens of plants have alerted us to the possibility of extreme departure from conventional virus structures.

The delta antigen of infectious hepatitis, a defective replicating particle with only 1700 bases on its genome (68,000 daltons) and requiring the infectious hepatitis B virus for its replication, offers further intriguing analogies to the unconventional virus.

Scrapie-Associated Fibrils of Merz and Somerville

In suspension of scrapie-affected brain sedimented in a density gradient, Merz and Somerville have demonstrated an amyloid-like 2- or 4-stranded fiber that increases in quantity with virus titer. We have found these structures in brains of CJD patients and in brains of primates with experimental CJD and kuru, and not in normal control brains or brains of patients with other neurodegenerative diseases [35,36]. It has been postulated that these structures may represent the scrapie or CJD or kuru infectious agent. Such structures bring to mind the filamentous plant viruses and filamentous phage fd, which are of approximately the same diameters [35-40].

These scrapie associated fibrils which may be the infectious agents are distinguishable from the paired helical filaments of neurofibrillary tangles and the fibrils of brain amyloid [35]. However, their similarity is sufficient to demand close discrimination, and our discovery that the autoimmune antibody to 10 nm neurofilament which often appears in these diseases also reacts with the neurofibrillary tangles of Alzheimer's disease and the accumulations of 10 nm neurofilament in the brains of 3,3'-iminopropionitrile treated rats, leads us to the conjecture that the scrapie associated fibrils may be related to normal 10 nm neurofilament, to the paired helical filament in neurofibrillary tangles, and to amyloid fibrils in the brain. We await our results in using hybridoma and other immunological techniques to search for the possibility of shared idiotopes indicating common sequences in these different brain fibrils.

References

1. Gajdusek DC, Gibbs Jr. CJ (1975) Slow virus infections of the nervous system and the Laboratories of Slow, Latent and Temperate Virus Infections. *In* Tower DB (ed) The Nervous System, Vol 2; Chase TN (ed) The Clinical Neurosciences, p 113
2. Gajdusek DC, Gibbs Jr CJ, Alpers M (1966) Experimental transmission of a kuru-like syndrome in chimpanzees. Nature 209:794-796
3. Gajdusek DC, Gibbs Jr CJ, Alpers M (eds) (1965) Slow, Latent and Temperate Virus Infections; NINDB Monograph No. 2, National Institutes of Health. PHS Publication No. 1378, US Govt Printing Office, Washington DC
4. Masters CL, Gajdusek DC, Gibbs Jr CJ (1981) The familial occurrence of Creutzfeldt-Jakob disease and Alzheimer's disease. Brain 104:535-558
5. Masters CL, Gajdusek DC, Gibbs Jr CJ (1981) Creutzfeldt-Jacob disease virus isolations from the Gerstmann-Sträussler syndrome, with an analysis of the various forms of amyloid plaque deposition in the virus-induced spongiform encephalopathies. Brain 104:559-588
6. Williams ES, Young S (1980) Chronic wasting disease of captive mule deer: A spongiform encephalopathy. J Wildl Dis 16:89-98
7. Williams ES, Young S (1982) Spongiform encephalopathy of Rocky Mountain elk. J Wildl Dis 18:465-471
8. Williams ES, Young S, March RF (1982) Preliminary evidence of the transmissi-

bility of chronic wasting disease of mule deer. Abstract No. 22 in Proceedings of the Wildlife Disease Associate Annual Conference, Madison, Wisconsin
9. Beck E, Daniel PM, Alpers M, Gajdusek DC, Gibbs Jr CJ, Hassler R (1975) Experimental kuru in the spider monkey. Histopathological and ultrastructural studies of the brain during early stages of incubation. Brain 98:592–620
10. Beck E, Daniel PM, Davey A, Gajdusek DC, Gibbs Jr CJ (1982) The pathogenesis of spongiform encephalopathies: An ultrastructural study. Brain 105:755–786
11. Klatzo I, Gajdusek DC, Zigas V (1959) Pathology of kuru. Lab Invest 8:799–847
12. Gajdusek DC, Zigas V (1957) Degenerative disease of the central nervous system in New Guinea. The endemic occurrence of "kuru" in the native population. N Engl J Med 257:974–978
13. Gajdusek DC, Zigas V (1959) Kuru: Clinical, pathological and epidemiological study of an acute progressive degenerative disease of the central nervous system among natives of the Eastern Highlands of New Guinea. Am J Med 26:442–469
14. Traub R, Gajdusek DC, Gibbs Jr CJ (1977) Transmissible virus dementias. The relation of transmissible spongiform encephalopathy to Creutzfeldt–Jakob disease. *In* Kinsbourne M, Smith L (eds) Aging and Dementia. Spectrum Publishing Inc., Flushing, New York, p. 91
15. Gajdusek DC (1977) Unconventional viruses and the origin and disappearance of kuru. Science 197:943–960
16. Prusiner SB (1982) Novel proteinaceous infectious particles cause scrapie. Science 216:136–144
17. Rohwer RG (1984) Virus-like sensitivity of the scrapie agent to heat inactivation. Science 223:600–602
18. Rohwer RG (1984) Scrapie infectious agent is virus-like in size and susceptibility to inactivation. Nature 308:658–662
19. Rohwer, RG, Gajdusek DC (1980) Scrapie-virus or viroid: The case for a virus. *In* Boese A (ed) Search for the Cause of Multiple Sclerosis and Other Chronic Diseases of the Central Nervous System. Proceedings of the First International Symposium of the Hertie Foundation, Frankfurt am Main, pp 333–355.
20. Rohwer RG (1984) Growth kinetics of hamster scrapie strain 263K: Sources of slowness in a slow virus infection. Virology, in press
21. Brown P, Gibbs CJ Jr., Amyx HL, Kingsbury DT, Rohwer RG, Sulima MP, Gajdusek DC (1982) Chemical disinfection of Creutzfeldt–Jakob disease virus. N Engl J Med 306:1279–1282
22. Brown P, Rohwer RG, Green EM, Gajdusek DC (1982) Effect of chemicals, heat and histopathological processing on high infectivity hamster adapted scrapie virus. J Infect Dis 145:683–687
23. Gibbs Jr CJ, Gajdusek DC, Latarjet R (1977) Unusual resistance to UV and ionizing radiation of the viruses of kuru, Creutzfeldt–Jakob disease and scrapie (unconventional viruses). Proc Natl Acad Sci USA 75:6268–6270
24. Haig DC, Clarke MC, Blum E, Alper T (1969) Further studies on the inactivation of the scrapie agent by ultraviolet light. J Gen Virol 5:455–457
25. Latarjet R, Muel B, Haig DA, Clarke MC, Alper T (1970) Inactivation of the scrapie agent by near-monochromatic ultraviolet light. Nature 227:1341–1343
26. Latarjet R (1979) Inactivation of the agents of scrapie, Creutzfeldt–Jakob disease and kuru by radiation. *In* Prusiner SB, Hadlow WJ (eds) Slow Transmissible Diseases of the Nervous System, Vol 2. Academic Press, New York, p 387
27. Sotelo J, Gibbs Jr CJ, Gajdusek DC, Toh BH, Wurth M (1980) Method for

preparing cultures of central neurons: Cytochemical and immunochemical studies. Proc Natl Acad Sci USA 77:653–657
28. Sotelo J, Gibbs Jr CJ, Gajdusek DC (1980) Autoantibodies against axonal neurofilaments in patients with kuru and Cruetzfeldt–Jakob disease. Science 210:190–193
29. Bahmanyar S, Moreau-Dubois MC, Brown P, Gajdusek DC (1983) Serum antineurofilament antibodies to neurofilament antigens in patients with neurological and other diseases and healthy controls. J. Neuroimmunol 5:191–196
30. Bahmanyar S, Liem RKH, Griffin JW, Gajdusek DC (1984) Characterization of antineurofilament autoantibodies in Creutzfeldt–Jakob disease. J Neuropath Exp Neurol, in press
31. Borras MT, Kingsbury DT, Gajdusek DC, Gibbs Jr CJ (1982) Inability to transmit scrapie by transfection of mouse embryo cells *in vitro*. J Gen Virol 58:263–271
32. Hunter GD, Collis SC, Millson GC, Kimberlin RH (1976) Search for scrapie-specific RNA and attempts to detect an infectious DNA or RNA. J Gen Virol 32:157–162
33. Diener T, Hadidi A (1977) Viroids. *In* Fraenkel-Conrat H, Wagner RR (eds) Comprehensive Virology. Plenum Press, New York
34. Sänger HL (1982) Biology, structure, functions and possible origins of plant viroids. *In* Nucleic Acids and Proteins in Plants. II. Encyclopaedia of Plant Pathology, New Series, 14B. Springer-Verlag, Berlin, pp 368–454
35. Merz PA, Rohwer RG, Kascsak R, Wisniewski HM, Somerville RA, Gibbs Jr CJ, Gajdusek DC (1984) Infection specific particle from unconventional slow virus diseases. Science, in press.
36. Merz PA, Rohwer RG, Somarville RA, Wisniewski HM, Gibbs Jr CJ, Gajdusek DC (1983) Scrapie associated fibrils in human Creutzfeldt–Jakob disease. J Neuropath Exp Neurol 42:327
37. Merz PA, Somerville RA, Wisniewski HM, Iqbal K (1981) Abnormal fibrils from scrapie-infected brain. Acta Neuropathol (Berlin) 54:63–74
38. Merz PA, Somerville RA, Wisniewski HM, Manuelidis L, Manuelidis EE (1984) Scrapie associated fibrils in Creutzfeldt–Jakob disease. Nature 306:474–476
39. Diringer H, Gelderblom H, Hilmert H, Ozei M, Edelbluth C, Kimberlin RH (1983) Scrapie infectivity, fibrils, and low molecular weight protein. Nature 306:476–478
40. Prusiner SB, McKinley MP, Bowman KA, Bolton DC, Benheim PD, Groth DF, Glenner GG (1983) Scrapie prions aggregate to form amyloid-like birefringent rods. Cell 35:349–358

Control of Viral Diseases

48. Antibodies to Synthetic Peptide Immunogens as Probes for Virus Protein Expression and Function
    THOMAS M. SHINNICK, J. GREGOR SUTCLIFFE, AND
    RICHARD A. LERNER 361

49. Recombinant DNA Vaccines
    LAURENCE A. LASKY AND JOHN F. OBIJESKI 366

50. Monoclonal Antibodies
    WALTER GERHARD AND HILARY KOPROWSKI 376

51. Antiviral Drugs
    GEORGE J. GALASSO 382

52. The Use of Interferons in the Control of Viral Diseases
    ANTHONY L. CUNNINGHAM AND THOMAS C. MERIGAN 389

53. Insect Viruses as Pesticides
    THOMAS W. TINSLEY 398

CHAPTER 48

Antibodies to Synthetic Peptide Immunogens as Probes for Virus Protein Expression and Function

THOMAS M. SHINNICK, J. GREGOR SUTCLIFFE, AND
RICHARD A. LERNER*

In general, the direction of research into the genetic structure of viruses has been that of tracing phenotype to genotype. The currently available recombinant DNA and nucleic acid sequencing techniques have somewhat reversed this process and we are beginning to find ourselves with genotypes in search of phenotypes, especially as research efforts move from questions about known viruses and viral proteins to questions about recently isolated or little-studied viruses. Indeed, from the nucleotide sequence of a virus, one can learn a great deal about the genetic structure of the viral genome, its coding capacity and its evolutionary history. What is needed, however, is a way to parlay the wealth of genetic information available in the nucleotide sequence into biological experiments. Furthermore, since proteins are the agents responsible for most of the biologically important activities and properties of viruses, one is faced with the not-so-trivial task of identifying the protein products of the viral genes and generating probes to study protein expression, processing, and biological activities. In other words, how does one identify a protein whose existence is merely inferred from the sequence of a string of nucleotides?

A solution to the problem of linking gene sequence and gene product is to elicit antibodies that specifically react with the putative gene product; but what can one use as immunogen? One approach is to construct by recombinant DNA technologies a bacterial plasmid or phage capable of expressing the desired gene into protein. The bacterially synthesized protein would be used as the immunogen. An alternative approach involves using a chemically synthesized portion of the deduced gene product as immunogen. That is, by using the genetic code, the nucleic acid sequence is translated into

---

* Scripps Clinic and Research Foundation, La Jolla, California.

amino acid sequence and then oligopeptides corresponding to portions of the deduced sequence are chemically synthesized, coupled to carrier proteins, and used as immunogen to elicit antibodies that react with the full-length gene product.

The usefulness of synthetic peptide immunogens requires that the elicited antibodies be capable of reacting with the full-length protein(s) carrying the peptide sequence and that the antibodies be specific for the peptide and appropriate full-length protein(s). Studies from a number of laboratories have shown that synthetic peptide immunogens often elicit antibodies that react with the appropriate full-length proteins (reviewed in [1–3]). For example, in our own hands 12 of 12 peptides corresponding to portions of the deduced amino acid sequence of the Moloney murine leukemia virus polymerase gene and 18 of 18 peptides corresponding to portions of the rabies glycoprotein elicited antisera that reacted with the appropriate gene products. The question of antibody specificity is relevant, especially when one is trying to detect a heretofore unidentified gene product. At very high antibody and antigen concentrations, antipeptide antibodies have been observed to react specifically with the target protein and occasionally with other normal cellular components [4]. This cross-reactivity occurs at 10^3–10^4 lower affinity than the specific reaction with the target protein. Such cross-reactivity is not surprising, since antibodies do obey mass action laws. Thus, the antibodies may bind to sites that resemble the target peptide sequence with low affinity, for example, a site that matches four of six amino acids. To avoid this complication the experimenter can simply assay reactivity at various dilutions of antisera and hence easily distinguish specific, high-affinity reactivity from the nonspecific, low-affinity reactivities.

The immunological probes generated by the synthetic peptide immunogens can be used to identify the protein product of a gene and to investigate its enzymatic activity. For example, the antisera to the 12 peptides that correspond to portions of the deduced protein sequence of the product of the Moloney murine leukemia virus polymerase gene were used to probe extracts of the virus-infected cells that had been labeled with ^{35}S-methionine. All 12 antisera detected the primary translation product of the *pol* gene, a 200,000-dalton protein. In addition, antisera to six of the peptides but none of the others detected an ~75,000-dalton protein in virions, which presumably represents the mature, processed *pol* product from this region of the gene. The coincident positive reactivities of these six peptides strongly argues that the ~75,000-dalton protein is a true product of the *pol* gene, rather than being the result of some nonspecific interaction. In general, the antibodies are useful in identifying the gene products, at least as bands on a gel.

If the deduced gene product has a suspected enzymatic activity, then the antisera can be used to investigate the correlation. In our example, one product of the polymerase gene is the reverse transcriptase enzyme, which consists of a single polypeptide chain of 70–80,000 daltons [5]. Antisera to the six peptides that recognized the ~75,000-dalton protein inhibited the

RNA-dependent DNA polymerase activity of reverse transcriptase in an *in vitro* assay as effectively as an antiserum raised against purified reverse transcriptase. None of the other antipeptide sera inhibited this reaction. Thus, the ~75,000-dalton protein is most likely the reverse transcriptase enzyme.

It is the biological activities of the viral proteins and not their *in vitro* enzymatic activities that are relevant for the understanding of the molecular details of the viral life cycle. By combining the knowledge of which peptides elicit antisera that react with which proteins with the nucleotide sequence of the genome, one can deduce rough limits for the coding capacity for each viral protein. Such a rough map can be used to direct *in vitro* mutagenesis experiments that could specifically mutate one viral protein at a time. The analysis of the phenotype of such mutants should reveal the biologically relevant role of each viral protein.

Besides analyzing the protein complement of a virus, the antibodies elicited by synthetic peptide immunogens may be useful in the detection, treatment, or prevention of disease. Obviously, if one has an antibody that specifically reacts with a protein, one can develop an assay to detect that protein, such as a radioimmune or solid-phase enzyme-linked immunosorbent assay. Here, one might take advantage of the observation that the synthetic peptide immunogens can elicit monoclonal hybridomas that produce antibodies that react with the full-length target protein [6]. Such doubly specific reagents can be prepared in large quantities by the standard hybridoma technologies and may be useful in large-scale diagnostic work.

The ability of the antibodies to inhibit enzymatic activity raises the possibility that peptide-elicited antibodies could be used in a passive immunization therapy. For example, the treatment for exposure to various toxins such as the one in rattlesnake venom is to inject an antitoxin that contains antibodies that will bind to and inactivate the toxin. Indeed, antibodies elicited by a tetradecapeptide corresponding to residues 188–201 of the 62,000-dalton polypeptide chain of diphtheria toxin can bind to the toxin, and when mixed with the toxin *in vitro* prior to intradermal injection into guinea pigs, can prevent the typical dermonecrotic reaction [7]. Such success *in vitro* provides promise for the development of a passive vaccine that will work *in vivo*.

The antipeptide antibodies can bind to viral proteins but can they neutralize virus? Several laboratories have shown that antipeptide antibodies can neutralize virus *in vitro* [1–3]. For example, mixing antibodies raised against the carboxyl terminal hexapeptide of the tobacco mosaic virus with the virus *in vitro* prior to application to tobacco leaves resulted in the neutralization of the ability of the virus to induce lesions on the leaves [8]. We have shown that immunization with synthetic peptide immunogens can protect animals from virus infection. For example, a single dose of a peptide corresponding to residues 141–160 of the coat protein (VP1) of foot-and-mouth disease virus is sufficient to protect animals from the disease [9].

When confronted with a novel pathogen, how can one generate a reagent for combating it? The following scenario, which combines immunology, recombinant DNA technology and the peptide immunogen approach, may produce antibodies that will neutralize the pathogenic organism. First, immunological and serological studies are undertaken to identify which proteins are the targets of a neutralizing immune response. For example, the selection of VP1 of foot-and-mouth disease virus for our initial studies was because it is the major target of neutralizing antibodies for this virus. In the absence of this information, it would be necessary to study all the proteins of a pathogen—obviously a formidable task. The next step is to determine the amino acid sequence of each protein identified by the immunological studies. Currently the most straightforward way to do this is to clone the gene that encodes the neutralizing target and then determine the nucleotide sequence of the gene. The amino acid sequence is deduced from the nucleotide sequence. For many viruses and cDNA clones of mRNA, the amino acid sequence is colinear with the nucleotide sequence. For genomic clones of eukaryotic DNA, coding regions (exons) are often interrupted by noncoding regions (introns) and hence deduction of an amino acid sequence is somewhat more difficult.

Using the amino acid sequence as a blueprint, peptides are chosen for synthesis and used as immunogens. The immunological or serological studies may be helpful in selecting neutralizing peptides. A comparison of the amino acid sequences of the protein from different serotypes may identify a particular region of the protein as the target of the neutralizing antibody elicited by a natural infection. Exactly this consideration led to the proper region of the hepatitis B virus surface antigen [10]. Alternatively, one could adopt a brute force approach and simply make peptides that represent many or all of the hydrophilic regions of the protein. This somewhat inelegant method is economically feasible and will eventually identify the appropriate peptide. In addition, this approach may identify a peptide that can cross-neutralize several serotypes as is the case for several of the influenza virus hemagglutinin peptides (Alexander, S; Alexander H; Green, N and Lerner, R, unpublished observations). Once an effective peptide is found, the optimal peptide corresponding to this region of the protein can be identified by analyzing a series of overlapping peptides. Finally, the appropriate peptide(s) or peptide-elicited antibodies must be developed for clinical applications. At this stage one must address questions regarding chemical purity, possible side-effects or cross-reactivities, carrier, adjuvant, immunization schedule, route of administration, dose, etc. Most of these questions have not yet been directly addressed and obviously will be intensively studied as a synthetic peptide vaccine approaches the marketplace.

References
1. Lerner RA (1982) Tapping the immunological repertoire to produce antibodies of predetermined specificity. Nature 299:592–596

2. Sutcliffe JG, Shinnick TM, Green N, Lerner RA (1983) Antibodies that react with predetermined sites on proteins. Science 219:660–666
3. Shinnick TM, Sutcliffe JG, Green N, Lerner RA (1984) Synthetic peptide immunogens as vaccines. Ann Rev Micro 37:425–446
4. Nigg EA, Walter G, Singer SJ (1982) On the nature of crossreactions observed with antibodies directed to define epitopes. Proc Natl Acad Sci USA 79:5939–5943
5. Verma IM (1975) Studies on reverse transcriptase of RNA tumor viruses. III. Properties of purified moloney murine leukemia virus DNA polymerase and associated RNase H. J Virol 15:843–854
6. Niman HL, Houghton RA, Walker LE, Reisfeld RA, Wilson IA, Hogle JM, Lerner RA (1983) Generation of protein-reactive antibodies by short peptides is an event of high frequency: Implications for the structural basis of immune recognition. Proc Natl Acad Sci USA 80:4949–4953
7. Audibert R, Jolivet M, Chedid L, Alouf JE, Boquet P, Rivaille P, Siffert O (1981) Active antitoxic immunization by a diphtheria toxin synthetic oligopeptide. Nature 289:593–594
8. Anderer FA (1963) Biochem Biophys Acta 71:246–250
9. Bittle JL, Houghten RA, Alexander H, Shinnick TM, Sutcliffe JG, Lerner RA, Rowlands PJ, Brown F (1982) Protection against foot-and-mouth disease by immunization with a chemically synthesized peptide predicted from the viral nucleotide sequence. Nature 298:30–33
10. Gerin JL, Shih JWK, Purcell RH, Dapolito G, Engle R, Green N, Alexander H, Sutcliffe JG, Shinnick TM, Lerner RA (1983) Chemically synthesized peptides of hepatitis B surface antigen duplicate the d/y specificities and induce subtype-specific antibodies in chimpanzees. Proc Natl Acad Sci USA 80:2365–2369

CHAPTER 49
Recombinant DNA Vaccines

LAURENCE A. LASKY AND JOHN F. OBIJESKI*

Introduction

Of the many promises of the biotechnological revolution, none appear to offer more immediate application for improving the welfare of mankind than the biosynthesis by recombinant DNA (R-DNA) methods of vaccines for the diseases of humans and animals. Conventional vaccine development research for the production of both live and killed vaccines for human and veterinary use has become rather static in the past few years. It is only recently that the emerging technology of R-DNA has become available for vaccine production. Coupled with the ability to produce abundant quantities of previously rare gene products [1-3] has also come a technology for the synthesis of important antigenic regions from disease agents using prokaryotic and eukaryotic organisms as microscopic production units. There are many obvious advantages of R-DNA-derived vaccines over older, more traditional vaccine-production methodologies. The use of isolated virus genes for subunit vaccines precludes the use of infectious agents during vaccine manufacture. The biosynthesis of antigens in innocuous organisms such as *E. coli* should allow a vaccine manufacturer to produce a homogeneous, pure, and safe product. Another important consideration is the potential for genetic modification of live-attenuated vaccines. In most instances, purified R-DNA vaccines should be more stable than comparable traditional vaccines, particularly with regard to temperature requirements. Finally, the costs for quality control of R-DNA vaccines should be reduced, which may ultimately allow developing countries access to the vaccines.

---

* Genentech, Inc., So. San Francisco, California.

Production of R-DNA Vaccines Using Microorganisms

Although many bacterial species have been characterized, *E. coli* is the bacteriuim most often used for industrial-scale production of R-DNA proteins. The reasons for its widespread use include its ease of growth, its well-characterized genetics, and, perhaps most importantly, a clear picture of the basic mechanisms for its replication [4]. In addition, our knowledge of *E. coli*-derived plasmids, extrachromosomal DNA elements with selectable markers, is extensive [5]. It is these extrachromosomal molecules that serve as vectors for the expression of foreign genes.

Perhaps the best-studied example of an *E. coli*-derived R-DNA vaccine is that for foot-and-mouth disease virus (FMDV) [6]. FMDV (a picornavirus) is the single most important virus disease of the world's livestock industries. Most FMDV vaccines now in use are inactivated whole virus preparations. In the past these vaccines have had imperfect records, particularly with respect to immunogenicity (protection) and safety. In 1965 it was shown that a surface component of the virus (the VP1 protein) could be isolated and that is could confer immunity in cattle [7,8]. Thus, it appeared that FMDV might be a likely candidate for an R-DNA vaccine.

The actual protocol followed for the production of an FMDV R-DNA vaccine has some similarities to that for the production of other R-DNA proteins. The RNA genome of the virus was first transcribed into DNA with reverse transcriptase and the resultant DNA form of the gene encoding the immunogenic capsid protein VP1 was cloned and its nucleic acid sequence was determined using traditional methods [9]. The deduced protein sequence was then utilized to determine an efficient method for expressing the antigen in *E. coli*. The expression of foreign proteins in *E. coli* may be accomplished either as a polypeptide fused to a naturally occurring bacterial protein or as a directly expressed protein that contains only the amino acid sequences encoded by the gene of interest. The most efficient expression system for the FMDV-VP1 protein was obtained by constructing a DNA sequence that encoded a fusion between the viral surface antigen (VP1) and the *E. coli* 190 amino acid trpLE fusion protein. The expression of this DNA construction was regulated by the inducible promoter from the tryptophan operon. In this system, the bacteria can be grown to high concentrations before expression of the viral protein, thereby eliminating the deleterious effects that some foreign gene products have on bacterial cell growth. Several key factors have been identified that influence the level of synthesis of any gene product. These include the number of gene copies, promoter efficiency, the stability and structure of the mRNA, the efficiency of ribosome binding, and the stability of the synthesized protein. In the case of the FMDV expression system, greater than 17% of the total cell protein was the fusion protein. Using appropriate purification procedures, it is possible to obtain large quantities of homogenous and pure protein, free of any contaminating bacterial substances [10]. Vaccination of cattle and swine with the purified fusion protein induced high titers of virus-neutralizing antibodies, which protected the animals from subsequent challenge with live FMDV. The immunity of these animals to infection by FMDV demonstrated for the first time the feasibility of the R-DNA approach for the production of subunit vaccines.

Several other viral proteins (antigens) have been expressed in the *E. coli* system. Hepatitis-B virus (HBV) core protein and surface antigen have been successfully synthesized, although the surface antigen was produced at extremely low levels [11]. The efficacy of the surface antigen as an immunogen was not reported. The gD glycoprotein from Herpes Simplex Virus type 1 (HSV-1) has also been expressed as a chimeric protein fused to the cro protein from the lambda bacteriophage and to *E. coli* β-galactosidase [12,13]. Although the protein contained *E. coli* sequences, which may make it unsatisfactory for human use, it nevertheless generated low levels of virus-neutralizing antibodies when injected into rabbits. The surface glycoprotein (G) from rabies virus has also been synthesized in *E. coli* as a directly expressed protein at fairly high levels [14]. While this protein appeared to contain all the naturally occurring antigenic determinants, its utility as a vaccine was not reported. The hemagglutinin (HA) protein from influenza has also been expressed as a fusion protein using various expression systems in *E. coli* [15–17]. The protein produced in *E. coli* was found to generate antibodies that recognized determinants found in native HA [16–18]. Thus, it appears that several more *E. coli*-produced vaccines may soon be forthcoming.

Although the *E. coli* system can synthesize several potential R-DNA vaccine antigens, it has the inherent problem that it is a gram negative microorganism that cannot secrete proteins into the extracellular medium. Thus, any antigens synthesized in *E. coli* must be purified from killed cells, a process that may be costly and, at times, difficult. As a result, attempts have been made to produce R-DNA vaccines in two microorganisms known to secrete proteins normally. In the first system, the bacterium *Bacillus subtilis* was engineered to produce the VP1 protein of FMDV [19]. Although the protein was expressed correctly, it was not secreted. Thus, the utility of this organism for the production of secreted R-DNA vaccines has yet to be demonstrated. A second organism, which has also been investigated for the production of an R-DNA vaccine, is the yeast *Saccharomyces cerevisae*. This lower eukaryotic microbe contains the biosynthetic pathways necessary for both the secretion and glycosylation of proteins. The hepatitis surface antigen mentioned previously was expressed in this system to determine if a glycosylated, secreted protein would be synthesized [20–23]. While the results demonstrated that the surface antigen gene was synthesized, and that this protein apparently formed a 22-nm particle that is characteristic of this antigen, it was observed that the particles were formed only upon lysis of the yeast cell. This result suggests that the surface antigen protein may be present intracellularly in a monomeric form and that monomer aggregation after cell lysis may result in the formation of surface antigen aggregates. The immunogenicity of the yeast-derived particles was not reported. The successful use of these systems for the production of other R-DNA vaccines remains to be determined.

Although the potential for vaccines manufactured in bacteria such as *E. coli* is attractive, there are inherent problems with the use of these bacteria as an R-DNA vaccine source. Of major concern is the potency of the immunogen and the secondary structure of the antigen. It is, of course, of paramount importance that the potency and structure of the bacterially-expressed antigen be similar, if not identical, to the native molecule. It becomes necessary, therefore, to examine, sometimes extensively, the isolation and purification conditions that give rise to a properly folded immunogenic protein. This can be, in many cases, difficult if not impossible. For example, in the case of hepatitis B virus surface antigen (HBsAg), the most immunogenic structure is the surface antigen protein in the form of a 22-nm particle [24]. The likelihood of assembling such a complex structure after protein isolation and purification appears extremely remote. *E. coli* is not able to glycosylate proteins and although it has not been conclusively demonstrated that sugar residues, which are found on virtually all surface antigens, are important to the immunogenicity of the antigen, it may nevertheless be necessary to investigate the role that carbohydrate may have in protein stability and immunogenic potency. Since bacterial contaminants may be intolerable in human vaccines, the increased costs of extensive purification procedures and quality control testing to ensure their removal may make some bacterially produced R-DNA vaccines uneconomical. It is for these reasons that an alternative expression system using mammalian cell culture has been investigated for the production of some R-DNA vaccines.

R-DNA Vaccines from Tissue Culture

There are now several examples where foreign gene products have been expressed in mammalian tissue culture cells [25–27]. Generally, by placing the desired gene under the control of a eukaryotic transcriptional promoter, usually of viral origin, one may express the protein product after the transfection of the DNA construction into cells. Although several different viral and cellular promoters have been isolated and characterized, the transcriptional signals thus far most commonly used in tissue culture expression systems are derived from the small DNA papovavirus, SV-40. The entire nucleotide sequence of this virus has aided in characterizing the genome regions that are involved in viral DNA replication and control of early and late viral gene transcription [28]. The SV-40 vectors can be grouped into two general categories: those that comprise most of the viral genome and result in a lytic infection of the cell [30–32], and those that use only the transcriptional signals of the viral genome [29,33–35]. The first class of vectors usually replaces a late region of the SV-40 genome with the gene of interest under viral transcriptional control. Upon infection of cells with a helper virus, the inserted nonviral gene is transcribed and translated as a viral sequence. As one might expect, this system eventually results in the virus-induced death of the host cell. The second class of vectors usually contains the viral promoter sequence and a bacterial origin of DNA replication and antibiotic resistance gene that allows the vector to be shuttled easily between prokaryotic and eukaryotic organisms. In addition, the vector lacks specific bacterial sequences that are inhibitory to transfection of the vector DNA into eukaryotic cells [36]. Thus, this vector would allow for both the cloning of the gene in bacteria as well as the expression of the gene by utilization of the viral promoter in tissue culture cells. To utilize this system efficiently, however, it is necessary to amplify the vector enough to obtain the high copy numbers normally found in virus-infected cells. This is accomplished by using COS cells derived from an established simian cell line (CV-1) that was transformed by an SV-40 origin-defective mutant that codes for wild type T antigen [37]. The resultant cell lines express sufficient quantities of T antigen to allow the efficient replication of molecules containing the SV-40 origin of replication. Thus, vectors containing the SV-40 origin of replication, which also functions to initiate transcription of the viral genome, will be replicated to high copy numbers (approximately 100,000-fold). In this way, the extent of transcription, and subsequent translation, of any gene contained in this vector and transcribed under the control of the SV-40 promoter is enhanced. Unfortunately cell death occurs as a result of the enormous DNA copy numbers. However, before this happens, considerable synthesis of the protein product takes place.

Several viral surface antigens have now been expressed in eukaryotic cells by using the SV-40-derived vector systems. These include the influenza [31,32] surface hemagglutinin gene (HA gene), the glycoprotein (G) from vesicular stomatitis virus (VSV) [34], and the hepatitis B virus surface antigen [29,30,35].

The HA gene of influenza virus has been expressed in cell culture by using lystic SV-40 expression vectors where either the late or the early region of the viral genome was replaced with the HA gene [31]. In both cases, synthesis of the HA protein was induced and the protein was found on the surface of the SV-40-infected cells. By several criteria, the expressed HA protein was identical to the native viral glycoprotein, even in its ability to hemagglutinate chicken erythrocytes. This work highlighted an important concept: namely, that the biosynthesis, transport, glycosylation, and cellular localization of the viral surface protein is determined by host-cell functions.

Although results obtained with the influenza HA suggested that surface glycoproteins could by synthesized from genetically engineered virus vectors in eukaryotic cells, the system was not practical for vaccine production because SV-40 viral genomes were used as vectors. Another system made use of a vector that was constructed from the SV-40 origin of replication, which contains viral transcriptional signals, and the bacterial vector pML [36], which contains a bacterial origin of replication and an ampicillin resistance gene. By placing the VSV glycoprotein G gene under the control of either the early or late SV-40 promoters, and subsequently transfecting this construction into COS cells, G protein was synthesized [34]. As was found with the influenza HA, the VSV G protein was also located on the cell surface. This study demonstrated that the SV-40 origin of replication alone was sufficient to regulate the expression of foreign genes in COS cells, particularly when the copy number of the vector was high.

The mammalian cell expression of the influenza HA and the VSV glycoprotein generated membrane-bound viral surface antigens. A more useful vaccine product might be obtained if the engineered mammalian cell cultures could transport the antigens extracellularly. As found with other membrane-bound glycoproteins, both the VSV glycoprotein and the influenza HA contain a carboxy-terminal hydrophobic domain that is thought to anchor the protein in the cell membrane. Removal of the transmembrane domain DNA sequences by restriction endonuclease digestion and subsequent expression of the truncated glycoprotein resulted in secretion of both proteins into the extracellular medium [32,34]. An obvious advantage of using a secretion system for vaccine manufacturing would be easier purification inasmuch as the producer cells would not have to be disrupted to release the antigen. In addition, isolation of the antigen might not result in cell death, so the immunogen could be produced continuously for a longer time.

The important immunogenic structure for a useful hepatitis B virus vaccine is the surface antigen protein complexed into a 22-nm particle. The HBsAg particle has been successfully induced in tissue culture cells using either lytic SV-40 vectors [30], or bacterial plasmid vectors containing SV-40 transcriptional signals [29,35], or by transfection of cells with plasmids that contain selectable markers as well as viral DNA encoding the surface antigen gene [38,39]. In each case, the expression of the surface antigen gene in the cells resulted in the production of particles apparently similar to serum-derived 22-nm HBsAg particles (Australia antigen). Thus, it appeared that the synthesis of this particular antigen within mammalian cells could direct its own self-assembly into a complex particle.

It is known that the plasma of hepatitis B carriers contains large numbers of 22-nm HBsAg particles that can be purified and used as a vaccine for hepatitis B [24]. These particles are, by all available criteria, identical to those produced by the tissue culture systems described above. Therefore, mammalian cell cultures that express the HBsAg particle may provide an acceptable alternative to the use of a plasma-derived vaccine.

Future R-DNA Vaccines

The concept of using genetically engineered viruses as carriers for immunogenic surface antigens has recently been reported in the vaccinia virus system. In this system, the gene for the hepatitis surface antigen was placed under the transcriptional control of a vaccinia promoter [40,41]. This DNA construction was recombined into the vaccinia virus genome by transfection of the cloned sequence into vaccinia virus-infected cells. Subsequent selection of recombinants made use of the loss of the vaccinia virus thymidine kinase gene. The resultant recombinant vaccinia virus genomes contained the HBsAg gene, which could be transcribed by the vaccinia RNA polymerase. The engineered virus produced secreted 22-nm HBsAg particles during virus replication *in vitro*. In addition, vaccination of rabbits with this virus induced HBsAg antibody. Although the efficacy of this system has not yet been tested, it is, in principle, a unique and entirely new method for obtaining live R-DNA-derived vaccines. Theoretically, any surface antigen could be incorporated and expressed in this system.

The techniques of R-DNA technology should allow the precise manipulation of viral genes for the production of attenuated live virus vaccines. For example, a normally virulent virus may be attenuated by cotransfection of the wild type viral genome with a cloned, mutated viral region that is necessary for virulence but not replication. *In vivo* recombination with this cloned region should result in incorporation of the mutated region into the wild type viral genome. Although this technique is limited to DNA viruses, it has apparently been successful with Herpes Simplex Virus [42]. The potential oncogenicity of the Herpes viral genome, along with its penchant for recombination, may render this type of vaccine unacceptable for human use. However, this work demonstrated that molecular attenuation is a possible alternative to random attenuation by virus passage. Similarly, the availability of complete nucleotide sequences of wild type and attenuated viral genomes may make it possible to create stable attenuated live virus vaccines by *in vitro* site-directed mutagenesis.

Summary

Genetic engineering methods have already provided a wide range of new developments with significant potential for the economical production of vaccines. Research is moving quickly and we can anticipate rapid progress in identifying and cloning many important immunogens and in the development of more efficient host–vector systems. An equally aggressive research effort must also be directed toward improved adjuvants, immuno-enhancers, and novel delivery systems (e.g., liposomes [43,44], slow release carrier compounds, mechanical slow release devices, etc.). The blending of both technologies will be needed to exploit fully the potential of purified R-DNA vaccines.

Although it is understandable that several approaches for future vaccine developments by R-DNA technology are being explored and many such developments are showing great promise, the ultimate test will be the ability of these vaccines to protect humans and animals against specific diseases. The technology is established and well characterized and we now look forward to its most appropriate application in the pursuit our common goal, containing infectious diseases.

References

1. Goeddel DV, Yelverton E, Ullrich A, Heyneker HL, Miozzari G, Holmes W, Seeburg PH, Dull T, May L, Stebbing N, Crea R, Maeda S, McCandliss R, Sloma A, Tabor JM, Gross M, Familletti PC, Pestka S (1980) Human leukocyte interferon produced by *E. coli* is biologically active. Nature 287:411–416
2. Pennica D, Holmes WE, Kohr WJ, Harkins RN, Vehar GA, Ward CA, Bennett WF, Yelverton E, Seeburg PH, Heyneker HL, Goeddel DV, Collen D (1983) Cloning and expression of human tissue-type plasminogen activator cDNA in *E. coli*. Nature 301:214–221
3. Hitzeman RA, Hagie FE, Levine HL, Goeddel DV, Ammerer G, Hall BD (1981) Expression of a human gene for interferon in yeast. Nature 293:717–722
4. Lewin B (1974) Bacterial genomes. Vol 1, Gene Expression. John Wiley and Sons, London
5. Lewin B (1977) Plasmids and phages. Vol 3, Gene Expression. John Wiley and Sons, London
6. Kleid DG, Yansura D, Smal B. Dowbenko D, Moore DM, Grubman MJ, McKercher PD, Morgan DO, Robertson BH, Bachrach HL (1981) Cloned viral protein vaccine for foot-and-mouth disease: Responses in cattle and swine. Science 214:1125–1129
7. Bachrach HL, Moore DM, McKercher PD, Polatnick J (1975) Immune and antibody responses to an isolated capsid protein of foot-and-mouth disease virus. J Immunol 115:1636–1641
8. Bachrach HL (1982) Recombinant DNA technology for the preparation and subunit vaccines. J Am Vet Med Assoc 181:992–999
9. Maniatis T, Fritsch E, Sambrook J (1982) Molecular cloning, a laboratory manual. Cold Spring Harbor
10. Johnson IS (1983) Human insulin from recombinant DNA technology. Science 219:632–637
11. Pasek M, Goto T, Gilbert W, Zink B, Schaller H, MacKay P, Leadbetter G, Murray K (1979) Hepatitis B virus genes and their expression in *E. coli*. Nature 282:575–579
12. Weis JH, Enquist LW, Salstrom JS, Watson RJ (1983) An immunologically active chimaeric protein containing herpes simplex virus type 1 glycoprotein D. Nature 302:72–74
13. Watson RJ, Weis JH, Salstrom JS, Enquist LW (1982) Herpes simplex virus type-1 glycoprotein D gene: Nucleotide sequence and expression in *Escherichia coli*. Science 218:381–384
14. Yelverton E, Norton S, Obijeski JF, Goeddel DV (1983) Rabies virus glycoprotein analogs: Biosynthesis in *Escherichia coli*. Science 219:614–620
15. Davis A, Nayak D, Ueda M, Hiti A, Dowbenko D, Kleid D (1981) Expression of the antigenic determinant of the hemagglutinin gene of a human influenza virus in *Escherichia coli*. Proc Natl Acad Sci USA 78:5376–5380
16. Emtage J, Tacon W, Catlin G, Jenkins B, Porter A, Carey N (1980) Influenza antigenic determinants are expressed from hemagglutinin genes cloned in *Escherichia coli*. Nature 283:171–174

17. Heiland I, Gething M (1981) Cloned copy of the hemagglutinin gene codes for human antigenic determinants in *E. coli*. Nature 292:851–852
18. Davis A, Bos T, Ueda M, Nayak D, Dowbenko D, Compans R (1983) Immune response to human influenza virus hemagglutinin depressed in *Escherichia coli*. Gene 21:273–284
19. Hardy K, Stahl S, Kupper H (1981) Production in *B. subtilis* of hepatitis B core antigen and of major antigen of foot and mouth disease virus. Nature 293:481–483
20. Valenzuela P, Gray P, Quiroga M, Zaldivar J, Goodman HM, Rutter WJ (1979) Nucleotide sequence of the gene coding for the major protein of hepatitis B virus surface antigen. Nature 280:815–819
21. Hitzeman RA, Chen CY, Hagie FE, Patzer EJ, Liu CC, Estell DA, Miller JV, Yaffe A, Kleid DG, Levinson AD, Oppermann H (1983) Expression of hepatitis B virus surface antigen in yeast. Nucl Acids Res 11:2745–2763
22. Valenzuela P, Medina A, Rutter WJ, Ammerer G, Hall BD (1982) Synthesis and assembly of hepatitis B virus surface antigen particles in yeast. Nature 298:347–350
23. Miyanohara A, Toh-e A, Nozaki C, Hamada F, Ohtomo N, Matsubara K (1983) Expression of hepatitis B surface antigen in yeast. Proc Natl Acad Sci USA 80:1–5
24. Szmuness W, Stevens CE, Harley EJ, Zang EA, Oleszko WR, William DC, Sadovsky R, Morrison JM, Kellner A (1980) Demonstration of efficacy in a controlled clinical trial in a high-risk population in the United States. N Eng J Med 303:833–841
25. Mulligan RC, Howard BH, Berg P (1979) Synthesis of rabbit β-globin in cultured monkey kidney cell following infection with a SV40 β-globin recombinant genome. Nature 277:108–114
26. Gruss P, Khoury G (1981) Expression of simian virus 40–rat preproinsulin recombinants in monkey kidney cells: Use of preproinsulin RNA processing signals. Proc Natl Acad Sci USA 78:133–137
27. Gray PW, Leung DW, Pennica D, Yelverton E, Najarian R, Simonsen CC, Derynck R, Sherwood PJ, Wallace DM, Berger SL, Levinson AD, Goeddel DV (1982) Expression of human immune interferon cDNA in *E. coli* and monkey cells. Nature 295:503–508
28. Tooze J (1980) *In* DNA Tumor Viruses: Molecular Biology of Tumor Viruses, Part 2. Cold Spring Harbor
29. Liu CC, Yansura D, Levinson AD (1982) Direct expression of hepatitis B surface antigen in monkey cells from an SV40 vector. DNA 1:213–221
30. Moriarty AM, Hoyer BH, Shih JW, Gerin JL, Hamer DH (1981) Expression of the hepatitis B virus surface antigen gene in cell culture by using a simian virus 40 vector. Proc Natl Acad Sci USA 78:2606–2610
31. Gething MJ, Sambrook J (1981) Cell-surface expression of influenza haemagglutinin from a cloned DNA copy of the RNA gene. Nature 293:620–625
32. Gething MJ, Sambrook J (1982) Construction of influenza haemagglutinin genes that code for intracellular and secreted forms of the protein. Nature 300:598–603
33. Mellon P, Parker V, Gluzman Y, Maniatis T (1981) Identification of DNA sequences required for transcription of the human alpha 1-globin gene in a new SV40 host–vector system. Cell 27:279–288
34. Rose J, Bergmann JE (1982) Expression from cloned cDNA of cell-surface se-

creted forms of the glycoprotein of vesicular stomatitis virus in eucaryotic cells. Cell 30:753–762
35. Crowley CW, Liu CC, Levinson AD (1983) Plasmid-directed synthesis of hepatitis B surface antigen in monkey cells. Mol Cell Biochem 3:44–55
36. Lusky M, Botchan M (1981) Inhibition of SV40 replication in simian cells by specific pBR322 DNA sequences. Nature 293:79–81
37. Gluzman Y (1981) SV40-transformed simian cells support the replication of early SV40 mutants. Cell 23:175–182
38. Christman JK, Gerber M, Price PM, Flordellis C, Edelman J, Acs G (1982) Amplification of expression of hepatitis B surface antigen in 3T3 cells cotransfected with a dominant-acting gene and cloned viral DNA. Proc Natl Acad Sci USA 79:1815–1819
39. Dubois MF, Pourcel C, Rousset S, Chany C, Tiollais P (1980) Excretion of hepatitis B surface antigen particles from mouse cells transformed with cloned viral DNA. Proc Natl Acad Sci USA 77:4549–4553
40. Mackett M, Smith GL, Moss B (1982) Vaccinia virus: A selectable eukaryotic cloning and expression vector. Proc Natl Acad Sci USA 79:7415–7419
41. Smith GL, Mackett M, Moss B (1983) Infectious vaccinia virus recombinants that express hepatitis B virus surface antigen. Nature 302:490–495
42. Roizman B, Warran J, Thuming C, Fanshaw M, Norrild B, Meignier B (1982) *In* Bonneau M, Hennessen W, (eds) Herpes Virus of Man and Animal: Standardization of Immunological Procedure. Karker S, p 287
43. North JR, Morgan AJ, Thompson JL, Epstein MA (1982) Purified Epstein–Barr virus M_r 340,000 glycoprotein induces potent virus-neutralizing antibodies when incorporated in liposomes. Proc Natl Acad Sci USA 79:7504–7508
44. Balcarova J, Helenius A, Simons K (1981) Antibody response to spike protein vaccines prepared from Semliki forest virus. J Gen Virol 53:85–92

CHAPTER 50
Monoclonal Antibodies

WALTER GERHARD AND HILARY KOPROWSKI*

The vast majority of antigens, when injected into an immunocompetent organism, induce a large number of distinct B-cell clonotypes (each committed to the production of a unique type of immunoglobulin) to divide and differentiate into antibody-secreting B-cell clones. Consequently, the antiserum will be composed of many distinct antibody populations which, depending on their relative concentrations, will determine the effector functions and overall antigenic specificity exhibited by the given antiserum. This polyclonality of antisera has been a major problem in viral immunology. It has been notoriously difficult and often impossible to generate antisera specific for individual viral proteins because the latter required purification procedures that, firstly, had to be stringent enough to produce a preparation of highest purity of a given viral protein and, secondly, did not result in the alteration of immunogenic and antigenic properties associated with the native structure of the given viral protein. More importantly, the polyclonality of antisera precluded, with rare exceptions, characterization of distinct antigenic regions on an individual viral protein and examination of the antiviral functions mediated by antibodies binding to these regions. Initial attempts to bypass the problems associated with polyclonal antisera by isolation, in tissue culture, of normal B-cell clones secreting antibodies of desired specificity were hampered by the fact that isolated precursor B cells gave rise to relatively small numbers of antibody-secreting progeny cells. The latter were rather short lived and thus provided only small quantities of monoclonal antiviral antibodies [1]. The situation changed drastically, however, when Köhler and Milstein [2] succeeded in transforming sheep erythrocyte-induced murine B cells by hybridization with murine myeloma cells into con-

---

* The Wistar Institute of Anatomy and Biology, Philadelphia, Pennsylvania.

tinuously growing and antibody-secreting (hybridoma) cell lines. Subsequently, transformation of virus-induced B cells into antibody-secreting cell lines [3] rapidly became the method of choice for production of antiviral antibody reagents because it provided virtually unlimited quantities of antibodies that were, for all practical purposes, homogeneous and therefore amenable to precise characterization with respect to physical properties, function, and specificity. In the last few years there have been numerous reports about production and various types of applications of monoclonal antibodies to many viral agents. In the following, we want to review two general findings that have come from those studies and discuss, on that basis, the potential of monoclonal antibodies to improve disease control as diagnostic or therapeutic reagents.

To date, the emphasis of many monoclonal antibody studies has been on the determination and detailed characterization of viral proteins that serve as target structures for virus-neutralizing antibodies. In this context, an interesting observation was made, originally with influenza viruses and subsequently with many other viruses: When monoclonal antibodies to the hemagglutinin molecule of influenza virus were tested for their efficacy to neutralize virus it was found that, with rare exceptions, a small fraction of virus (usually in the range of 10^{-4} to 10^{-6}) resisted neutralization by individual monoclonal antibodies [4]. Analysis of the neutralization-resistant virus fractions revealed that they consisted of variants of the parental virus in which the epitope recognized by the antibody used in the neutralization test was altered. This alteration prevented the given antibody from binding to and neutralizing the variants. Amino acid and nucleotide sequence studies showed that the majority of the variants were point mutants (in the HA1 polypeptide) of the parental virus [5,6]. Antigenically distinct sets of neutralization-resistant virus mutants could be selected from cloned parental virus seed in the presence of distinct antibodies. Analogous findings have been made (to our knowledge, without exception) with all viruses that have been studied so far with neutralizing monoclonal antibodies. These observations serve to demonstrate two important points: First, an extensive microheterogeneity exists within cloned virus stocks, which apparently results from spontaneous mutations occurring in viral genes in the course of virus replication. Selective conditions that favor the growth of a particular set of mutants may lead to a rapid change in the composition of a virus stock. Second, by virtue of their homogeneity, monoclonal antibodies can be exquisitely sensitive to minor alterations in the target antigen and can distinguish unequivocally viruses that differ from each other by only a single changed amino acid residue. Not surprisingly, therefore, monoclonal antibodies frequently detected antigenic differences among virus strains (of the same designation) that had been passaged independently from each other in different laboratories.

Another type of information that is relevant in the present context has come from the use of monoclonal antibodies in functional antiviral assays *in vitro* (such as prevention of virus adsorption to target cells, virus neutralization, prevention of cell fusion, modulation or lysis of virus-infected target cells, etc.) and *in vivo* (protection against infection, recovery from or modulation of virus-induced disease, etc.). In several virus systems, it was found that distinct antibodies, even though directed to the same viral protein, exhibited drastically different antiviral activities. An instructive example is the finding of Burstin et al. [7] that antibodies to the s1 protein of reovirus type 3 can be grouped, on the basis of virus neutralization (VN) and hemagglutination inhibition (HI), into four functionally distinct sets: antibodies that were either VN- and HI-positive, VN-positive and HI-negative, VN-negative and HI-positive and, lastly, both VN- and HI-negative. Other studies showed that antibodies of different function may act competitively with each other and exhibit, in combination, an overall reduction of a desired antiviral activity [8,9] or, by contrast, may act not only additively but even synergistically [10]. Also, individual antibodies have been found to differ from each other with respect to inhibition of cytotoxic antiviral T cells [11,12] and, obviously, isotype-related effector functions. Furthermore, using individual antibodies in adoptive transfer studies, it was found that assays *in vitro* (such as virus neutralization) did not always correlate with the antibody's protective efficacy against virus challenge *in vivo* [13–15]. In the long run, the use of monoclonal antiviral antibodies in functional assays *in vitro* and *in vivo* will help to unravel the complex interactions that occur *in vivo* between viral agents and antibody-dependent host defense mechanisms and may ultimately provide a rational basis for production of optimally active mixtures of antiviral hybridoma antibodies for immunotherapy.

If one considers now the diagnostic value of individual monoclonal antibodies, it is clear that their sensitivity to minor antigenic changes limits their usefulness as general diagnostic reagents. This limitation can be readily overcome, however, by using appropriate mixtures rather than individual antibodies for diagnostic purposes [16,17]. By contrast, for scientific purposes, the sensitivity of individual antibodies provides extraordinary opportunities. It may make it possible, for instance, to trace the spread of antigenically similar virus strains. Also, it may provide evidence for relations existing between antigenic, biologic, and epidemiologic properties of viruses. The serologic surveys performed with monoclonal antibodies on a large number of attenuated strains and field strains of rabies virus exemplify these points. Thus, it was found that rabies virus isolates from different geographic regions could be differentiated antigenically from each other [17]. In addition, certain field strains could be shown to differ antigenically from the standard vaccine strains, a situation that might have been responsible for the occasional failure of post-exposure vaccination treatment and that pointed to the need for producing vaccine strains "tailored" to the specific field strains circulating in different geographic areas [18]. Lastly, in the course of the serologic comparison of rabies virus strains it was noticed that two anti-glycoprotein antibodies reacted with strains of high neurovirulence but failed to react with strains of low neurovirulence, suggesting a relationship between the corresponding glycoprotein epitopes and neurovirulence of rabies virus. This was subsequently confirmed by antibody-mediated selection of rabies virus variants of low neurovirulence from rabies virus strains of high neurovirulence [19] and by comparison of peptide maps and amino acid sequence of rabies virus strains of high and low neurovirulence [20]. An analogous observation was made with reovirus type 3 [21] and evidence suggesting a relationship between antigenicity and virus virulence has been obtained also with influenza type A [22]. Thus, selected monoclonal antibodies might become useful for prognostication of viral diseases.

There are a number of reports describing the use of monoclonal antibodies in virus protection experiments *in vivo*. Essentially, these studies showed that, under suitable experimental conditions, individual monoclonal antibodies could protect against various virus diseases [13–15,23–26]. However, analogous to what has been said with respect to the usefulness of individual antibodies as general diagnostic reagents, their use for therapeutic purposes may not be advantageous: It might promote the selection of mutant viruses (perhaps with altered tropism, virulence [27], or susceptibility to T-cell-mediated lysis [28]) and result in antigenic drift away from the predominant wild type virus strain to which part of the host population has acquired immunity. By contrast, appropriate mixtures of antibodies that act noncompetitively at different stages of virus infection and through different effector mechanisms (neutralization of free virus, prevention of cell-to-cell spread, complement- or cell-mediated lysis of virus-infected cells) may be vastly superior to antisera. Using this approach allows one to decrease the therapeutically effective dose of antibodies and thus minimize adverse reactions resulting from sensitization against foreign immunoglobulin determinants or from fortuitous cross-reactions (not recognized in screening assays) of individual monoclonal antibodies with normal components of the treated host organism. Another obvious advantage is the feasibility of generating species- and allotype-matched antibodies by *in vitro* immunization and fusion with suitable species-specific myelomas, thus performing antiviral immunotherapy in a completely physiological manner.

So far, the use of monoclonal antibody reagents as research tools has provided much new information about epidemiologic, structural, and biologic properties of individual proteins of many different viral agents. More detailed understanding of the mechanisms by which viruses induce disease is likely to improve, indirectly, disease control. To date, there are only a few examples of direct applications of monoclonal antibodies to disease control, be they as diagnostic or therapeutic reagents. There is no reason, however, to doubt that these reagents will, in the long run, totally replace antisera and vastly improve control of viral diseases.

References
1. Gerhard W, Braciale TJ, Klinman NR (1975) The analysis of the monoclonal immune response to influenza virus. I. Production of antiviral antibodies *in vitro*. Eur J Immunol 5:720–725

2. Köhler G, Milstein C (1975) Continuous cultures of fused cells secreting antibody of pre-defined specificity. Nature 256:495–497
3. Koprowski H, Gerhard W, Croce CM (1977) Production of antibodies against influenza virus by somatic cell hybrids between mouse myeloma and primed spleen cells. Proc Natl Acad Sci USA 74:2985–2989
4. Gerhard W, Yewdell JW, Lopes D, Caton A, Brownlee GG (1984) Point mutations in the hemagglutinin molecule of influenza virus: Their frequency, phenotypic characteristics and biological relevance. Med Virol, in press
5. Laver WG (1982) The use of monoclonal antibodies to investigate antigenic drift in influenza virus. In Hurrell JGR (ed) Monoclonal Hybridoma Antibodies: Techniques and Applications. CRC Press, Boca Raton, p 103
6. Caton AJ, Brownlee GG, Yewdell JW, Gerhard W (1982) The antigenic structure of the influenza virus A/PR/8/34 hemagglutinin (H1 subtype). Cell 31:417–427
7. Burstin SJ, Spriggs DR, Fields BN (1982) Evidence for functional domains on the reovirus type 3 hemagglutinin. Virology 117:146–155
8. Massey RJ, Schochetman G (1981) Viral epitopes and monoclonal antibodies: Isolation of blocking antibodies that inhibit virus neutralization. Science 213:447–449
9. Yewdell J, Gerhard W (1982) Delineation of four antigenic sites on a paramyxovirus glycoprotein via which monoclonal antibodies mediate distinct antiviral activities. J Immunol 128:2670–2675
10. Webster RG, Hinshaw VS, Berton MT, Laver WG, Air G (1981) Antigenic drift in influenza viruses and association of biological activities with the topography of the hemagglutinin molecule. In Nayak DB (ed) Genetic Variation among Influenza Viruses. Academic Press, New York, p 309
11. Effros RB, Frankel ME, Gerhard W, Doherty PC (1979) Inhibition of influenza-immune T cell effector function by virus-specific hybridoma antibody. J Immunol 123:1343–1346
12. Lefrancois L, Lyles DS (1983) Cytotoxic T lymphocytes reactive with vesicular stomatitis virus: Analysis of specificity with monoclonal antibodies directed to the viral glycoprotein. J Immunol 130:1408–1412
13. Schmaljohn AL, Johnson ED, Dalrymple JM, Cole GA (1982) Non-neutralizing monoclonal antibodies can prevent lethal alphavirus encephalitis. Nature 297:70–72
14. Balachandran N, Bacchetti S, Rawls WE (1982) Protection against lethal challenge of BALB/c mice by passive transfer of monoclonal antibodies to five glycoproteins of herpes simplex virus type 2. Infect Immun 37:1132–1137
15. Rector JT, Lausch ARN, Oakes JE (1982) Use of monoclonal antibodies for analysis of antibody-dependent immunity to ocular herpes simplex virus type 1 infection. Infect Immun 38:168–174
16. Nowinski RC, Tam MR, Goldstein LC, Stong L, Kuo C, Corey L, Stamm WE, Handsfield H, Knapp JS, Holmes KK (1983) Monoclonal antibodies for diagnosis of infectious diseases in humans. Science 219:637–644
17. Wiktor TJ, Flamand A, Koprowski H (1980) Use of monoclonal antibodies in diagnosis of rabies virus infection and differentiation of rabies and rabies-related viruses. J Virol Meth 1:33–46
18. Wiktor TJ, Koprowski H (1980) Antigenic variants of rabies virus. J Exp Med 152:99–112
19. Coulon P, Rollin PE, Flamand A (1983) Molecular basis of rabies virus virulence.

II. Identification of a site on the CVS glycoprotein associated with virulence. J Gen Virol 64:693–696
20. Dietzschold B, Wunner WH, Wiktor TJ, Lopes AD, Lafon M, Smith CL, Koprowski H (1983) Characterization of an antigenic determinant of the glycoprotein that correlates with pathogenicity of rabies virus. Proc Natl Acad Sci USA 80:70–74
21. Spriggs DR, Fields BN (1982) Attenuated reovirus type 3 strains generated by selection of hemagglutinin antigenic variants. Nature 297:68–70
22. Kilbourne ED (1978) Genetic dimorphism in influenza viruses: Characterization of stably associated hemagglutinin mutants differing in antigenicity and biological properties. Proc Natl Acad Sci USA 75:6258–6262
23. Doherty PC, Gerhard W (1981) Breakdown of the blood cerebrospinal fluid barrier to immunoglobulin in mice injected intracerebrally with a neurotropic influenza A virus. J Neuroimmunol 1:227–237
24. Dix RD, Perreira L, Baringer JR (1981) Use of monoclonal antibody directed against herpes simplex virus glycoproteins to protect mice against acute virus-induced neurological disease. Infect Immun 34:192–199
25. Mathews JH, Roehrig JT (1982) Determination of the protective epitopes on the glycoproteins of Venezuelan equine encephalomyelitis virus by passive transfer of monoclonal antibodies. J Immunol 129:2763–2767
26. Letchworth GI, Appleton JA (1983) Passive protection of mice and sheep against bluetongue virus by a neutralizing monoclonal antibody. Infect Immun 39:208–212
27. Spriggs DR, Bronson RT, Fields BN (1983) Hemagglutinin variants of reovirus type 3 have altered central nervous system tropism. Science 220:505–507
28. Ertl HC, Greene MI, Noseworthy JH, Fields BN, Nepom JT, Spriggs DR, Finberg RW (1982) Identification of idiotypic receptors on reovirus-specific cytolytic T cells. Proc Natl Acad Sci USA 79:7479–7483

CHAPTER 51
Antiviral Drugs

GEORGE J. GALASSO*

Until recently, when it was shown that vidarabine (adenine arabinoside) was effective in treating serious viral infections, most virologists questioned whether antiviral agents would be clinically useful. It was believed to be impossible to block viral replication without inhibiting normal cell metabolism. Viruses are, after all, obligate intracellular parasites, dependent on host cells for energy, nucleotide and amino acid building blocks. Viruses utilize the metabolic pathways of the cell. As more has been learned about the molecular biology of viruses, it has become evident that there are viral and virus-specified components that are sufficiently unique in the virus-infected cell that they can be inhibited without adversely affecting the host cell.

In the past, antiviral agents have been developed by serendipity, primarily through large screening programs. There is now little doubt that molecular biologists can identify virus-coded components for which teams of biochemists, organic chemists, and physical chemists can design specific inhibitors. There are basically three stages of viral replication to be considered in designing inhibitors: (1) viral attachment, penetration, and uncoating; (2) synthesis of viral components; and (3) assembly and release.

Viral attachment is highly specific, usually involving the interaction of a protein on the viral surface and a particular site on the cell membrane.

The presence of the attachment (receptor) site on the cell membrane determines the susceptibility of the cell to infection. Thus, if the nature of the receptor site can be determined, blocking agents can be developed to prevent infection. Viruses penetrate the cell either by engulfment (viropexis) or fusion with the cell membrane, as is the case with enveloped viruses. The

---

*National Institutes of Health, Bethesda, Maryland

viruses are then uncoated, usually by cellular enzymes. Since penetration and uncoating usually does not involve virus-specified components, it would appear that they would not be potential target areas for antiviral agent development. However, some drugs such as amantadine and arildone appear to inhibit viral uncoating.

Once the virus is uncoated, mRNA synthesis is initiated with subsequent viral genome and protein replication. The viral mRNA synthesis will differ from that of the normal cell and require enzymes not normally present. In some instances an RNA-dependent RNA polymerase may be a component of the virion. It is at this stage that the greatest potential for specific antiviral agent development exists. Detailed studies of viral or virus-specified enzymes may yield ideal sites for development of inhibitory agents either to block the enzyme or to foster faulty assemby of viral nucleic acid.

The final stage of assembly and release is less likely to yield promising sites for effective antivirals. Viruses are either packaged into a capsid and released through cell lysis or, as in the case of some enveloped viruses, through a budding process at the cell membrane where they acquire their envelope.

It appears quite likely, therefore, that by profiting from the work of molecular biologists and appropriate chemists, we should be able to develop specific, nontoxic antiviral agents. However, there are other difficulties that must be overcome if antivirals are to be clinically useful. As stated, most of the target sites are within the cell; therefore, the inhibitor must traverse the cell membrane in sufficient quantity to be effective. The viral disease itself presents another obstacle in that viral replication usually peaks during the incubation period and is already in decline when symptoms become apparent. If antivirals are to be maximally effective, we need to develop better methodologies for rapid viral diagnosis. Another factor that cannot be overlooked as a hindrance is the high cost, not only for the identification of a suitable compound but also for the extensive toxicology, drug distribution, mutagenicity, teratogenicity, oncogenicity, pharmacology, and clinical efficacy studies needed prior to drug approval.

Of the agents listed (Table 1), only a few are widely accepted, approved by the FDA, and available to the clinician; the others are promising compounds currently under study. Those approved by the FDA are IDU, Ara-A, and F_3T ointments as topical agents for the treatment of herpes ocular infections; amantadine for prophylaxis and therapy of influenza A; Ara-A for herpes encephalitis and zoster in the immunocompromised; and ACV for topical treatment of primary herpes genitalis and severe initial genital herpesvirus infection, and systemic treatment of serious mucocutaneous herpes in the immunosuppressed.

Influenza A remains one of the last uncontrolled viral diseases capable of causing yearly epidemics and pandemics. Because of viral antigenic drift and shift, the current vaccine is not always as protective as desirable. Fortunately, an effective antiviral, amantadine, is available. Its importance has

Table 1. Antiviral agents

Drug	Virus	Specific viral target	Toxicity	Efficacy
Acyclovir (Zovirax, ACV)	herpes (HSV-1, HSV-2, EBV, VZ, and CMV)	DNA polymerase DNA inhibition	Drug highly specific, minimal toxicity.	Clinical efficacy generally accepted topical treatment of primary genital herpes and systemic use in severe mucosal and cutaneous herpes infections of the immunosuppressed. Other clinical studies including oral application under study.
Adenine arabinoside (Vidarabine, Vira-A, Ara-A)	herpes hepatitis B*	DNA polymerase DNA inhibition	Minimal and acceptable toxicity. Systemic application side effects include nausea, vomiting, diarrhea; some neurological side effects reported.	Clinically effective for systemic treatment of herpes encephalitis, neonatal herpes and zoster; topical treatment of ocular herpes.
Adenine arabinoside monophosphate (Ara-AMP)	herpes, hepatitis B*	DNA polymerase DNA inhibition	As with adenine arabinoside	Clinical studies in herpes encephalitis, neonatal herpes, and chronic active hepatitis B underway.
Amantadine (Symmetrel)	influenza A	Attachment and uncoating	Minimal. Side effects in less than 6% of patients: insomnia, dizziness, difficulty in concentration, and drowsiness.	Clinically effective for prevention and treatment of influenza A.
Arildone	DNA, RNA viruses including: herpes, picorna, and respiratory syncytial virus (RSV)	Uncoating	Under study.	Clinical studies ongoing with topical treatment of genital herpes.

Drug	Virus	Mode of action	Toxicity	Clinical efficacy
Bromovinyldeoxyuridine (BVDU)	herpes	DNA polymerase DNA inhibition	Under study.	Beneficial effect reported with topical drug for ocular herpes and oral drug for zoster.
2-Deoxy-D-glucose	Enveloped viruses, herpes, influenza	Assembly	Minimal.	No definitive clinical efficacy demonstrated
Enviroxime	picorna	RNA synthesis	Systemic toxicity, gastric symptoms, nasal spray acceptable.	Some reduction of symptoms in challenged volunteers with rhinovirus type 4.
2'-fluoro-5-iodoaracytosine (FIAC) and analogues FMAU, FIAU	herpes	DNA polymerase DNA inhibition	Under study.	Early reports indicate positive results against zoster.
Iododeoxyuridine (IDU)	herpes	DNA inhibition	Minimal when applied topically.	Clinical efficacy for topical treatment of ocular herpes.
Isoprinosine (methisoprinol)	RNA, DNA viruses	immune potentiation (?)	None reported.	No definitive clinical efficacy demonstrated.
Phosphonoformate (PFA, Foscarnet)	herpes, retro, hepatitis B	DNA polymerase	Bone retention, topically 3% ointment tended to irritate mucosal membranes.	Some clinical benefit seen in topical treatment of labial herpes; genital studies and studies against CMV underway.
Ribavirin	DNA, RNA viruses, including arena, herpes, RSV, and influenza	mRNA capping, under study	Dose related systemic hematopoietic toxicity, some increase in serum uric acid and iron; not seen with aerosol.	Aerosol application appears to have beneficial effects in influenzal pneumonia and RSV infections; oral application under study for Lassa fever.
Rimantadine	influenza A	Uncoating	None seen.	Clinically effective in prevention and treatment of influenza A.
Trifluorothymidine (F_3T, viroptic)	herpes simplex	DNA inhibition	Minimal. Applied topically, mild burning, palpebral edema in 3%–5% studied.	Clinically effective against ocular herpes infections.

*Some effect in clinical chronic active hepatitis B observed, mode of action not fully understood been underestimated, largely because of the overrated side effects. These are usually mild and occur in less than 6% of the recipients. Currently an analogue, rimantadine, is proving to be equally effective with no apparent side effects [1,2].

Ara-A is of particular interest because of the role it has played in overcoming the skepticism shown to antivirals. It was the first systemic drug demonstrated to be efficacious against a life-threatening infection. It has been successful in reducing the mortality resulting from brain biopsy-proven herpes encephalitis from 70% to 28% [3]. It has been equally effective against neonatal herpes infections, reducing mortality from 74% to 38% [4]. Studies in immunocompromised patients with zoster have shown that the drug is useful when administered early in infection by accelerating healing, decreasing cutaneous dissemination, zoster-related visceral complication, and duration of post-herpetic neuralgia [5].

ACV has received considerable attention because of its effectiveness against genital herpes and its strong viral specificity. It selectively inhibits replication of herpes simplex virus after being converted to the active form by the viral thymidine kinase. The ointment has been reported to be effective in reducing the duration and severity of the initial episode but has no clinical efficacy in recurrent genital herpes. Systemic application has proved effective therapeutically [6] and prophylactically [7] against serious herpes infections of the immunocompromised. Current studies with ACV are focused on a comparison to Ara-A and evaluation of an oral formulation.

Of the remaining compounds there are some deserving special mention. BVDU [8] is a relatively new compound, with great specificity for herpes viruses, and is effective in animal studies when given orally even after viral symptoms appear. Preliminary clinical studies in Europe against zoster are very promising. PFA is being tested extensively in Europe as a topical agent against labial and genital herpes with favorable, although not dramatic, results. Ribavirin is proving to be an interesting compound. Although it was initially disappointing as an oral drug against influenza, current studies using it as an aerosol against both influenzal pneumonia and RSV infections in infants show promise [9]. It also has potential, administered systemically, against arenaviruses such as Lassa fever. The pyrimidine analogues FIAC, FMAU, and FIAU, show excellent promise. Preliminary clinical studies show good activity against zoster infections. There have been recent reports on a new nucleoside analogue of ACV referred to as 9-2[[hydroxy-1-(hydroxymethyl)ethoxy]methyl]-guanine, 2′nor-2′deoxyguanosine, and 9-(1,3-dihydroxy-2-propoxymethyl)guanosine, which appears to be quite interesting. In addition to its good activity against herpes simplex, it appears to have excellent activity against CMV. The mode of action has not been clarified but, if it is different from ACV, it may be active against some viruses that become resistant to ACV. There has been an extensive search for compounds active against the common cold, to no avail; however, enviroxine does show some effect. Considerable work needs to be done to maximize this, but at least it shows that there is some possibility for an effective agent against rhinoviruses.

There are a number of agents whose efficacy has not been demonstrated but they need to be mentioned because of the extensive visibility that they have been accorded. 2-Deoxy-D-glucose has been reported to have clinical efficacy against genital herpes but this report has not been substantiated. The study has been widely criticized and animal studies in several laboratories have shown no effect of the drug. Isoprinosine has received some acceptance outside the United States, but again, there is no clear evidence of efficacy. It was originally presented as an antiviral agent and is currently advocated as an immune potentiator.

The search for clinically suitable antiviral agents is a continuing one and, fortunately, one that is now achieving success. Knowledge gained from research on the molecular biology of viral replication is useful for identifying targets for attack and will facilitate further development of antiviral agents. Another largely unexplored area is drug combinations. It is conceivable that two minimally effective drugs can act synergistically in overcoming an infection. This has been demonstrated by the combination of interferon and F_3T, which appear to be more effective than either drug alone against ocular herpes infections. However, care must be taken to assure that these various drug combinations do not increase the toxicity of either or both drugs.

Another concern is development of resistance. Herpesviruses have shown acquired resistance in the laboratory as well as *in vivo*. Thus far the resistant strains have been less virulent than wild strains and have not occurred too readily. However, if the drugs are administered prophylactically over prolonged periods, concern for resistant virulent strains must be raised.

Superior drugs are obviously needed; techniques are now available for their development. The final caution is that once such compounds are identified, there is no substitute for a carefully conducted double-blind, placebo-controlled study involving proven cases of the disease to demonstrate efficacy properly. The research momentum in the development of antiviral agents is gaining speed; we have every reason to expect that several new and effective agents will be available in the next few years.

References

1. Dolin R, Reichman RC, Madore HP, Maynard R, Linton PN, Webber-Jones J (1982) A controlled trial of amantadine and rimantadine in the prophylaxis of influenza A infection. N Engl J Med 307:580–584
2. Van Voris LP, Betts RF, Hayden FG, Christmas WA, Douglas Jr RG (1981) Successful treatment of naturally occurring influenza A/USSR/77 H1N1. J Am Med Assoc 245:1128–1131
3. Whitley RJ, Soong SJ, Hirsch MS, Krachmer AW, Dolin R, Galasso GJ, Dunnick JK, Alford Jr CA (NIAID Collaborative Antiviral Study Group) (1980) Herpes simplex encephalitis—vidarabine therapy and diagnostic problems. N Engl J Med 304:313–318

4. Whitley RJ, Nahmias AJ, Soong SJ, Galasso GJ, Fleming CL, Alford Jr CA (NIAID Collaborative Antiviral Study Group) (1980) Vidarabine therapy of neonatal herpes simplex virus infection. Pediatrics 66:495–501
5. Whitley RJ, Hilty M, Haynes R, Bryson Y, Connor JD, Soong SJ, Alford Jr CA (NIAID Collaborative Antiviral Study Group) (1982) Vidarabine therapy of varicella in immunocompromised patients. J Pediat 101:125–131
6. Wade JC, Newton B, McLaren C, Flournay N, Keeney RE, Meyers JD (1982) Intravenous acyclovir to treat mucocutaneous herpes simplex virus infections after marrow transplantation. Ann Int Med 96:265–269
7. Saral R, Burns WH, Laskin OK, Santos GW, Lietman PS (1981) Acyclovir prophylaxis of herpes simplex virus infections: A randomized, double-blind controlled trial in bone marrow transplant recipients. N Engl J Med 305:63–67
8. DeClercq E, Descamps J, Verhelst G, Walker RT, Jones AS, Torrence PF, Shugar D (1980) Comparative efficacy of antiherpes drugs against different strains of herpes simplex virus. J Infect Dis 141:563–574
9. Taber LH, Knight V, Gilbert BE, McClung HV, Wilson SZ, Norton HJ, Thurston JM, Gordon WH, Atmar RL, Schaudt WR (1983) Ribavirin aerosol treatment of bronchiolitis associated with respiratory syncytial virus infection in infants. Pediatrics 72:613–618

CHAPTER 52

The Use of Interferons in the Control of Viral Diseases

ANTHONY L. CUNNINGHAM AND THOMAS C. MERIGAN*

Introduction

In the quarter century since the discovery of interferon by Isaacs and Lindenmann an enormous literature describing its effects on biologic systems has accumulated. Most of the early clinical trials were conducted with leukocyte interferon prepared by Sendai virus stimulation of human buffy coats by Kari Cantell at the Finnish Red Cross. This material, now known to be a mixture of ten or more types of alpha interferon, was relatively impure (0.1%–1%) with a maximum specific activity of 10^6 units/mg of protein. Recombinant DNA technology has revolutionized the approach to interferons as therapeutic substances. The cloning of interferons alpha (1980) [1], beta (1980) [2], and gamma (1982) [3] not only allowed independent confirmation of the primary structure of the first two and revealed the great diversity of the alpha interferon gene family (28 genes or pseudogenes, of which at least 13 are functional and nonallelic) but has also led to their production in commercial quantities as pure preparations (10^8 units/mg of protein). Restriction endonucleases, organic oligonucleotide synthesis, and mutagenic techniques are currently being used to produce hybrid and altered molecules with varying biologic actions and potencies. A difference of only three amino acids between hybrids may result in marked differences in metabolism, antiviral activity, and immunomodulating effects [4]. Synergistic combinations of interferons alpha (or beta) and gamma, which combine with different receptor sites, may lead to enhanced efficacy in viral and neoplastic diseases in animal model systems and in humans [5]. The number of potential combinations defies independent clinical trials. Selection of the best molecules and

---

* Stanford University, Stanford, California.

their combinations undoubtedly will be made by extrapolation from biologic activity *in vitro*. Thus, a background knowledge of the structure and functions of interferons is necessary for an understanding of this rapidly advancing field and is briefly reviewed here.

Structure and Functions of the Interferons

Currently there appear to be at least ten different functional alpha interferons but one species each of beta and gamma interferon. The alpha interferons, 165–166 amino acids in length, exhibit more than 70% homology, whereas the 165 amino acid sequence of interferon-beta demonstrates at most 40% homology with the alpha interferons. These interferons, formerly grouped as Type I interferons, share a common gene structure (without introns), a common cell receptor, and are similar in their resistance to acid and heat inactivation. Interferon-gamma, 146 amino acids long, shares closest homology with interferon-beta. It is more distantly related to interferon-alpha. Both interferon-beta and interferon-gamma are glycosylated, whereas the alpha interferons are not. Interferon-gamma previously named Type II or immune interferon, differs from the others in its segmented gene structure (three introns), a different cell receptor, its susceptibility to acid and heat inactivation, and its mode of induction. Whereas interferon-alpha and interferon-beta may be induced by intact viruses, double-stranded RNA, or synthetic polynucleotides and other chemical inducers from most human cells *in vivo* or *in vitro,* interferon-gamma is induced nonspecifically by mitogens from T lymphocytes or from sensitized T lymphocytes by antigens. Monocytes or macrophages (and perhaps B cells) provide essential accessory functions [6,7].

Interferons may exert their antiviral effects *in vivo* by direct inhibition of cellular synthesis of viral mRNA or by augmentation of host defenses. Comparative antiviral and immunoregulatory potencies as defined by specific molecular activity (i.e., molecules per cell) should soon be available now that the pure substances are being tested. This is the only way to resolve differences in cell and virus sensitivity to the different interferons. The lack of human (or murine) interferon-gamma standards also hampers such comparisons.

Interferons probably exert their major antiviral (and antiproliferative) activity through the induction of synthesis of three groups of proteins: (1) enzymes that regulate translation, such as 2,5A synthetase and a (eIF-2) protein kinase [8]; (2) cell surface antigens such as the class I (A,B) and II (Ia-like) histocompatibility antigens and beta 2 microglobulin [6]; and (3) unidentified proteins detected by gel electrophoresis. Interferon-gamma, besides inducing the same set of peptides as interferon-alpha and beta, apparently induces an additional unique set of 12 polypeptides [9]. Their function is yet to be determined. The relative importance of the two main enzyme systems (2,5A synthetase and eIF-2 protein kinase) in the antiviral action of interferon is also yet to be established. All interferon types induce both systems. In addition, both enzymes require the presence of double-stranded (ds) RNA for full activation. After enzyme induction (probably via a cyclic nucleotide second messenger, following combination of interferon with its membrane receptor) 2,5 oligo A synthetase is activated by ds RNA and converts ATP to a series of 2,5' linked oligoriboadenylates, which have a 5'-terminal triphosphate. These oligomers can activate a latent cellular endoribonuclease, which cleaves RNA preferably after UA or UU sequences, thereby inhibiting protein synthesis. 2-5 As are unstable, and therefore their activating effects are transient [8]. Currently, several groups are attempting to use 2-5A synthetase activity in peripheral leukocytes as an index of viral infections, endogenous interferon activity, or response to interferon action [10]. eIF-2 protein kinase phosphorylates the alpha subunit of eukaryotic initiation factor 2 (eIF-2), decreasing its ability to form an initiation complex with a 40S ribosomal subunit, GTP, and Met-tRNA. Recently, use of inhibitors of 2-5A synthetase (analogues of 2-5A) in a cell-free system have suggested this may be the most important pathway for interferon inhibition of mRNA translation [8]. Application of these techniques to intact cells and, finally, *in vivo* observation is required. The specificity of these mechanisms for selective degradation of viral and cellular mRNAs is still unexplained. Currently, isolation and cloning of the genes responsible for the interferon-inducible polypeptides is underway and may lead to new methods of monitoring interferon effects. Interferons may inhibit viral replication in several other ways, e.g., by inhibiting uncoating or budding of enveloped viruses from the cell surface [11].

The immunomodulating effects of interferons are more clearly defined *in vitro* than *in vivo*. All interferons inhibit cell growth, enhance natural killer (NK) cell activity and antibody-dependent cytotoxicity, have time- and dose-dependent effects on antibody production by B cells, and inhibit mitogen-activated lymphocyte proliferation and leukocyte migration inhibition. All interferons enhance the expression of cell surface antigens, including Class I and II HLA antigens. Interferon-gamma is more potent in enhancing DR expression on monocytes than the others, a factor that may be important in viral antigen presentation [12]. The unique role of interferon-gamma as an immunoregulatory lymphokine is becoming increasingly apparent [7]. Together with interleukin 2, it is important in the induction of cytotoxic T lymphocytes and is released by these cells on contact with virus-infected target cells. It may also stimulate expression of IL-2 receptors and is probably the main macrophage-activating factor. Exactly how this affects function *in vivo* is complex. Interferon may enhance NK activity in several ways, but also enhances resistance of target cells to cytolysis. However, both increased and decreased NK activities have been reported in patients undergoing therapy with interferon [13].

Sources of Interferon for Clinical Use

Currently two new sources of purified interferons, lymphoblastoid and recombinant DNA expression in *E coli,* are replacing the less pure (0.1%–1%) leukocyte and fibroblast interferons in clinical trials. Lymphoblastoid interferon derived from Sendai virus stimulation of Namalwa cells contains a small amount (5%–10%) of beta interferon and at least eight species of interferon-alpha [14]. Recombinant alpha and beta interferons are pure preparations of the one species. This mode of production in *E. coli* precludes glycosylation, unimportant for interferon-alpha, but possibly important for activity of interferon-beta and gamma. Methods of producing recombinant interferon in cultured eukaryotic cells are being developed [1].

Fibroblast (mainly beta) interferon prepared by "superinduction" of fibroblasts has been used in several clinical trials not requiring intramuscular administration.

Spectrum of Antiviral Activity *in Vitro*

In vitro leukocyte (alpha) interferon is active against most of the major viral pathogens of humans, including both DNA and RNA viruses (e.g., herpes simplex types 1 and 2, cytomegalovirus, varicella zoster, EB virus, adenovirus, papovaviruses, poxviruses, rabies, rubella, arboviruses, enteroviruses, influenza, rhinoviruses, and most other respiratory tract viruses). There is considerable variability in susceptibility, not only between viruses but also depending upon culture systems, use of cell-free and cell-associated virus, and input multiplicity [11].

Administration, Pharmacokinetics, and Toxicity

Interferons may be applied topically to conjunctiva (drops), upper respiratory tract (sprays or pledgets), skin (hydrophilic gels), injected intravenously, subcutaneously, intramuscularly, or intrathecally via ventricular reservoir or lumbar puncture. The use of both cell cultures and radioimmunoassays has confirmed that the pharmacokinetics of interferon-alpha 2 are similar to those of leukocyte interferon [11,15]. Intravenous administration of leukocyte or alpha 2 interferon is followed by biphasic clearance with a rapid phase having a half-life of 10 minutes. Early experiences of systemic reactions, including shock, have made intravenous injection unpopular. Intramuscular or subcutaneous administration has been most widely used in clinical trials in humans. Peak serum levels of leukocyte interferon are usually achieved by 4 hours with serum levels still detectable at 24 hours. Hence, once-daily dosage may be used. The metabolism of interferon is still poorly understood, but probably most is eliminated from blood by adsorption to and catabolism by a variety of tissues and cells. The role of hepatic metabolism of alpha interferons is unclear, but may be important for glycosylated interferon-beta. Glycosylation may also increase tissue binding which, together with low *in vivo* stability, leads to its poor absorption from intramuscular injection sites [16]. Leukocyte interferon diffuses poorly into the cerebrospinal fluid and brain (30-fold gradient in animals and humans), aqueous and vitreous humors, and the placenta and foetus. These limitations are clearly relevant to the treatment of diseases such as viral encephalitis, rabies, CMV retinitis, and congenital viral infections.

Recent phase I trials with recombinant interferon-alpha and lymphoblastoid interferon have shown that the dosage tolerated may be increased from $30–50 \times 10^6$ units daily for leukocyte interferon to single doses of 198×10^6 units and 100×10^6 units for recombinant and lymphoblastoid interferons, respectively. These trials clearly demonstrated that the previously described side-effects of fever, fatigue, myalgias, malaise, and headache are intrinsic properties of the molecule and are not due to impurities [11]. With increasing dosage, the prevalence of these side-effects increases, with fever and chills being almost invariable above 30×10^6 u/day. Fever is most prominent with the first few injections. With prolonged treatment, mild hair loss is also common. Reversible granulocytopenia ($<1000/mm^3$) or thrombocytopenia ($<50,000/mm^3$) is common and hematologic monitoring is essential. Elevated serum transaminases are occasionally observed. Numbness and parasthesia are very infrequent with interferon alone, but occur more frequently in combination with adenine arabinoside [6]. This may be related to the depression of the drug-metabolizing activity of cytochrome p450 by interferon.

Clinical Efficacy of Interferons

In general, for success, interferon therapy needs to be administered prophylactically or early in the course of the disease. This has been well demonstrated in animal models with both interferon and other antivirals. The earliest clinical trials with leukocyte interferon demonstrated topical efficacy against herpetic and vaccinial keratitis and rhinoviral upper respiratory infections [17]. Interferon is unique as an antiviral agent in having a broad specificity against the 100 or more viruses capable of causing upper respiratory infections. Recently double-blind controlled trials of prophylactic recombinant interferon-alpha 2 against rhinovirus challenge confirmed earlier studies showing successful prevention of infection [18]. Slow-release delivery systems to counteract rapid interferon clearance from the nose may be required for this to be practical in prophylaxis or therapy. High-potency interferon alone was insufficient for the topical therapy of herpetic keratitis, but in combination with minimal wiping debridement or trifluorothymidine, a synergistic effect in hastening healing and reducing recurrence has been demonstrated. A controlled trial has also shown therapeutic efficacy of fibroblast interferon in adenoviral epidemic keratoconjunctivitis [17].

Randomized double-blind controlled trials have demonstrated the efficacy of leukocyte interferon in the prophylaxis and treatment of herpes simplex and varicella zoster infections in normal and immune compromised hosts [17]. In herpes zoster in immunocompromised patients, a one-week course of intramuscular leukocyte interferon reduced local new vesicle formation, decreased cutaneous and visceral dissemination and associated morbidity, and also the duration of post-herpetic neuralgia. Early therapy with leukocyte interferon for varicella in children with neoplasia significantly reduced new vesicle formation and *tended* to reduce life-threatening visceral dissemination [19]. Further trials with recombinant interferon in the prophylaxis and therapy of the immunocompromised with zoster and varicella are underway or planned. Prophylactic leukocyte interferon also reduced virologic and, to a lesser degree, clinical reactivation of herpes simplex after surgical section of the trigeminal nerve for intractable neuralgia. Subsequent recurrent rate was unaffected [17]. Clinical trials of recombinant alpha-interferon therapy for genital herpes are underway.

Although treatment of cytomegalic inclusion disease in neonates with leukocyte interferon has not been clinically successful, it does appear to diminish urinary viral shedding during therapy [6]. Treatment of bone marrow recipients afflicted with established severe CMV pneumonia with interferon alone or in combination with vidarabine did not alter the high mortality [17]. Prophylaxis may be more successful, and trials to examine this with recombinant interferons are underway. Prophylactic administration of interferon to renal transplant patients has been successful in delaying urinary excretion of CMV and reducing the incidence of viremia, especially in patients not receiving antithymocyte globulin. Viremia is associated with significant clinical syndromes and renal dysfunction. Interferon prophylaxis did not diminish the incidence of EBV reactivation [17].

Open and controlled trials have demonstrated the capacity of leukocyte and recombinant alpha interferons to depress and occasionally (in female patients) eradicate markers for viral replication in chronic hepatitis B [6]. Combination of interferon with adenine arabinoside increases the frequency of both responses, especially in males. Randomized-placebo controlled trials have been commenced to evaluate interferon therapy carefully in comparison with adenine arabinoside monophosphate. Neurotoxicity may restrict the combined use of these antivirals.

Interferon therapy has been attempted in many other viral diseases in which no other therapy is available. The evaluation of efficacy claimed from single case reports or small open studies is difficult, but promising clinical responses have been noted in papovaviral infections such as warts [20] and juvenile laryngeal papillomatosis [17] and the viral hemorrhagic fever Ebola [21]. No efficacy has been demonstrated in the treatment of Creutzfeld–Jakob disease [22], subacute sclerosing panencephalitis [23], well-established rabies (TC Merigan et al., submitted), or measles encephalopathy in children with leukemia [24]. Earlier therapy may be useful in rabies. A transient reduction in pharyngeal viral shedding was the only response observed in three infants treated for congenital rubella [6]. Other diseases with a putative viral origin, such as multiple sclerosis or juvenile-onset diabetes, are the objects of future or ongoing trials. Apart from varicella, herpes simplex, and CMV infection (see above), currently trials with recombinant interferons are underway or being planned with condylomata acuminata and juvenile laryngeal papillomatosis. Other possible targets include viral hemorrhagic fevers (Ebola, Marburg, Lassa, Junin), arboviral encephalitis, enteroviral conjunctivitis, myocarditis, and meningoencephalitis, acute viral hepatitis, and severe EBV infections.

Future directions of interferon therapy may also be guided by discovery of absolute or relative interferon deficiencies: Severe influenza infection has been associated with mucosal interferon-alpha deficiency [25], chronic EBV infection in a child was associated with an absolute deficiency of interferon-gamma production [26], and relative deficiencies of interferon-gamma production may be associated with frequent recurrences of herpes simplex [27]. Interferons will almost certainly be used in combination with other antivirals in a wider range of diseases. They may also be used for increasing the efficacy of vaccines, as demonstrated with rabies in experimental animals. With the supply problems almost solved, an exciting era for the clinical investigation of interferon therapy has begun.

Acknowledgments. This work was supported by a grant from the US PHS (AI-05629-20). AL Cunningham is supported by an Applied Health Sciences Travelling Fellowship from the National Health and Medical Research Council of Australia.

References
1. Weissman C, Nagata S, Boll W, Fountoulakis M, Futisawa A, Fujisawa JI, Haynes J, Henco K, Mantei N, Ragg H, Schein C, Schmid J, Shaw G, Streuli M, Taira H, Todokor K, Weidle U (1982) Structure and function of human alpha-interferon genes. *In* Merigan TC and Friedman RM (eds) Interferon, (UCLA Symposia on Molecular and Cellular Biology, Vol 25). Academic Press, New York, p 295
2. Taniguchi T, Guarente L, Roberts TM, Kimelman D, Douhan J, Ptashne, M (1980) Expression of the human fibroblast interferon gene in *Escherichia coli.* Proc Soc Natl Acad Sci USA 77:5230
3. Gray PW, Leung DW, Pennica D, Yelverton E, Najarian R, Simonsen CC, Derynck R, Sherwood PJ, Wallace DM, Berger SL, Levinson AD, Goeddel DV (1982) Expression of human immune interferon cDNA in *E. coli* and monkey cells. Nature 295:503–508
4. Lee SH, Weck PK, Moore J, Chen S, Stebbing N (1982) Pharmacological comparison of two hybrid recombinant DNA-derived human leucocyte interferons. *In* Merigan TC and Friedman RM (eds) Interferons (UCLA Symposia on Molecular and Cellular Biology, Vol 25). Academic Press, New York, p 295

5. Merigan TC (1982) Interferon—the first quarter century. J Am Med Assoc 248:2513–2516
6. Stiehm ER, Kronenberg LH, Rosenblatt HM, Bryson Y, Merigan TC (1982) Interferon: Immunobiology and clinical significance. Ann Int Med 96:80–93
7. Epstein LB (1981) Interferon-gamma: Is it really different from the other interferons? In Gresser I (ed) Interferon 3. Academic Press, New York, p 13
8. Torrence PF, Imai T, Lesiak K, Johnston MI, Jacobson H, Friedman RM, Sawai H, Safer B (1982) Double stranded RNA and 2',5' oligoadenylates: Companions in interferon action? In Merigan TC and Friedman RM (eds) Interferons (UCLA Symposia on Molecular and Cellular Biology, Vol 25). Academic Press, New York, p 295
9. Weil J, Epstein CJ, Epstein LB, Sedmak JJ, Sabran JL, Grossberg SE (1983) A unique set of polypeptides is induced by gamma interferon in addition to those induced in common with alpha and beta interferons. Nature 301:437–439
10. Schattner A, Wallach D, Merlin G, Hahn T, Levin S, Revel M (1981) Assay of an interferon-induced enzyme in white blood cells as a diagnostic aid in viral diseases. Lancet 2:497–499
11. Stewart II WE (1977) The interferon system. Springer-Verlag, New York
12. Basham T, Merigan TC (1983) Recombinant interferon γ increases HLA-DR synthesis and expression. J Immunol 130:1492–1494
13. Herberman RB, Ortaldo JR, Riccardo C, Timonen T, Schmidt A, Maluish A, Djeu J (1982) Interferon and NK (natural killer) cells. In Merigan TC and Friedman RM (eds) Interferons (UCLA Symposia on Molecular and Cellular Biology, Vol 25). Academic Press, New York, p 295
14. Finter NB (1982) Large scale production of human interferon from lymphoblastoid cells. In Baron S, Dianzani F, Stanton GJ (eds) Texas Reports on Biology and Medicine, Vol 41. The interferon system: A review to 1982, Part I. The University of Texas Medical Branch, Galveston, Texas, p 175
15. Gutterman JU, Fine S, Quesada J, Horning SJ, Levine JF, Alexanian R, Bernhardt L, Kramer M, Spiegel H, Colburn W, Trown P, Merigan TC, Dziewanowski Z (1982) Recombinant leucocyte A interferon: Pharmacokinetics, single-dose tolerance, and biologic effects in cancer patients. Ann Int Med 96:549–555
16. Bocci V (82) Pharmacokinetics of interferon—a reappraisal. In Baron S, Dianzani F, Stanton GJ (eds) Texas Reports on Biology and Medicine, Vol 41. The interferon system: A review to 1982, Part I. The University of Texas Medical Branch, Galveston, Texas, p 336
17. Attallah AM, Petricciani JC, Galasso GJ, Rabson AS (1980) Update on clinical trials with exogenous interferon. J Infect Dis 142:293–301
18. Scott GM, Philpotts RJ, Wallace J, Gauci CL, Greiner J, Tyrell DAJ (1982) Prevention of rhinovirus colds by human interferon alpha-2 from *Escherichia coli*. Lancet 2:186–187
19. Arvin AM, Kushner JH, Feldman S, Baehner RL, Hammond D, Merigan TC (1982) Human leukocyte interferon for the treatment of varicella in children with cancer. N Engl J Med 306:761–765
20. Pazin GJ, Ho M, Haverkos HW, Armstrong JA, Breinig MC, Wechsler HL, Arvin AM, Merigan TC, Cantell K (1982) Effects of interferon-alpha on human warts. J Interferon Res 2:235–243

21. Edmond RTC, Evans B, Bowen ETW, Lloyd G (1977) A case of Ebola virus infection. Br Med J 2:541–544
22. Kovanen J, Maltia M, Cantell K (1980) Failure of interferon to modify Creutzfeldt–Jakob disease. Br Med J 280:902
23. Bartram CR, Henke J, Treuner J, Basler M, Esch A, Mortier W (1982) Subacute sclerosing panencephalitis in a brother and sister. Therapeutic trial of fibroblast interferon. Eur J Pediatr 138:187–190
24. Olding-Stenkvist E, Forsgren M, Hanley D, Kreuger A, Lundmark KM, Nilsson A, Wadell G (1982) Measles encephacopathy during immunosuppression: Failure of interferon treatment. Scand J Infect Dis 14:1–4
25. Isaacs D, Clarke JR, Tyrrell DAJ, Webster ADB, Valman MB (1981) Deficient production of leucocyte interferon (interferon-alpha) *in vitro* and *in vivo* with recurrent respiratory tract infections. Lancet 2:950–952
26. Lipinski M, Virelizier JL, Tursz T, Griscelli C (1980) Natural killer and killer cell activities in patients with primary immunodeficiencies or defects in immune interferon production. Eur J Immunol 10:246–249
27. Cunningham AL, Merigan TC (1983) Gamma interferon appears to predict time of recurrence of herpes labialis. J Immunol 130:2397–2400

CHAPTER 53
Insect Viruses as Pesticides

THOMAS W. TINSLEY*

The concept of the use of insect pathogenic viruses as *biological control* agents is very topical, inasmuch as the potential of these systems has been adequately demonstrated. The reasons are straightforward and reflect simple economics overlaid with a concern and a conscience over the deterioration of the environment. The rise in crude oil prices is a familiar ingredient in the story of inflation and is relevant to the conventional methods of insect control. The increased costs of petrochemicals and labor have contributed in large measure to the increased costs of control by chemical insecticides. An additional problem arose with the development of resistance in many insect pests to a wide range of insecticides. Farmers and growers always respond to such events by increasing the quantities of insecticides applied and by demanding new and more effective compounds. This can lead to serious misuse of the chemicals and the eventual contamination of the environment, as shown by the increasing level of *pesticide residues* in the soil, water supplies, and in various wildlife such as predatory birds and fish. All of these factors have considerable sociopolitical consequences and, in combination, produce mental attitudes receptive to systems based on biological control methods. This state of mind can be seen in the use of various euphemistic synonyms by the media and the conservationists, such as "ecological," "organic," "natural," "integrated," and "nature's own remedies." However, apart from a few isolated examples, insect pathogenic viruses are not yet commercially available and their application in the field is still on a very small scale indeed. Why should this be, in view of the claims made for the potential of virus control of insect pests and the obvious advantage of popular support? Unfortunately, as for most apparently simple questions, a sim-

---

* Natural Environment Research Council, Oxford, England.

ple and uncomplicated answer is not available. The word virus has undertones that rank it second only to cancer in causing public anxiety and concern. Further, the idea of deliberately releasing a virus to kill its host has only one good historical precedent and that is *myxomatosis* virus against the common rabbit (*Oryctolagus cunninculus*) [1]. It is sometimes argued that the use of attenuated or killed viruses as vaccines (e.g., smallpox, yellow fever, and poliomyelitis), is in fact an example of biological control systems. Technically, this is true but the actual control is exerted by the immunological response of the vaccinated animals. Nevertheless, vaccination is an example of the widespread release of viruses for biological purposes.

There is no doubt that the irresponsible release of an insect-pathogenic virus would be just as reprehensible as, and possibly more dangerous than, the irresponsible use of chemical insecticides, fungicides, and herbicides. Unlike noxious chemicals, viruses are replicating systems and, if infection and replication occurred in the wrong host, i.e., in nontarget organisms, then serious consequences could result if debilitating and/or lethal diseases were caused. However, it would be necessary for such viruses to adapt to a new environment and also to develop methods of long-term survival, particularly if the death of the exotic hosts occurred. The chances of these events taking place are remote, particularly if candidate viruses are selected from groups known to affect only insects and if adequate safety tests are undertaken before application of the virus.

On the other hand, what of the insect pests themselves? Insects, particularly those that cause defoliation or loss of harvested crops in storage, are the human population's greatest competitor for food. Further, biting flies carry pathogens capable of causing serious diseases in humans and in domestic animals. The world population is steadily rising, despite the intensive efforts of the family-planning organizations. Malnutrition is present in most countries of the Third World and this is overshadowed by the ever-present threat of famine. More food crops must be grown, but perhaps just as importantly, a fairer and more equitable distribution of existing food would relieve a great deal of suffering. Further, an answer must be found to the conundrum of the 20th century: Do we grow crops to feed animals for indirect production of protein or do we grow high-protein crops and thereby reduce our dependence on animal products? In tropical countries, many peasant farmers operate at the subsistence level and so their existence is very finely balanced. A heavy infestation of a defoliating pest at a crucial stage of crop development can lead rapidly to a state of famine. In this event, the demand for absolute safety from insect viruses may seem a little academic. It would also follow that advanced pest management techniques involving frequent applications of insecticides could not be undertaken unless substantial government subsidies for aid programs from wealthier countries were available. Therefore, it is in this context that biological control programs must be considered, as they can provide an alternative technology to chemical insecticides.

The idea of using insect viruses to control their hosts is not new and *natural epizootics* of insect viruses are well documented in the literature. The high level of control that can be exerted is also well known. However, the processes that are involved in the initiation and development of such epizootics are only now being investigated. This is unfortunate because such information is vital to the satisfactory application of insect viruses in the field.

It is interesting to find, in the majority of the early attempts to use viruses to control insect pests, that the viruses to be applied were selected from diseased individuals in the target pest population. Viruses found in other host species were evidently not considered as candidate control agents. The rationale for this policy probably arose from the prevalent belief that insect viruses in general, and baculoviruses in particular, were very specific in their host ranges. We know now that varying degrees of *specificity* occur and that it is unwise to generalize. These procedures had the effect of preventing the release of viruses at random either at the species level or on a geographical basis. Even so, it is still considered advisable to search for suitable viruses in the original host populations first before looking elsewhere.

There are at least ten *groups of viruses* known from insects at the moment. However, insects comprise 75% of the known animal species and only a small fraction of the insect species have been examined to date. Further, the methods of virus extraction currently in use very probably eliminate the more unstable, lipid-containing groups of viruses. Therefore, all in all, our knowledge of the viruses harbored by insects is extremely fragmentary and reflects only the tip of the iceberg. The groups listed here must be regarded very much as a minimum number.

The groups are shown in Table 1 and the physicochemical affinities to viruses found in vertebrates and plants are indicated where known. The system followed is based on that published by Matthews [2]. Perhaps only the Baculoviridase are truly unique to invertebrate animals, as this group does not seem to have any similarities either in physicochemical or biological properties to any other virus group. This *restricted host range* of the baculoviruses to invertebrates was an obvious and compelling reason for their selection as potential biological control agents.

These particular viruses were in fact among the first to be recorded in insects [3]. The large size of the virus-associated inclusion bodies or polyhedra (0.5 μm–5 μm) make it possible for them to be seen in stained preparations with the light microscope. Their isolation and recognition is therefore easy and requires only relatively simple equipment. The present *classification* of baculoviruses is shown in Table 2. Viruses have been isolated principally from diseased Lepidoptera and Hymenoptera, though there are reports of baculoviruses in Diptera, Coleoptera, and Crustacea. The association of baculoviruses with their hosts in relation to control systems has been treated in detail by Tinsley [4] and Burges [5]. There seems to be little doubt that the large number of baculoviruses reported from defoliating caterpillars is a

Table 1. Groups of viruses found in insects.

Family	Genus	Nucleic acid	Particle symmetry	Association with inclusion body	Biochemical and biophysical similarities to viruses found in Vertebrates	Biochemical and biophysical similarities to viruses found in Plants
Baculoviridae	Baculovirus	DNA	Rod	+	None	None
Poxviridae	Entomopoxvirus	DNA	Ovoid	+	Poxviruses	None
Reoviridae	Cytoplasmic polyhedrosis virus	RNA	Isometric	+	Reovirus Blue-tongue	Rice dwarf, Wound tumor
Iridoviridae	Iridovirus	DNA	Isometric	–	African swine fever Frog viruses 1–3	None
Parvoviridae	Densovirus	DNA	Isometric	–	Parvoviruses	None
Picornaviridae	Unassigned	RNA	Isometric	–	Picornaviruses	Many small RNA viruses
Caliciviridae	*Amyelois* virus	RNA	Isometric	–	Caliciviruses	None
Nodaviridae	Nodamura virus group	RNA	Isometric	–	?*	None
Unassigned	*Nudaurelia* β virus group	RNA	Isometric	–	None	None
Rhabdoviridae	Sigmavirus	RNA	Bullet-shaped bacilliform	–	Rhabdoviruses	Rhabdoviruses

* Nodamura virus replicates in various vertebrate cells and in suckling mice.

Table 2. Classification of the Baculoviridae.

Family	Genus	Subgroup	Type species
Baculoviridae	Baculovirus	"A": Nuclear Polyhedrosis Virus (NPV)	*Autographa californica* (NPV)
		"B": Granulosis Virus (GV)	*Trichoplusia ni* (GV)
		"C": Non-occluded rod-shaped nuclear virus	*Oryctes rhinoceros* virus
		"D": Non-occluded nuclear viruses with a polydisperse DNA genome	"D1" *Hyposoter exiguae* virus Nucleocapsids surrounded by 2 envelopes
			"D2" *Apanteles melanoscelus* virus. Nucleocapsids surrounded by a single envelope.
		Unassigned Virus particles with similar general structure to baculoviruses have been isolated from mites, crustaceans, and fungi. Further subgroups will be delineated as more data become available.	

direct reflection on the pest status of the host species and the consequent intensified search for pathogenic viruses.

The purpose of releasing these viruses is to create an epizootic within the pest population and thereby to reduce the degree of infestation. Control of larvae as soon as possible after hatching would be the ideal. Significantly, first instar larvae are very susceptible to baculoviruses, whereas the fifth and sixth instar larvae are virtually immune to infection. This means that very small doses of virus are required to kill hatching larvae. Naturally, the frequency of spraying will depend on the number of generations produced by each pest species.

The basic question to be answered is, if insect pathogenic viruses have such a high control potential why are they not in everyday use? The stumbling blocks appear to be the economics of large-scale production and purification of viruses and their subsequent registration for field use. Many baculoviruses can be grown in tissue culture systems, usually, but not invariably, derived from the target host. Unfortunately, the yields of virus are frequently low and do not compare with the amounts that can be obtained from *in vivo* systems. Also, the cost of tissue culture media is high and the scale of production is still at the laboratory or initial "pilot" level. Nevertheless, these technical problems can be solved. Baculoviruses have only recently been cloned and so mixed populations of virus particles have been used routinely to initiate infection both *in vitro* and *in vitro*.

The development of protocols for assessing the possible *ecological hazards* arising from the release of insect viruses has occupied the time of regulatory authorities of many governments and United Nations bodies such as Food and Agriculture and World Health Organizations. The latest memorandum from WHO on the mammalian safety of microbial agents for vector control [6] has gone a long way to simplify the levels of testing and the procedures to be employed. However, such tests will continue to be expensive and may act as a deterrent to full commercial exploitation. Nevertheless, it is very probable that as more and more baculoviruses are tested without detectable hazard, then the restrictions on their use may be reduced.

Methods of virus release other than by spraying have not been adequately tested. One exception was the release of adults of the coconut beetle artificially infected with a baculovirus (*Oryctes rhinoceros*), which proved to be effective in the South Pacific in the dissemination of the virus to the beetle larvae, resulting in a significant decrease in beetle populations. Ignoffo [7] has made some intriguing and pertinent suggestions for such an *alternative strategy*. The use of virus-containing *baits, contamination* of adult moths by virus dusts in light traps, the release of infected or virus-contaminated parasites and predators are methods that could either supplement virus release by spraying or even be sufficiently effective in their own right. Insect pest populations are frequently infected at an inapparent level with insect viruses; this is the so-called *latency phenomenon*. This state is so common that it has been suggested that latency could be an important factor in the insects' defensive mechanisms against viruses [8]. These inapparent or subclinical infections can be activated to produce frank expression of the disease by a wide range of physical and chemical stimuli. The processes involved in such responses are not understood, but could be associated with the probability of entry of virus particles into the blood system. It is very probable that natural epizootics arise from the *activation* of *inapparent* or latent *infections*. If such universal low-level infections could be activated at will, this would closely mirror natural events and thereby present less risk of ecological hazard. Unfortunately, the literature shows that no serious experimentation along these lines has been attempted.

It is very probable that insect pathogenic viruses will continue to have only theoretical potential unless and until their undoubted usefulness is taken more seriously. It is possible that the agrochemical industry as a whole does not want an alternative to chemical insecticides. Therefore, perhaps the commercial development should come from the more biologically oriented firms producing vaccines or from the recent growth of companies interested in biotechnology. Perhaps state-controlled organizations could also play a part, particularly in the less-developed countries of the world. Certainly the technology is available on an experimental level; it is development on a full scale that is required.

References

1. Fenner F, Ratcliffe FN (1965) *Myxomatosis*. Cambridge University Press, Cambridge
2. Matthews REF (1982) Classification and nomenclature of viruses. Intervirology 17:1–199
3. Tinsley TW, Harrap KA (1978) Viruses of invertebrates. *In* Fraenkel-Conrat H, Wagner RR (eds) Comprehensive Virology, Vol 12. Plenum Press, New York, p 1
4. Tinsley TW (1979) The potential of insect pathogenic viruses as pesticidal agents. Ann Rev Entomol 24:63–87
5. Burges HD (1981) Microbial control of pests and plant diseases 1970–80. Academic Press, London
6. Mammalian safety of microbial agents for vector control: A WHO memorandum (1981) Bull WHO 59:857–863
7. Ignoffo CM (1978) Strategies to increase the use of Entomopathogens. J Invertebr Pathol 31:1–3
8. Tinsley TW (1975) Factors affecting virus infection of insect gut tissue. *In* Maramorosch K, Shope RE (eds) Invertebrate Immunity—Mechanisms of Invertebrate Vector–Parasite Relations. Academic Press, New York, p 55

Index

Antibody
  affinity, 32
  classes, 32
  complement, 32
  dependent cellular cytotoxicity, 11, 58
  enhancement of virus infectivity, 17
  enhances C3b deposition on virus infected cells, 41
  FAB'2 fragment in lysis, 39
  hypothesis of neutralization, 35, 36
  initiate virus persistence, 187
  monoclonal, 158, 376
  neutralization, 32, 34, 325
  non-neutralized fraction, 35
  sensitization of virus, 35
Antibody-dependent cellular cytotoxicity, 58
Antigenic domains
  Coxsackie B4, 158
  reovirus type 3 HA, 104
Antigenic drift
  genetic basis, 144, 154
  immune selection, 152
Antigen specific proliferation
  activation of cytotoxic and suppressor T cells, 225
  polyclonal B cell activation and autoimmunity, 231
  retrovirus induced lymphomagenesis, 216
Antiviral drugs, 382
Antiviral therapy
  drugs, 382
  interferon, 389
  monoclonal antibody, 376
  vaccines
    recombinant DNA, 366
    synthetic peptides, 361
Arthritis
  virus induced, 254
Arthropod vectors, 194, 338
Atherosclerosis
  virus induced, 248
Avian leukosis virus, 178, 216
Autoimmunity
  molecular mimicry, 187, 201, 211, 231
  multi-organ reactive antibodies, 213
  polyclonal B cell activation, 211, 231
  virus induced, 210

Bats, rabies infected, 344
B cell growth factors
  soluble, nonspecific and differentiation factors, 231
B cell lymphomas, 216

B cell mitogens
  viruses, 231
B lymphocytes, 32, 49
  bystander activation, 234
  Epstein–Barr virus, 307
  polyclonal activation, 231
Burkitt's lymphoma, 307

Canine distemper, 26
Caprine arthritis-encephalitis virus (CAEV), 254
C-*mos*, 167–168, 179
C-*myc*, 167–168, 179
Complement
  alternative pathway, 21, 44
  classical pathway, 21
  deficient states in animals, 43
  deficient states in man, 44
  enhanced absorption on cells, 117
  lysis of virus infected cells, 20, 39
  neutralization of antibody, 20, 32
  purified components in lysis of virus infected cells, 39
Coxsackie B4
  antigen variants and clinical disease, 158
Creutzfeldt–Jakob disease, 350
C3, 39
C2 and C4 depleated sera, 39
C5 deficient mice
  influenza infection, 43
  sindbis virus infection, 43

Delta agent, 293
Demyelination
  virus induced, 260
Dengue virus
  antibody enhanced tissue tropism, 117
Diabetes
  virus induced, 241
Differentiation
  virus effects, 130, 269
DI particles, 71, 137
Disseminated intravascular coagulation (DIC), 325
Duck hepatitis B virus (DHBV), 288

EBNA
  complement activation, 43
Ebola virus, 338
Encephalomyelitis
  post infectious, 260
Encephalomyocarditis (EMC) virus, 98–99
Endocrine disorders virus induced
  diabetes, 241
  growth hormone synthesis, 269
Endothelial injury
  hemorrhagic fever viruses, 325
Epstein–Barr virus
  C3 receptor, 97
  infection of epithelial cell, 308
  pathogenesis, 312
F(ab)$_2$ fragments
  role in lysis of virus infected cells, 42
F$_c$ fragment
  role in ADCC, 60
  role in hemorrhagic fever viruses, 325
  role in virus absorption, 117
Flaviviruses
  genetic control (host), 74, 75
Fusion protein
  paramyxoviruses, 26, 27
Fv-1 genes, 76

Gastroenteritis
  diagnosis, 318
  vaccine strategy, 320, 321
  virus induced, 315
Genes
  identifying the protein products by antibody to predetermined sequences, 361
Genetic control (host genes)
  immune complexes, 201
  immune response, 79
  murine leukemia viruses, 75
  orthomyxoviruses, 73
  virus infection, 71
Genital herpetic infection, 300
Genome RNA viruses
  mutation frequency, 137
  varients, 141
Genomes
  segmented, 101

Glycoprotein
  expression in membranes, 124
  insert in membrane, 29
  reduced expression in virus persistence, 273
  stripping off by antibody, 187
  transport, 125
Ground squirrel hepatitis virus (GSHV), 288

Hemagglutinating activity, 26
Hepatitis B surface antigens (HBsAg), 288
Hepatitis B virus
  associated with cancer, 178, 288
  disease, 288
  immune balance, 56
  immune complex, 288
Hepatocellular carcinoma (HCC), 178, 288
Herpes simplex virus
  latency
    gene expression, 174
    gene state, 173
Herpes simplex virus type 2 (HSV-2)
  association with carcinoma, 300
  transformation, 300
Host protease, 26
Human T cell leukemia virus, 165, 216, 279
  association with AIDS, 279
  association with leukemia, 279

Immune complexes
  host gene control, 203
  tissues injured, 201
  viral gene control, 205
  virus induced, 201
Immune response
  genetic control, 79
  turned off by virus, 274
Immune surveillance
  escape by antibody induced modulation, 187
  escape by virus mutation, 141
Immune tolerance
  absence in virus persistence, 187
  split tolerance, 225
Immunodeficiency syndrome
  acquired, 279
Immunoglobulin A, 32, 33
Immunoglobulin G, 32, 33
  enhances C3b deposition, 41
Immunoglobulin M, 32, 33
Immunosuppressive diseases, 269, 279
Inflammation, 46
Influenza virus
  cytotoxic T lymphocyte recognition, 55
  reassortments, 144
Insects
  ovarian infection, 194
  transmission of viruses, 194
  viruses as pesticides, 398
Interferon (IFN), 3
  activity, 7
  classes, 3, 389
  control of virus infection, 8, 389
  induction, 5, 6, 389
  system, 5
Interleukin 2 (IL-2), 11, 65, 231

Kaposi sarcoma, 279
K-cells, 11, 58
Kuru, 350

Lassa fever, 325
Leukemia gene, 86, 165, 178, 279
Lymphocytes
  B, 32
  cloning, 65
  interferon, 8-9
  large granular (LG-L), 11
  receptors for virus, 97
  subsets, 13, 47, 68, 227
  T, 46, 53
Lymphocytic choriomeningitis virus (LCMV)
  cytotoxic T cells, 53
  polyclonal B cell activation, 231
  RNA and proteins, 271
Lymphokines, 3, 46, 65, 231
Lysis, 20

Major histocompatibility region
  control antibody responses, 79, 203
  control cytotoxic T cells, 47, 53
  HLA organization, 82
  H-2 organization, 81
  restriction, 53, 225
  susceptibility in virus disease, 71, 203
Marburg virus, 338
Mast cells, 46
Measles, 26, 187
Measles virus
  atypical, 26
Membrane domains
  virus budding, 123
Molecular regulation of virus tropism, 130
Monoclonal antibodies
  detection of viral variants, 158, 376
  therapeutic use, 376
Monocyte/Macrophage
  absorb virus by Fc and C3 receptors, 49
  in inflammation, 49
Mosquito vectors, 194
Mumps, 26
Murine leukemia virus
  gene products, 86
  genetic control (host), 75, 76
  host range, 86
  lymphomagenesis, 216
  receptors, 216
  recombination, 86, 88
  sequence and structure, 86
Mutation frequencies
  RNA viruses, 137
Myxovirus, 28

Nasopharyngeal carcinoma
  association EBV, 311
Natural killer cells, 11, 50, 58
  makers, 13
Neutralization
  antibody, 32
New Castle disease, 26
Non-neutralized fraction, 35

Oncogenes
  capture by retroviruses, 168
  relationship to cancer, 178
  transcriptional control, 182
Orthomyxoviruses
  host genetic control, 73

Parainfluenza type 1, 26
Paramyxovirus
  F protein, 26
  hemagglutinin, 26
  membrane fusion, 26
Perturbed homeostasis
  virus induced, 269
Promoter insertion, 165, 178
Properdin, 39
Proteolytic cleavage viruses, 26

Rabies, 344
ras oncogene, 178
Reassortments
  arenaviruses, 144
  bunyavirus genes, 113
  influenza virus genes, 144
  in nature, 146
  mechanism, 145
  reovirus, 103
  tool for analysis of virulence, 148
Receptors viral
  attachment to nonviral receptors, 117
  causing retrovirus induced lymphomas, 216
  EBV, 307
  enhancement by antibody, 117
  host proteins, 20, 97, 109
  tropism, 97
Recombinant DNA vaccines, 366
Recombinants, viral genes
  murine leukemia virus, 86, 165
  RNA viruses, 139
Reovirus type 1, 102
Reovirus type 3, 102
Retroviruses
  association with human disease, 169, 178
  cellular oncogenes, 168, 178
  chronic leukemia, 279

classification, 165
interaction with cellular genes, 167, 178
mode of replication, 86, 166, 178
murine leukemia virus
  genetic control, 86
  recombination, 75, 167, 178
RFv-1, RFv-2 genes, 77
Rift valley fever, 325
Rotaviruses, 315

Scrapie, 350
Slow viruses, 350
src gene, 178
Subacute sclerosing panencephalitis, 187
Synthetic peptide vaccines, 361

T cell growth factor (TCGF), 65, 279
T cell helper responses, 231
T cell hybridomas, 65
T cell suppressor responses, 225, 231
T cells, cytotoxic (T_c)
  genetic control, 53, 58, 65
  in inflammation, 46
  mapping to viral gene, 103
T cells, delayed type hypersensitivity (T_D)
  in inflammation, 46
  mapping to viral genes, 103
  regulated by virus, 225
T lymphocytes
  clones, 65
  control of virus infection, 46, 53
  cytotoxicity, 46, 47, 53, 58, 65
  delayed hypersensitivity, 47

Tropism, tissue, 26, 97
  acetylcholine receptor, 117
  antibody influence, 117
  lymphocyte receptor, 97
  molecular basis, 130
  nervous system, 106

Unconventional viruses, 350

Vaccines, new strategies
  recombinant DNA, 366
  synthetic peptides, 361
Viral hemorrhagic fever (HVF), 325
Viroids, 350
Virus models
  arthritis, 254
  atherosclerosis, 248
  autoimmunity, 210, 231
  demyelination, 260
  diabetes mellitus, 241
  immune complex disease, 201
Virus variants
  amino acid sequence, 344
  generation in nature, 137, 144
  immune selection, 152
  relevance to disease, 158, 344
  typing by monoclonal antibody, 158
  typing peptide maps, 338
Visna virus
  antigenic drift, 153
  immune selection, 152

Woodchuck hepatitis B virus (WHV), 288